INTERNATIONAL CENTRE FOR MECHANICAL SCIENCES

COURSES AND LECTURES - No. 287

# ABELIAN GROUPS AND MODULES

PROCEEDINGS OF THE UDINE CONFERENCE

UDINE APRIL 9-14, 1984

EDITED BY

R. GÖBEL
UNIVERSITÄT ESSEN

C. METELLI - A. ORSATTI
UNIVERSITA' DI PADOVA

L. SALCE
UNIVERSITA' DI UDINE

Springer-Verlag Wien GmbH

VOLUME STAMPATO CON IL CONTRIBUTO DEL CONSIGLIO NAZIONALE DELLE RICERCHE – COMITATO PER LE SCIENZE MATEMATICHE


Originally published by CISM, Udine in 1984.

ISBN 978-3-211-81847-3 ISBN 978-3-7091-2814-5 (eBook)
DOI 10.1007.978-3-7091-2814-5

*Dedicated to Laszlo Fuchs*

*on his 60th Birthday*

## PREFACE

*The Udine '84 Conference on Abelian Groups and Modules fits into a by now long list of International Meetings on the subject; the recent ones: Las Cruces '76, Roma '77, Trento '80, Oberwolfach '81, Honolulu '82; it will have a continuation in Oberwolfach '85, and, hopefully, Perth '87.*

*The special feature of this, as of all Italian Conferences on the subject, has been the blending of Abelian Group Theory with Module Theory: thus supplying stimulating outlooks between bordering fields.*

*Many of the greatest names in the subject were present. The approximately 70 participants, both attending and contributing the 11 Main Lectures (see list on page IV) and 43 Communications, provided the highest participation so far; their interest was rewarded by the wide spectrum of subjects treated, the generally good, and sometimes outstanding, quality of mathematical contributions, and the vitality witnessed by both classical and recent research trends throughout the field.*

*Most of this appears in the present Volume; regrettably, due to the physical bound on the number of pages, we had to make some choices: thus, except for the Main Lecturers, the number of pages allotted to each contributor was restricted and no survey article was accepted; publication was subjected to favourable reports by Referees appointed by the Editors; no contribution was accepted from non-participants.*

*Nevertheless, we trust this Volume represents a faithful and exhaustive photograph of the state of the field as of today, and will provide many a challenging suggestion for the work of tomorrow.*

*Last but not least, we like to recall that the Conference caught the opportunity supplied by Laszlo Fuchs' 60th birthday to throw a party in his honour, and dedicate this Volume to him: in thankful acknowledgement of the great contribution his work and presence are giving to the lively and promising state of our subject.*

*The Editors*

**ACKNOWLEDGEMENTS**

*The Conference on Abelian Groups and Modules held in Udine on April 9-14 1984 was organized by the Istituto di Algebra e Geometria of the Padova University and the Istituto di Matematica, Informatica e Sistemistica of the Udine University.*

*The prestigeous location at Palazzo del Torso, and the excellent organizational services, were supplied by C.I.S.M., the International Centre of Mechanical Sciences at Udine.*

*Financial support was provided by the Research Funds of the two above mentioned Institutes, of the National Research Group "Teoria dei Gruppi ed Algebra non Commutativa", and, for the publication of this Volume, by the National Research Council – CNR. Smaller, but very important, contributions to the success of the meeting were given by the Presidency of the Regional Council of Friuli-Venezia Giulia, by the Banca Popolare Udinese, and by the Travel-Agency ACITUR.*

*Special thanks are due to Miss Elsa Venir, for her invaluable collaboration to the actual organization and her help to the participants, and to Mrs. Rosanna D'Andrea, Miss Lisetta Del Bianco and Mr. Umberto Modotti for the typing and editing of some of the manuscripts.*

## LIST OF GENERAL LECTURES

1. A.L.S. Corner: Endomorphism rings I
2. R. Göbel: Endomorphism rings II
3. L. Fuchs: Divisible modules
4. S. Shelah: A combinatorial principle and modules with few endomorphisms
5. A. Orsatti: Sulla struttura degli anelli linearmente compatti e loro dualità
6. E.A. Walker: Value-bounded valuated p-groups
7. B. Osofsky: Cardinality and projective dimensions
8. B. Müller: Morita duality

9 P. Hill: On the classification of abelian groups

10. J.D. Reid: Abelian groups as modules over their endomorphism rings
11. R. Wiegand: Direct-sum cancellation of finitely generated modules

## LIST OF PARTICIPANTS

U. Albrecht, Department of Mathematics, Auburn University, Auburn, Alabama 36849, USA.

T. Albu, Facultatea de Matematica, Str. Academiei 14, Bucuresti 1, Romania.

D. Arnold, Department of Mathematical Sciences, New Mexico State University, Las Cruces, New Mexico 88003, USA.

G. Baccella, Istituto Matematico, Università de L'Aquila, Via Roma 33, 67100 L'Aquila, Italy.

S. Bazzoni, Istituto di Algebra e Geometria, Università di Padova, Via Belzoni 7, 35100 Padova, Italy.

D. Beers, Department of Mathematics, Connecticut College, New London, Ct. 06320, USA.

K. Benabdallah, Département de Mathématiques et Statistique, Université de Montreal, Montreal, Quebec, H3C3J7, Canada.

L. Bican, Matem. Fyz. Fakulta KU, Sokolovska 83, Praha 8 (Karlin), Czechoslovakia.

W. Brandal, Department of Mathematics and Applied Statistics, University of Idaho, Moscow, Idaho 83843, USA.

R. Burkhardt, Mathematisches Institut, Universität Würzburg Am Hubland, 87 Würzburg, Western Germany.

A.L.S. Corner, Worcester College, Oxford, Great Britain.

D. Cutler, Department of Mathematics, University of California at Davis, Davis, California 95616, USA.

G. D'Este, Seminario Matematico, Università di Padova, Via Belzoni 7, 35100 Padova, Italy.

R. Dimitric, 29 Novembra 108, Beograd, Yugoslavia.

M. Dugas, Department of Mathematics, University of Colorado, Colorado Springs, Co. 80933, USA.

P. Eklof, Department of Mathematics, University of California at Irvine, Irvine, California 92717, USA.

A. Facchini, Istituto di Matematica, Informatica e Sistemistica, Università di Udine, Via Mantica 3, 33100 Udine, Italy.

T. Fay, Department of Mathematics, University of Southern Mississippi, Hattiesburgh, Ms. 39406, USA.

B. Franzen, Fachbereich 6, Mathematik, Universität Essen, 43 Essen 1, Western Germany.

L. Fuchs, Department of Mathematics, Tulane University, New Orleans, Louisiana 70118, USA.

R. Göbel, Fachbereich 6, Mathematik, Universität Essen GHS, 43 Essen 1, Western Germany.

B. Goldsmith, Department of Mathematics, Dublin Institute of Technology, Kevin Street, Dublin 8, Ireland.

E. Gregorio, Istituto di Algebra e Geometria, Università di Padova, Via Belzoni 7, 35100 Padova, Italy.

J. Hausen, Department of Mathematics, University of Houston, Houston, Texas 77004, USA.

P. Hill, Department of Mathematics, Auburn University, Auburn, Alabama 36849, USA.

K.Y. Honda, Department of Mathematics, St. Paul's University, Nishi-Ikebukuro, Toshima-Ku, Tokyo 171, Japan.

S. Khabbaz, Department of Mathematics, Lehigh University, Bethlehem, Pennsylvania 18015, USA.

J. Irwin, Department of Mathematics, Wayne State University, Detroit, Michigan 48202, USA.

H. Lausch, Technische Universität Clausthal, Institut für Mathematik, Erzstr. 1, D-3392 Clausthal-Zellerfeld, Western Germany.

A. Letizia, Dipartimento di Matematica, Università di Lecce, 73100 Lecce, Italy.

W. Liebert, Mathematisches Institut, Technische Universität München, 8000 München 2, Western Germany.

A. Mader, Department of Mathematics, University of Hawaii, 2565 The Mall, Honolulu, Hawaii 96822, USA.

F. Menegazzo, Seminario Matematico, Via Belzoni 7, 35100 Padova, Italy.

C. Menini, Istituto di Matematica, Università di Ferrara, Via Macchiavelli 35, 4100 Ferrara, Italy.

C. Metelli, Istituto di Algebra e Geometria, Università di Padova, Via Belzoni 7, 35100 Padova, Italy.

R. Mines, Department of Mathematical Sciences, New Mexico State University, Las Cruces, New Mexico 88003, USA.

E. Monari Martinez, Seminario Matematico, Università di Padova, 35100 Padova, Italy.

B. Müller, Mathematics Department, McMasters University, Hamilton, Ontario L8S4KI Canada.

M.G. Murciano, Dipartimento di Matematica, Università di Lecce, 73100 Lecce, Italy.

O. Mutzbauer, Mathematisches Institut, Universität Würzburg Am Hubland, 8700 Würzburg, Western Germany.

C. Nastasescu, Facultatea Matematica, University of Bucharest, Str. Academiei 14, 70109 Romania.

J. Ohm, Department of Mathematics, LSU. Baton Rouge, Louisiana 70803, USA.

A. Orsatti, Istituto di Algebra e Geometria, Università di Padova, Via Belzoni 7, 35100 Padova, Italy.

B. Osofsky, Department of Mathematics, Rutgers University, New Brunswick, N.J. 08903 USA.

R. Pierce, Department of Mathematics, University of Arizona, Tucson, Arizona 85271, USA.

L. Prochazka, Matem. Fyz. Fakulta KU, Sokolovska 83, Praha 8, Karlin, Czechoslovakia.

G. Regoli, Dipartimento di Matematica, Università di Perugia, Via Vanvitelli 1, 06100 Perugia.

J.D. Reid, Department of Mathematics, Wesleyan University, Middletown, Ct. 06457, USA.

N. Rodinò, Istituto Matematico "U. Dini", Università di Firenze, Viale Morgagni 67/A, 50134 Firenze, Italy.

W. Roselli, Istituto di Matematica, Università di Ferrara, Via Macchiavelli, 44100 Ferrara.

L. Salce, Istituto di Matematica, Informatica e Sistemistica, Università di Udine, Via Mantica 3, 33100 Udine, Italy.

A. Sands, Mathematics Department, The University of Dundee, Dundee DDI 4HN, Great Britain.

P. Schultz, Department of Mathematics, University of Western Australia, Nedlands, WA 6009, Australia.

S. Shelah, Institute of Mathematics, The Hebrew University, Jerusalem, Israel.

A. Skowronski, Institute of Mathematics, Nicholas Copernicus University, ul. Chopina 12/18, 87100 Torun, Poland.

D. Simson, Institute of Mathematics, Nicholas Copernicus University, ul. Chopina 12/18, 87100 Torun, Poland.

A. Soifer, Department of Mathematics, University of Colorado, P.O.B. 7150, Colorado Springs, Col. 80933, USA.

B. Thomé, Mathematisches Institut, Universität Freiburg, Alberstrasse 23b, 7800 Freiburg, Western Germany.

E. Toubassi, Department of Mathematics, University of Arizona, Tucson, Arizona 85721, USA.

H.P. Unseld, Mathematisches Institut, Albert Ludwigs Universität, 7800 Freibrug, Western Germany.

P. Vamos, Department of Mathematics, University of Exeter, North Park Road, Exeter EX4 4QE, Great Britain.

C. Vinsonhaler, Department of Mathematics, University of Connecticut, Storrs, Connecticut 06268, USA.

B. Wald, Institut für Mathematik II, Freie Universität Berlin, Arnimallee 3, 1000 Berlin 33, Western Germany.

E.A. Walker, Department of Mathematical Sciences, New Mexico State University, Las Cruces, NM 88003, USA.

R. Wiegand, Department of Mathematics and Statistics, University of Nebraska, Lincoln, Nebraska 68588, USA.

S. Wiegand, Department of Mathematics & Statistics, University of Nebraska, Lincoln, Nebraska 68588, USA.

G. Zacher, Seminario Matematico, Università di Padova, Via Belzoni 7, 35100 Padova, Italy.

P. Zanardo, Seminario Matematico, Università di Padova, Via Belzoni 7, 35100 Padova, Italy.

W. Zimmermann, Mathematisches Institut, Universität München, Theresienstrasse 39, 8 München 2, Western Germany.

**CONTENTS**

# The Classification Problem

Paul Hill

Auburn University

Having reached the full measure of a half century since countable abelian p-groups were classified, it seems appropriate that we consider this occasion (a survey talk on the classification of abelian groups presented to the Udine Conference) as a kind of golden anniversary of that event, which still ranks as one of the great achievements in the history of abelian groups. We wish to honor here especially the work of Prüfer, Ulm, Zippin, Baer and other pioneer researchers who were able to determine completely the structure of important classes of abelian groups [1, 23, 27, 30]. It is proper and fitting that we review at this time what has been accomplished in classifying abelian groups over the last fifty years and that we reflect on what we have done to carry on and complete the work they started. Moreover, it may be beneficial for us to examine methods and techniques that have developed over this period and to analyse those in current use. Finally, we consider a few open problems and discuss briefly directions for future research. In presenting the above program we will place special emphasis on torsion (primary) groups, but we will not neglect entirely the corresponding results on the classification of mixed groups, which in recent years have appeared to overshadow somewhat the torsion case. Our development will begin formally with countable p-groups and principally culminate with a new class of groups that I discovered recently called A-groups.

The primary A-groups form an additive class that encompasses both totally projective groups and S-groups. But we will find that they, too, can be classified with numerical invariants. Let me emphasize at the outset that this is a survey, and virtually all of the results mentioned have either appeared or will appear elsewhere. Therefore proofs are generally omitted. Naturally, all groups considered are abelian.

To initiate our survey, we begin with a time line that attempts to mark when the most important classes of groups have been classified. Events marked above the line designate torsion (primary) groups, while those below correspond to torsion-free or mixed groups. Dates may not be precise. For example, the cornerstone [27] of the classification of countable p-groups reads 1933, but the classification was not complete until the existence theorem was provided two years later [30]. In a very broad sense the classification of countable p-groups actually took thirty years to complete. Prüfer [23] laid the foundation in 1921, and the finish-work was performed by Kaplansky and Mackey in 1951 [20].

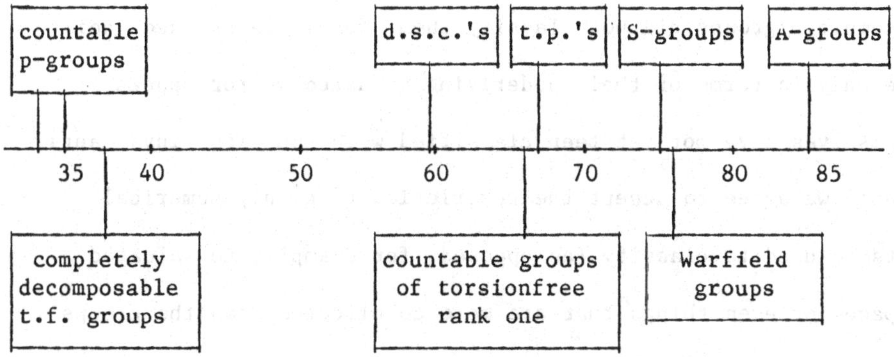


There have been several other classes, some of which we will discuss, not listed on the chart, but those listed constitute in my opinion (unless there is an inadvertent omission) the major classes of groups that have been classified with numerical invariants. However, one is invited to add, if he wishes, additional classes. We mention also that some very recent work (some of which is not yet in final form) may prove to be as significant as any of the results listed.

It was almost two decades ago that I first began work on the classification of groups. The first groups that I was able to classify were coproducts of torsion-complete groups; in those days I called them direct sums of closed groups [5]. Well, at least in 1965 I thought that I had classified coproducts of torsion-compelete groups, but now I would probably say that I did not. It all depends on what is meant by a classification (or as some would say a complete classification) of abelian groups. There are those who tend to favor the narrow definition of classification that requires numerical invariants. If this clause is accepted in the definition, then my result on coproducts of torsion-complete groups did not classify them; for it determined their structure only in terms of their underlying valuated vector spaces, which themselves have not yet been classified with numerical invariants.

Even if we agree to accept the restriction that only numerical invariants be used to classify (as opposed, for example, to valuated vector spaces or even things that are more complicated than the groups themselves), questions remain concerning the precise meaning of classification in the context of abelian groups. Let me hasten to add that the question whether or not a theorem satisfies some formal classification criteria may not be relevant to its real worth. Nevertheless, it would serve a useful purpose to have a universally accepted definition of what the classification of abelian groups should mean. At the very least such a definition would provide a guide and a goal to work toward. We now offer a provisional definition. First, a few preliminaries are needed. By an additive class of groups we mean a class closed with respect to arbitrary direct sums. Of course, it is to be understood that a class is well defined in the sense that inclusion is independent of notation. Denote the class of ordinals (with infinity adjoined) by $\mathcal{O}$ and the class of cardinals by $\mathcal{C}$. Let $\mathcal{O}_\lambda$ be the set of ordinals less than $\lambda$ union $\infty$. If $\lambda < \mu \in \mathcal{O}$ and $m_\alpha \in \mathcal{C}$ for each $\alpha$ between $\lambda$ and $\mu$, we denote $\sum_{\lambda \leq \alpha < \mu} m_\alpha$ by $\int_\lambda^\mu m_\alpha$. Likewise, $\int_\lambda^{\mu +} m_\alpha$ denotes the summation subject to $\lambda \leq \alpha \leq \mu$.

Definition. An additive class $C$ of groups is classified by $\Phi$ if the following conditions are satisifed.

(1) There is a class $\mathcal{F}$ of cardinal-valued functions $f(\alpha_1,\alpha_2,\ldots,\alpha_n,\ldots)$ with a finite or infinite sequence of ordinal variables $\alpha_i$. In other words, if $f \in \mathcal{F}$, then $f\colon O^n \to C$ where n is an ordinal; for our present purpose, we can take $n \leq \omega$. Moreover, we require $\operatorname{supp}(f) \subseteq O^n_{\lambda(f)}$ for some ordinal $\lambda(f)$.

(2) The inclusion of a function $f\colon O^n \to C$ in $\mathcal{F}$ can be determined by one or more inequalities involving sums of the values of f indexed by a certain range of its variables; we call these integral inequalities. The range of each variable in the summations will be a point or an interval. For example, if $n = 1$ and $f = f(\alpha)$ is a function of a single variable, the inequality $\int_\lambda^\infty f(\alpha) \leq \int_\lambda^{\lambda+\omega} f(\alpha)$, for all $\lambda$, is typical.

(3) There is an additive bijection $\Phi$ from $\mathcal{C}$ to $\mathcal{F}$. In particular, if $f_G$ denotes $\Phi(G)$ when $G \in \mathcal{C}$, then $f_G = \sum f_{G_i}$ whenever $G = \bigoplus G_i$.

The preceding definition of classification is compatible with the major existing classifications, and perhaps more importantly it seems appropriate for future ones. Certainly the Ulm-Zippin-Kaplansky-Mackey classification of countable p-groups satisfies the above definition and has served as a model for it, except that countable groups are not closed with respect to arbitrary direct sums. As is well known, Kolettis [21] classified direct sums of countable p-groups (d.s.c.'s) in

1960. In 1966, I gave a very short (two-page) proof of the uniqueness part of the classification of d.s.c.'s [6]. Aside from its brevity this paper had another noteworthy feature. The proof involved a technique that I learned from Kaplansky [19] that would have a significant impact on the theory of abelian groups. In the late sixties Megibben and I used it as if it were going out of style and called it the back-and-forth procedure. Needless to say, it did not go out of style but became fashionable. Richman and Walker [24] went one step further. They observed that the extension of Ulm's theorem from countable p-groups to d.s.c.'s involved no group theory at all; in other words, an application of the back-and-forth procedure was all that was ever required.

Now we move on to totally projective groups, but to do so is to leave out a chapter, actually one of the most action-filled chapters of the whole classification story - from d.s.c.'s to totally projectives. There are at least three papers, written in a short period of time that classify different classes of groups somewhere between d.s.c.'s and totally projectives. Each of these classes is determined by a bound on the length of the groups belonging to the class. In order of increasing generality the papers are by Crawley [2], Hill and Megibben [13], and Parker and Walker [22]. For a different approach to the classification of totally projective groups, see Crawley and Hales [3].

It has now been clearly revealed that totally projective groups, whatever characterization is used, were destined to become an important part of the theory of groups. One manifestation of this is the trinity for totally projectives, three distinctly different but yet equivalent and natural characterizations: the totally projective, simply presented, and Axiom 3 characterizations. Even though Axiom 3 (in more detail, the third axiom of countability) is the most useful for many purposes - in particular, for the purpose of classification - it appears not to have been, to this point, for the abelian group public the most appealing of the three concepts. Certainly its name is well known but has not been preferred to the more popular terms "totally projective" and "simply presented". Nevertheless, it appears now that Axiom 3 will not only endure but may receive more recognition and gain favor in the future. One reason for this is that the original axiom lends itself readily to modifications and generalizations. For example, L. Fuchs and I have recently used a version of Axiom 3 to characterize the p-groups having a given balanced-projective dimension [4]. Recall, too, that the third axiom concept was the means by which totally projectives were classified [7]. In this classification, as with d.s.c.'s, $\Phi(G)$ is the Ulm-Kaplansky function and

$$F = \{ f: O \to C \mid \int_{\lambda}^{\lambda+\omega} f(\alpha) \geq \int_{\lambda}^{\infty} f(\alpha) \} .$$

Since totally projectives are closed with respect to summands, it is immediate that the class of totally projective groups is maximal with respect to $\Phi$. Hence a classification of any additive class larger than totally projectives must be different from the mere assignment of the Ulm-Kaplansky function of a group to the group. This one-dimensional barrier was broken and we entered a new dimension in the classification of p-groups with Warfield's S-groups in 1975 [28].

Warfield regarded S-groups as torsion subgroups of local balanced-projectives; he first called them KT-modules [29]. But one can flip the coin and just as naturally construct KT-modules from S-groups. Indeed, we can (and do) take for the definition of a $\mu$-elementary S-groups, where $\mu$ is a limit ordinal not cofinal with $\omega = \omega_0$, the following. The p-group $H$ is a $\mu$-elementary S group if it appears as an isotype and $p^{\mu}$-dense subgroup of a totally projective group $G$ of length $\mu$ so that $G/H = Z(p^{\infty})$. An S-group is the direct sum of a totally projective group and $\mu$-elementary S-groups for various limit ordinals $\mu$ not cofinal with $\omega$. For another description of S-groups, see [18]. If $H$ is a $\mu$-elementary S-group and $G$ is the containing totally projective, the pullback $W$ in the diagram

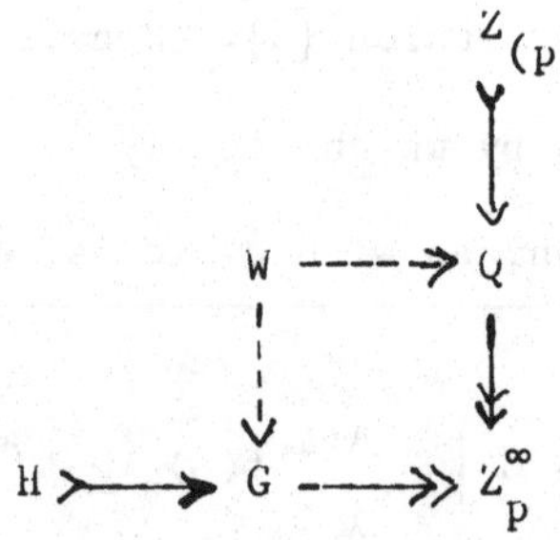


is in fact, a $\mu$-elementary KT-module. Thus with S-groups and KT-modules it is like the chicken-and-egg argument as to which comes first.

We next observe that S-groups are swallowed by a larger species of groups that we call A-groups.

If $\mu$ is a limit ordinal and $G$ is a p-group of length $\mu$, a subgroup $H$ of $G$ is said to be *almost balanced* if $H$ is isotype in $G$ and $p^{\lambda}(G/H) = \langle p^{\lambda}G,H\rangle/H$ when $\lambda < \mu$.

Definition. Let $\mu$ be a limit ordinal not cofinal with $\omega$. A p-group $H$ is called a $\mu$-elementary A-group if $H$ appears as an almost balanced subgroup of a totally projective group $G$ of length $\mu$ so that $G/H$ is totally projective (not necessarily reduced). A p-group $H$ is an A-group if it is the direct sum of $\mu$-elementary A-groups for various limit ordinals $\mu$ not cofinal with $\omega$.

We will now show how to classify A-groups (but proofs are not included).

If $H$ is a $\mu$-elementary A-group and $G$ is a containing totally projective group (that demonstrates that $H$ is a $\mu$-elementary A-group), we call $(H,G)$ an $A_{\mu}$-pair. Now, let $H$ be an arbitrary A-group. We can write $H = \oplus H_i$ where $H_i$ is $\mu(i)$-elementary and $(H_i,G_i)$ is a $\mu(i)$-pair for distinct limit ordinals $\mu(i)$ not cofinal with $\omega$. Set $G = \oplus G_i$ and

$$E_{\mu} = \bigcap_{\lambda<\mu} \langle p^{\lambda}G,H\rangle/\langle p^{\mu}G,H\rangle .$$

Although it appears that $E_{\mu}$ depends on $G$ and consequently on the choice of $G_i$, it in fact does not.

Theorem [12]. $E_{\mu} = (H/p^{\mu}H)^{c}/H/p^{\mu}H$, where $(H/p^{\mu}H)^{c}$ is the completion of $H/p^{\mu}H$ in the $p^{\mu}$-topology. Moreover, $E_{\mu} = p^{\mu}(G_i/H_i)$ if $\mu = \mu(i)$, and otherwise $E_{\mu} = 0$.

Associate with $H$ a cardinal-valued function $f_H$ of two ordinal variables as follows. If $\mu \neq 0$ and $\mathrm{cof}(\mu) \leq \omega$, then $f_H(\mu,\alpha) = 0$ for all $\alpha$. If $\mu = 0$ or $\mathrm{cof}(\mu) > \omega$, define

$$f_H(\mu,\alpha) = \begin{cases} \dim(p^{\alpha}H[p]/p^{\alpha+1}H[p]) & \text{if } \mu = 0 \text{ and } \alpha \neq \infty. \\ \dim(p^{\infty}H[p]) & \text{if } \mu = 0 \text{ and } \alpha = \infty. \\ \dim(p^{\alpha}E_{\mu}[p]/p^{\alpha+1}E_{\mu}[p]) & \text{if } \mu \neq 0 \text{ and } \alpha \neq \infty. \\ \dim(p^{\infty}E_{\mu}[p]) & \text{if } \mu \neq 0 \text{ and } \alpha = \infty. \end{cases}$$

The major result on A-groups is the following uniqueness theorem.

Theorem [12]. If $H$ and $H'$ are A-groups and $f_H = f_{H'}$, then $H \cong H'$.

The proof of the uniqueness theorem evolved over a series of papers including [8], [9], [10], and [14]. The method of proof involves embedding both $H$ and $H'$ in a single totally projective group $G$ and then constructing an automorphism of $G$ that maps $H$ onto $H'$ (without looking directly at $H$ and $H'$ but rather at $G/H$ and $G/H'$). This method of proof was perfected in [15].

To complete the classification of A-groups an existence theorem needs to be formulated. Consider the following integral inequalities imposed on a function $f\colon O^2 \to C$.

(1) $\int_0^{\infty+} f(\mu,\alpha) \leq 0$ if $\mathrm{cof}(\mu) \leq \omega$ and $\mu \neq 0$.

(2) $\int_{\lambda}^{\infty} f(\mu,\alpha) \leq \int_{\lambda}^{\lambda+\omega} f(\mu,\alpha)$; $\mu$ fixed, $\mu \geq 0$.

(3) $\int_{\mu>\lambda} \int_0^{\infty+} f(\mu,\alpha) \leq \int_{\lambda}^{\lambda+\omega} f(0,\alpha)$.

Let $F = \{f: O^2 \to C \mid f$ satisfies (1), (2) and (3)$\}$. Finally, let $C$ be the class of A-groups and define $\Phi$ from $C$ to $F$ by $\Phi(H) = f_H$.

Theorem [12]. $\Phi$ is a classification of A-groups, that is, $\Phi$ is an additive bijection from $C$ to $F$.

An illustration, provided by S. Brown, of the function $f_H(\mu,\alpha)$ for a typical A-group $H$ is given in figure 1. If $H$ is an A-group, then $H$ is totally projective if and only if $f_H(\mu,\alpha) = 0$ when $\mu \neq 0$. In other words when $f_H(\mu,\alpha)$ has no projection in the large cofinality dimension $L$. Likewise, $H$ is an S-group when the projection of $f_H(\mu,\alpha)$ in the large cofinality dimension is restricted to the line at infinity.

Although A-groups are classified, an important open question remains.

Problem 1. Are A-groups closed with respect to direct summands?

The answer to the preceding problem for S-groups is in the affirmative [25], [18].

Another natural question is the following.

Problem 2. Are there subclasses of A-groups other than S-groups that serve as the torsion for interesting and natural classes of mixed groups?

We remark that the classification theories of primary groups and local mixed groups seem to be beginning to merge into a more unified theory. In fact we can generalize the definition of A-groups to obtain (not necessarily torsion) A-modules over a discrete valuation ring by

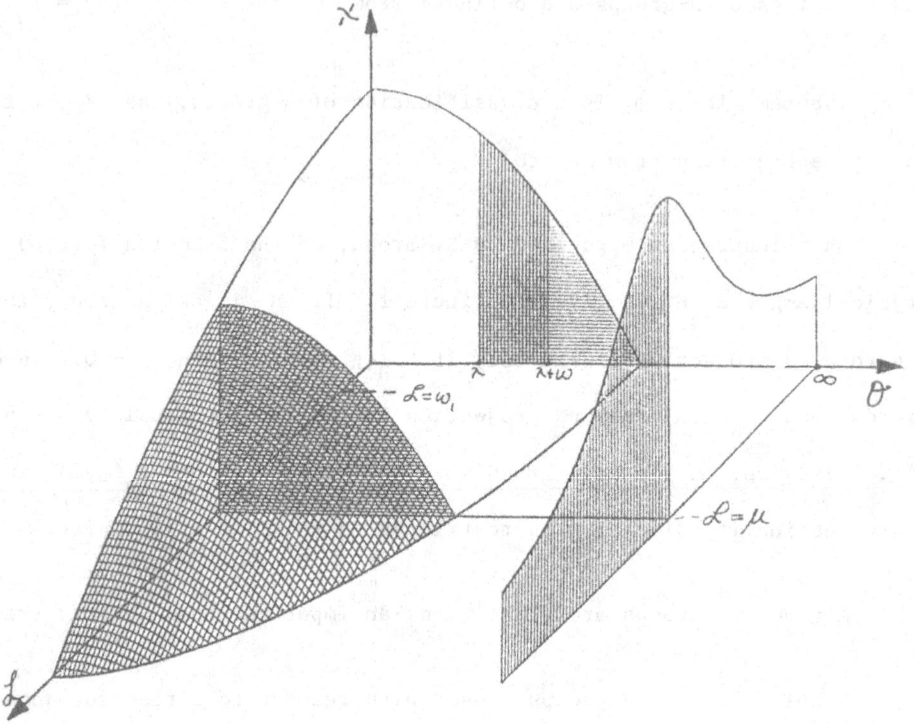


$f_H(\mu, \alpha)$

A typical function $f_H(\mu,\alpha)$ associated with an A-group H.

The "volume" $\int_{\mu>\lambda} \int_0^{\infty+} f_H(\mu,\alpha)d\alpha\, d\mu$ does not exceed the "area" $\int_\lambda^{\lambda+\omega} f(0,\alpha)d\alpha$. Moreover, $\int_\lambda^{\lambda+\omega} f(\mu,\alpha)d\alpha \geq \int_\lambda^{\infty} f(\mu,\alpha)d\alpha$ for each $\mu$.

Fig. 1

letting the containing totally projective group $G$ be replaced by a local Warfield module. In this case, it is desirable to replace $p^{\mu}G = 0$ by the weaker condition that $p^{\mu}G$ is torsion. Thus, if $cof(\mu) > \omega$, a $\mu$-elementary A-module (over a discrete valuation ring with prime p) is a module $H$ that appears as a submodule of a Warfield module $G$ so that the following conditions are satisfied.

(a) $p^{\mu}G$ is torsion.

(b) $H$ is isotype in $G$.

(c) $p^{\lambda}(G/H) = \langle p^{\lambda}G,H\rangle/H$ if $\lambda < \mu$.

(d) $G/H$ is a torsion, totally projective module (not necessarily reduced).

An A-module of course is a coproduct of $\mu$-elementary A-modules for various $\mu$ not cofinal with $\omega$. Observe that A-modules are more general than Warfield modules. We claim (see [11]) that an A-module $H$ is uniquely determined by the function $f_H(\mu,\alpha)$ and the Warfield-Stanton function [26]

$$g_H(\overline{\alpha}) = \sup_{n<\omega} \{\dim(H(p^n\overline{\alpha})/H^*(p^n\overline{\alpha}))\} .$$

The two functions can easily be joined together to produce a single function $\hat{f}_H\colon O^{\omega} \to C$ that remains additive in the group variable $H$ and alone determines $H$. However, it may not be desirable to go through the mechanics of this step, for it may be more natural to work with the ordered pair $(f_H,g_H)$. To complete the classification of A-modules (in the technical sense of our definition), we need a description of the class of functions $\hat{f}_H$ (or of the pairs $(f_H,g_H)$) in terms of integral inequalities, where $H$ ranges over the class of A-modules. We suspect that the Hunter-Richman-Walker [16] existence theorem for local Warfield modules can be extended and translated to obtain this result.

Problem 3. Formulate and prove the existence theorem for A-modules.

Finally, we remark that a class of modules apparently larger than the class of A-modules is the class of modules M over a discrete valuation ring that have a representing sequence

$$F \rightarrowtail M \twoheadrightarrow H,$$

where F is a valuated coproduct of infinite cyclics, is nice in M, and H is a torsion A-module. This is the analogue of the Hunter-Richman-Walker representation of Warfield modules. It would be desirable to classify these modules, which would extend in a natural fashion the classification of local Warfield modules from the view of [17]. C. Megibben and I are currently working in this direction although our approach is somewhat different from [17].

To summarize, much progress has been made on classifying different kinds of groups, and this gives us a wealth of knowledge about the groups. However, much work is left to be done, but the prospects look bright that more results will be obtained shortly with current and improved techniques.

BIBLIOGRAPHY

[1] R. Baer, Abelian groups without elements of finite order. Duke Math Journal 3 (1937), 68-122.

[2] P. Crawley, Abelian p-groups determined by their Ulm sequences, Pacific J. Math. 22 (1967), 235-239.

[3] P. Crawley and A. Hales, The structure of abelian p-groups given by certain presentations, J. Algebra 12 (1969) 10-23.

[4] L. Fuchs and P. Hill, Balanced-projective dimension of abelian p-groups, to appear.

[5] P. Hill, A classification of direct sums of closed groups, Acta Math. Acad. Sci. Hungar. 17 (1966), 263-266.

[6] P. Hill, Sums of countable primary groups, Proc. Amer. Math. Soc. 17(1966), 1469-1470.

[7] P. Hill, On the classification of abelian groups, photocopied manuscript, (1967).

[8] P. Hill, On the equivalence of high subgroups, Proc. Amer. Math. Soc. 88 (1983), 207-211.

[9] P. Hill, Balanced subgroups of totally projectives, J. Algebra, to appear.

[10] P. Hill, On the classification of N-groups, Houston J. Math., to appear.

[11] P. Hill, Toward the classification of mixed groups, Arch. Math., to appear.

[12] P. Hill, On the structure of abelian p-groups, to appear.

[13] P. Hill and C. Megibben, Extending automorphisms and lifting decompositions in abelian groups, Math. Ann. (1968), 159-168.

[14] P. Hill and C. Megibben, On the congruence of subgroups of totally projective groups, LNM 1006, 513-518, Berlin-Heidelberg-New York, 1983.

[15] P. Hill and C. Megibben, On the theory and classification of abelian groups, to appear.

[16] R. Hunter, F. Richman, E. Walker, Existence theorems for Warfield groups, Trans. Amer. Math. Soc. 235 (1978),345-362.

[17] R. Hunter, F. Richman, E. Walker, Warfield modules, LNM 616, 87-123, Berlin-Heidelberg - New York 1977.

[18] R. Hunter and E. Walker, S-groups revisited, Proc. Amer. Math. Soc. 82 (1981), 13-18.

[19] I. Kaplansky, Projective modules, Ann. Math. 68 (1958), 372-377.

[20] I. Kaplansky and G. Mackey, A generalization of Ulm's theorem. Summa Brasil Math. 2 (1951), 195-202.

[21] G. Kolettis, Direct sums of countable groups, Duke Math. J. 27 (1960), 111-125.

[22] L. Parker and E. Walker, An extension of the Ulm-Kolettis theorem, in Studies on Abelian groups, 309-325, Paris, 1968.

[23] H. Prüfer, Unendliche abelsche Gruppen von Elementen endlicher Ordnung, Dissertation, Berlin, 1921.

[24] F. Richman and E. Walker, Extending Ulm's theorem without group theory. Proc. Amer. Math. Soc. 21 (1969), 194-196.

[25] R. Stanton, S-groups, to appear.

[26] R. Stanton, An invariant for modules over a discrete valuation ring, Proc. Amer. Math. Soc. 47 (1975), 51-54.

[27] H. Ulm, Zur Theorie der abzählbar-unendlichen abelschen Gruppen, Math. Ann. 107 (1933), 774-803.

[28] R. Warfield, A classification theory for abelian p-groups, Trans. Amer. Math. Soc. 210 (1975), 149-168.

[29] R. Warfield, Classification theory of Abelian groups, I: Balance projectives, Trans. Amer. Math. Soc. 222 (1976), 33-63.

[30] L. Zippin, Countable torsion groups, Ann. Math. 36 (1935) 86-99.

# SUBGROUPS OF BOUNDED ABELIAN GROUPS

by

Roger Hunter, Fred Richman, and Elbert Walker[1]

New Mexico State University, Las Cruces N.M.

**0. Introduction.** Valuated groups are a topic of central interest in Abelian group theory. On one hand, they provide a viewpoint for classical Abelian theory problems, and on the other hand are of interest in their own right. In this latter regard, there has been some progress in getting structure theorems for certain valuated groups. Complete sets of invariants have been provided for finite direct sums of cyclic valuated p-groups [HRW1], for finite simply presented valuated p-groups [AHW], and for direct sums of torsion-free cyclic valuated groups [AHW]. A general discussion of simply presented valuated p-groups, with an aim toward a structure theory, is presented in [HW]. In [BHW], a basis for a general study of finite valuated p-groups is suggested. However, structure theories for simply presented valuated p-groups and for finite valuated p-groups are only in the initial stages.

This paper is concerned with the structure of valuated p-groups with a finite bound on the values that may be assumed by non-zero

---

[1] These authors were supported by NSF grant #MCS 8301606.

elements. Each bounded abelian group decomposes into a direct sum of cyclic groups of prime power order, the number of cyclic groups of each order in the decomposition being an invariant of the group. Thus the structure theory of bounded abelian groups may be considered complete. The picture is far murkier when we attempt to classify subgroups of bounded groups, that is, pairs $A \subset B$ where $B$ is bounded and $A$ is a bounded subgroup of $B$. With no loss of generality we restrict ourselves to groups that are p-primary for a fixed prime $p$. If $A$ is a valuated p-group and $\alpha$ is an ordinal, let $A(\alpha) = \langle a \in A : v(a) \geq \alpha \rangle$. The valuated group $A$ is of length $\alpha$, denoted $\lambda(A) = \alpha$, if $\alpha$ is the smallest ordinal such that $A(\alpha) = 0$. The study of pairs $A \subset B$ with $p^n B = 0$ is equivalent to the study of valuated groups $A$ such that $A(n) = 0$, that is, those $A$ with $\lambda(A) \leq n$. This equivalence is made precise in Theorem 1.2.

The valuated groups of length $\leq 1$ consist of p-bounded groups, and these are direct sums of cyclic groups. The structure of valuated groups of length $\leq 2$ is not obvious, although finite ones are direct sums of cyclics [HRW1, Theorem 4]. It turns out that the valuated groups of length $\leq 3$ are direct sums of cyclics, and one of our main theorems (Theorem 4.2) classifies those valuated groups of length $\leq 4$. To this end we construct, for each finite cyclic valuated group $C$, a functor $F_C$ such that if $A$ is a direct sum of copies of $C$, and $A$ is a subgroup of a valuated group $B$, then $A$ is a summand of $B$ if and only if $A \cap F_C(B) = 0$ (Theorem 2.5).

The functors $F_C$ also enable us to show that any $p^2$-bounded valuated group of finite length is a direct sum of cyclics (Theorem 3.2);

this generalizes [HRW1; Theorem 4]. As limiting examples to this theorem we construct a valuated group $A$ with $p^2A = A(\omega) = 0$ that is not a direct sum of cyclics (Example 3.3), and a valuated group $B$ with $p^2B = B(\omega + \omega) = 0$ that has no nonzero cyclic summand (Example 3.4). The last example is best possible in that any valuated group $G$ such that $p^2G = G(\omega + n) = 0$ has a nonzero cyclic summand (Lemma 3.1). We also give an easy construction of an infinite indecomposable valuated group $G$ such that $G(\omega) = 0$ (Example 6.2).

Berman and Zilinskaja [BZ; Theorem 6] announced the construction of arbitrarily large finite indecomposable valuated groups that can be imbedded in a direct sum of cyclic groups of orders $p^4$ and $p^8$. We show how to construct arbitrarily large finite indecomposable, as well as infinite, $A$ such that $p^3A = A(7) = 0$ (Example 6.1). Our construction is based on a categorical equivalence (Corollary 5.3) between vector spaces with distinguished subspaces over a field $F$, and modules over $F[x,1/x,1/(1-x)]$. The categorical equivalence is reminiscent of, and indeed may follow from, [BUT; 3.1].

**1. A Category of Pairs.** In order to study subgroups, we consider the category $P$ whose *objects* are pairs $A \subset B$ of bounded p-groups. A *map* from a pair $A \subset B$ to a pair $A' \subset B'$ is a homomorphism $f : A \to A'$ that can be extended to a homomorphism from $B$ to $B'$. Isomorphism in $P$ captures the notion of isomorphism of subgroups in the following sense.

**THEOREM 1.1.** Two pairs $A \subset B$ and $A' \subset B'$ are isomorphic in $P$ if and only if there are summands $S$ and $S'$ of $B$ and $B'$ containing

$A$ and $A'$, and an isomorphism from $S$ to $S'$ that takes $A$ onto $A'$.

**PROOF.** The "if" part is clear. Let $f : B \to B'$ and $g : B' \to B$ be homomorphisms so that $fg$ is the identity on $A'$ and $gf$ is the identity on $A$. Then $f$ induces the required isomorphism between the summands

$$S = \langle x \in B : gfx = x \rangle \text{ and } S' = \langle x \in B' : fgx = x \rangle. \ \square$$

The study of the category $P$ is the same as the study of the category of value-bounded valuated $p$-groups. We call a valuated group $A$ *value bounded* if $\lambda(A)$ is finite.

**THEOREM 1.2.** The functor $F( A \subset B ) = A$ is an equivalence between the category $P$ and the category of value-bounded valuated $p$-groups.

**PROOF.** As $A$ inherits its values from the height function on $B$, it is clear that $F$ is a functor. We know from [RW; Theorem 1] that every value-bounded $p$-group is isomorphic to $F$ applied to some pair in $P$. What remains to show is that if $A \subset B$ and $A' \subset B'$ are in $P$, then every homomorphism of valuated groups $A \to A'$ extends to a homomorphism from $B$ to $B'$. But this is true because $B'$ is an algebraically compact group, hence injective in the category of valuated groups [RW; Theorem 29]. $\square$

The fact that algebraically compact groups are injective in the category of valuated groups also plays a role in constructing complementary summands of pure subgroups of bounded groups.

**THEOREM 1.3.** Let $A$ be a subgroup of a bounded group $B$. If $S$ is a subgroup of $B$, then $A$ is a summand of $B$ and $S$ is contained in a complementary summand of $A$ if and only if $ht_B(a + s) \leq ht_A(a)$ for each $a$ in $A[p]$ and $s$ in $S$.

**PROOF.** The "only if" part is clear. To prove the "if" part we first show that the inequality on heights holds for arbitrary $a$ in $A$. Suppose that $a$ is not in $A[p]$ and that $ht_B(a + s) > ht_A(a)$. We may assume that $A$ is isomorphic to a direct sum of copies of a fixed cyclic group. Then $ht_B(pa + ps) > ht_B(a + s) > ht_A(a) = ht_A(pa) - 1$, so $ht_B(pa + ps) > ht(pa)$, an impossibility by induction on the order of $a$.

The inequality on heights now says that the projection map from $A + S$ onto $A$ is a map of valuated groups. As $A$ is algebraically compact, the map can be extended to $B$. The kernel of the extended map is a complementary summand of $A$ that contains $S$.

2. **A Generalization of a Theorem of Szele.** Let $C$ be a cyclic group of order $p^m$, and $A$ a direct sum of copies of $C$. A theorem of Szele [FCHS; Prop. 27.1] states that if $A$ is a subgroup of a group $B$, and $A \cap p^mB = 0$, then $A$ is a summand of $B$. In this section we generalize Szele's theorem by allowing $B$ and $C$ to be valuated groups.

Let $m$ be a nonnegative integer, $\alpha$ an ordinal, and $B$ a p-valuated group. We will be concerned with the functors that take $B$ to $p^mB(\alpha)$. If $B$ is a group, then $p^mB(\alpha) = B(\alpha + m) = p^{\alpha+m}B$, so there is no reason to consider the combination $p^mB(\alpha)$. For valuated groups it is essential.

**LEMMA 2.1.** The following are equivalent:

i) $p^mB(\alpha) \subset p^nB(\beta)$ for each p-valuated group $B$

ii) $m \geq n$ and $\alpha + m \geq \beta + n$.

Moreover for each $m$ and $\alpha$ there is a valuated p-group $B$ such that

$p^mB(\alpha) \neq 0$ and, if $p^nB(\beta) \neq 0$, then ii) holds.

**PROOF.** If ii) holds, then $p^mB(\alpha) = p^np^{m-n}B(\alpha) \subset p^nB(\alpha+m-n) \subset p^nB(\beta)$ so i) holds. Now let B be cyclic of order $p^{m+1}$ with generator x such that $vp^ix = \alpha + i$; in particular $p^mB(\alpha) \neq 0$. If $p^nB(\beta) \neq 0$, then $p^mx \in p^nB(\beta) \subset B(\beta + n)$, so ii) holds. Thus i) implies ii), and the valuated p-group B satisfies the last statement of the theorem. □

Let I be a collection of pairs $(m,\alpha)$ where m is a nonnegative integer and $\alpha$ is an ordinal. Let $B_I = \Sigma_{(m,\alpha)\in I}\, p^mB(\alpha)$, and call I and J *equivalent* if $B_I = B_J$ for all p-valuated groups B.

**THEOREM 2.2.** Each I contains a finite set J such that

i) J is equivalent to I

ii) if $(m,\alpha)$ and $(n,\beta)$ are distinct elements of J, then $m \neq n$ and, if $m > n$, then $\alpha + m < \beta + n$.

Moreover, there is only one set J satisfying these two conditions.

**PROOF.** That such a set J satisfying i) and ii) exists is clear from Lemma 2.1 by eliminating redundancies. That it is finite follows from the fact that the ordinals decrease as the integers increase. That it is unique follows from the last statement of Lemma 2.1. □

**THEOREM 2.3.** The functor taking B to $B_I$ is an additive subfunctor of the identity. If $B \to C$ is a nice cokernel, then $B_I$ maps onto $C_I$.

**PROOF.** The first statement is obvious. If $B \to C$ is a nice cokernel, then $B(\alpha)$ maps onto $C(\alpha)$, so $p^nB(\alpha)$ maps onto $p^nC(\alpha)$, and thus $\Sigma_I\, p^nB(\alpha)$ maps onto $\Sigma_I\, p^nC(\alpha)$. □.

We say that $A \subset B$ is *I-pure* if $A_I = A \cap B_I$. (* Note: $A/A(\alpha)$ is pure in $B/B(\alpha)$ precisely when $A \subset B$ is I-pure for all $I$ of the form $\langle (n,0), (0,\alpha) \rangle$. *)

**COROLLARY 2.4.** The I -pure nice exact sequences of valuated groups form a proper class.

**PROOF.** Let $0 \to A \to B \to C \to 0$ be an I-pure, nice, short exact sequence. We must show that its pullbacks and pushouts are also I-pure. The pullback argument is an easy diagram chase, so we prove only the pushout case. Consider the pushout diagram

$$\begin{array}{ccccc} A & \longrightarrow & B & \longrightarrow & C \\ \downarrow & & \downarrow & & \| \\ A' & \longrightarrow & B' & \longrightarrow & C \end{array}$$

The map $f$ from $A' \oplus B$ onto $B'$ is nice [RW; Lemma 4], so $A'_I \oplus B_I$ maps onto $B'_I$ (Theorem 2.3). Suppose $a' \in A' \cap B'_I$. Then $a'$ is the image of $(a'_I, b_I)$ in $A'_I \oplus B'_I$. Since $(a'_I, b_I) - (a', 0) = (a'_I - a', b_I)$ is in the kernel of $f$, we have $b_I \in A$, and $-b_I$ maps to $a'_I - a'$. As $A$ is I-pure in $B$, we have $b_I \in A_I$. Thus $a'_I - a' \in A'_I$ and hence $a' \in A_I$. □

If $C$ is a cyclic valuated p-group, we set

$$I(C) = \langle (n,\alpha) : p^n C(\alpha) = 0 \rangle$$

and let $F_C(B) = B_{I(C)}$. If $C = \langle x \rangle$ is of order $p^m$, then $I(C)$ is equivalent to $\langle (m,0) \rangle \cup \langle (i, vp^{m-i-1}x + 1) : 0 \le i \le m \rangle$. If, in addition, $C$ is a group, then $F_C(B) = p^m B$.

**THEOREM 2.5.** Let $B$ be a valuated group, $C$ a cyclic valuated p-group, and $A$ a subgroup of $B$ which is a direct sum of copies of $C$. Then $A$ is a summand of $B$ if and only if $A \cap F_C(B) = 0$. Moreover if

$A + K = B$ and $A \cap K = 0$ and $K \supset F_C(B)$, then $K$ is a complementary summand of $A$.

**PROOF.** If $B = A \oplus K$, then $F_C(B) = F_C(A) \oplus F_C(K) = F_C(K) \subset K$, so $A \cap F_C(B) = 0$. Conversely, suppose $A \cap F_C(B) = 0$. If $a \in A[p]$ and $x \in F_C(B)$, then $ht_B(a + x) \leq ht_A a$, for if $ht_B(a + x) \geq n$, then either $p^n A(0) = 0$, whence $p^n C(0) = 0$ and $a + x \in p^n B(0) \subset F_C(B)$ so $a = 0$, or $p^n A(0) \neq 0$, and since $a \in A[p]$, $ht_A a \geq n$. By Theorem 1.3 we can write $B = A + K$ with $K \supset F_C(B)$ and $A \cap K = 0$. Suppose $a \in A$ and $x \in K$. For any $n$, if $p^n A(v(a + x)) = 0$, then $p^n C(v(a + x)) = 0$, so $p^n(a + x) \in p^n B(v(a + x)) \subset F_C(B) \subset K$, so $p^n a = 0$. But if $v(a + x) > v(a)$, then since $A$ is a direct sum of copies of $C$, there is an $n$ such that $p^n A(v(a + x)) = 0$ and $p^n(v(a)) \neq 0$, whence $p^n a \neq 0$. Hence $v(a + x) \leq va$. □

COROLLARY 2.6. Let $B$ be a valuated group and $C$ a finite cyclic valuated group. Then $B = A \oplus K$ where $A$ is a direct sum of copies of $C$, and $K$ has no summand isomorphic to $C$. Moreover if $A' \oplus K'$ is another such decomposition, then $B = A \oplus K'$.

**PROOF.** A standard Zorn's lemma argument gets the first part. To get the second, let $0 \neq x \in (A \cap K')[p]$. Then $x$ is in the socle of a copy of $C$ in $K'$. As $F_C(B) \subset K'$, this copy of $C$ is a summand, which is impossible. Thus $A \cap K' = 0$, whereupon $B = A \oplus K''$ where $K'' \supset K'$ as $K' \supset F_C(B)$. As $K''$ is isomorphic to $K$ which has no summand isomorphic to $C$, we have $K'' \cap A = 0$ so $K'' = K'$. □

**THEOREM 2.7.** Let $C$ be a cyclic valuated $p$-group. If $A$ is a $p$-valuated group, then $F_C(A) = 0$ if and only if there is a one-to-one map of $A$ into a direct sum of copies of $C$.

**PROOF.** Clearly if such a map exists, then $F_C(A) = 0$. Conversely, suppose $F_C(A) = 0$. By 2.6, any direct product of copies of $C$ is a direct sum of copies of $C$; thus it suffices to show that if $x \in A$ is nonzero, then there is a map $f$ from $A$ to $C$ such that $f(x) \neq 0$. As $C$ is $I(C)$-pure injective (Theorem 2.5), it suffices to get a nonzero map from $\langle x \rangle$ to $C$. If $vx = \alpha$, then $C(\alpha) \neq 0$ because $A(\alpha) \neq 0$. Thus we can map $x$ to any element in $C$ of order $p$. □

**COROLLARY 2.8.** Let $C$ be a cyclic valuated $p$-group, and $A$ a direct sum of $p$-valuated groups with finite value sets. Then there is a map $\varphi$ from $A$ to a direct sum of copies of $C$ such that $F_C(A) = \operatorname{Ker} \varphi$.

**PROOF.** As every subroup of a $p$-valuated group with a finite value set is $p$-nice, the subgroup $F_C(A)$ is nice in $A$. Theorem 2.3 says that $F_C(A/F_C(A)) = 0$. Theorem 2.7 says that $A/F_C(A)$ can be mapped one-to-one into a direct sum of copies of $C$. □

**3. $p^2$-Bounded Valuated Groups.** Every $p^2$-bounded finite valuated group is a direct sum of cyclics [HRW1; Theorem 4]. We extend this result to $p^2$-bounded valuated groups with finitely many values.

**LEMMA 3.1.** If $G$ is a nonzero valuated group such that $p^2 G = 0$ and $G(\omega + n) = 0$ for some $n < \omega$, then $G$ has a nonzero cyclic summand.

**PROOF.** If there is an element of finite value in $G[p]$, let $x$ in $G[p]$ have smallest value $m$. If $x \notin F_{\langle x \rangle}(G) = G(m+1) + pG$, then $\langle x \rangle$ is a nonzero cyclic summand. Otherwise there is $y$ in $G$ such that $vpy = m$. We may choose such $y$ of maximum value. Then $py \notin F_{\langle y \rangle}(G) =$

$G(m+1) + pG(vy + 1)$ so $\langle y \rangle$ is a summand.

If, on the other hand, $G[p] \subset G(\omega)$, choose nonzero $x$ in $G[p]$ of maximum value. If $x \notin F_{\langle x \rangle} = pG$, then $\langle x \rangle$ is a summand. Otherwise write $x = py$ where $vy < \omega$ or $vy$ is maximal among $\{vz : pz = x\}$. We must show that $x \notin F_{\langle y \rangle}(G) = pG(vy + 1)$. Suppose $x = pz$. Then $z - y \in G[p]$ so $v(z - y) \geq \omega$. If $vy < \omega$, then $vz = vy$ so $z \notin G(vy + 1)$. If $vy \geq \omega$, then $z \notin G(vy + 1)$ by the maximality of $y$. □

**THEOREM 3.2.** If $B$ is a $p^2$-bounded valuated group with finitely many values, then $B$ is a direct sum of cyclics.

**PROOF.** It suffices to prove the theorem for the case $B(n) = 0$. As there are only finitely many cyclics $C$ with $C(n) = 0$, we may assume that $B$ has no cyclic summands (Corollary 2.6). But then $B = 0$ by Lemma 3.1. □

**EXAMPLE 3.3.** A $p^2$-bounded countable valuated group $G$ of length $\omega$ that is not a direct sum of cyclics.

Let $x_1$ be of order $p$ and value $0$. For $n > 1$ let $x_n$ have order $p^2$ and value sequence $(n-1,n)$. Let $G$ be the subgroup of $\Sigma\langle x_n \rangle$ generated by $y_n = x_n + x_{n+1}$

· 0 · 1 · 2 · 3

· 2 · 3 · 4 . . .

The $n^{th}$ Ulm invariant of $G$ is 1 for $n \neq 1$, and $G[p] \subset G(2)$. We shall show that $G$ has no finite summand with a nonzero $0^{th}$ Ulm invariant. Suppose that $G = A \oplus B$, and that the $0^{th}$ Ulm invariant of $A$ is nonzero. Let $\pi$ be the projection on $A$. If $vx = 0$, then $x$ represents the $0^{th}$ Ulm invariant of $G$, so $v\pi x = 0$. We shall show that if $vx = 0$, then there is $y$ such that $vy = 0$ and $vp\pi y > vp\pi x = n$, so $A$ is infinite. Now $py_n$ represents the $n^{th}$ Ulm invariant of $G$, as does $p\pi x$. Thus there is a unit $u$ so that $v(p\pi x - upy) > n$. Let $y = x - uy_n$. Then $vy = 0$ and $vp\pi y = v(p\pi x - up\pi y_n) \geq v\pi(px - upy) > n$. □

Note that $\langle y_2, y_4, y_6, \cdots \rangle \oplus \langle y_2-y_1, y_4-y_3, y_6-y_5, \cdots \rangle$ is a decomposition of $G$ into infinite summands. That $G$ necessarily has nonzero cyclic summands, even though $G$ is not a direct sum of cyclics, follows from Lemma 3.1. If we allow length $\omega + \omega$, then we can get an example of a $p^2$-bounded valuated group with no nonzero cyclic summands

**EXAMPLE 3.4.** A nonzero valuated group $G$ such that $p^2G = G(\omega + \omega) = 0$, but $G$ has no nonzero cyclic summand.

Let $G_n$ be a direct sum of cyclics of order $p^2$ with basis $a_{n0}, \ldots, a_{nn}$ such that $va_{nj} = j$ and $vpa_{nj} = \omega + (n - j)$. We may view $G_n$ as a subgroup of $G_{n+1}$ under the identification

$$a_{nj} = a_{(n+1)j} - a_{(n+1)(j+1)}.$$

Let $G$ be the union of the groups $G_n$. Then

$$pG_n(j) \cap G_n(\alpha) \subset G_{n+1}(\alpha+1) + pG_{n+1}(j+1)$$

so $G$ has no nonzero cyclic summands. □

**4. The Smallest Tree.** Not every valuated $p$-group $B$ with $B(4) = 0$ is a direct sum of cyclics. In fact the tree

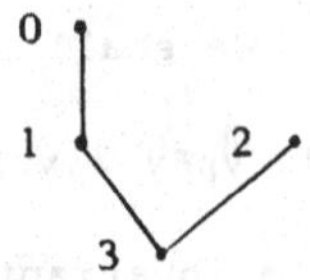


which we write linearly as $(0-1,2)-3$, gives a simply presented valuated $p$-group $T$ that is not a direct sum of cyclics. We shall show that $T$ is, up to isomorphism, the only indecomposable valuated $p$-group, with no values exceeding $3$, that is not cyclic.

To construct a functor that tells when $T$ is a summand, we need to go beyond $I$-purity. To see this let $x$ have order $p$ and value $2$, let $y$ have order $p^3$ and value sequence $(0,2,3)$, and let $z$ bave order $p^2$ and value sequence $(0,1)$. If $B = \langle x\rangle \oplus \langle y\rangle \oplus \langle z\rangle$ and $A$ is generated by $x + py$ and $y + z$, then $A \subset B$ is $I$-pure for all $I$, and $A$ is isomorphic to $T$. But $A$ is not a summand of $B$, as every summand of $B$ is a direct sum of cyclics.

Recall that $G(0,2) = \langle x \in G : vpx \geq 2 \rangle$.

**LEMMA 4.1.** Let $F(G) = G(4) + p^2G(0,2)$. Let $A$ be a subgroup of a valuated group $B$ such that

$$F(A) = 0$$

$$A(3) = pA(2) = A(2)[p]$$

$$A(1) = A(2) + pA = A(0,2)$$

Then $A$ is a summand of $B$ if and only if $A \cap F(B) = 0$.

**PROOF.** If $B = A \oplus K$, then $F(B) = F(A) \oplus F(K) = F(K)$ so $A \cap F(B) = 0$. So suppose $A \cap F(B) = 0$. As $p^2B(2) \subset B(4) \subset F(B)$, and $A(2)$ is a direct sum of cyclics of order $p^2$, we can find $H$

containing $F(B)$ such that

$$B(2)/F(B) = H/F(B) \oplus (A(2) + F(B))/F(B).$$

In particular, $B(2) = A(2) + H$ and $A \cap H = 0$.

In order to apply Theorem 1.3 we shall show that if $a \in A$ and $h \in H$, then $ht_B(a + h) \leq ht_A(a)$. If $va = ht_A(a) \leq 1$, then $ht_A(a + h) \leq v(a + h) = va = ht_A(a)$. If $va > ht_A(a) = 0$ and $ht_B(a + h) > 0$, then $a + h = pb$ for some $b$ in $B$. But $a = a_2 + pa'$ where $a_2 \in A(2)$ and $a' \in A$, so $a_2 + h = p(b - a') \in B(2)$, whence $p^2(b - a) \in p^2B(0,2) \subset H$. Thus $pa_2 = p^2(b - a') - ph \in H$, so $pa_2 = 0$ and therefore $va_2 \geq 3$, so $ht_A(a_2) > 0$, a contradiction. Thus we may assume that $ht_A(a) \geq 2$. If $ht_B(a + h) > 2$, then $a + h \in p^3B \subset H$ so $a \in A \cap H = 0$.

By Theorem 1.3 we can find $K \supset H$ so that $B = A + K$ and $A \cap K = 0$. We shall show that the sum $A \oplus K$ respects values. Suppose $a \in A$ and $k \in K$. If $v(a + k) \geq 4$, then $a \in K$ so $a = 0$. If $va \geq 2$ and $v(a + k) \geq 3$, then $pa \in K$ so $pa = 0$ and thus $va \geq 3$. If $v(a + k) \geq 2$, then $a \in A(2) + K$ so $a \in A(2)$. If $v(a + k) \geq 1$, then $pa \in A(2)$, so $a \in A(1)$. □

**THEOREM 4.2.** Let $B$ be a valuated p-group with $B(4) = 0$. Then $B$ is a direct sum of cyclic valuated groups and simply presented valuated groups of type (0-1,2)-3.

**PROOF.** We may assume that $B$ has no nonzero cyclic summands. Thus $p^3B = 0$ as any element of order $p^4$ would have value sequence (0,1,2,3) and so would be a summand. Similarly no element of $B$ can have value sequence (0,1,2). If $x$ in $B$ had value sequence (1,2,3), then $F_{\langle x\rangle}(B) = B(4) + p^3B = 0$, so $\langle x\rangle$ would be a summand of $B$. Hence no such $x$ is possible, whence $p^2B(1) = 0$. If $x$ in $B$ had value sequence $(0,2,3)$, then $F_{\langle x\rangle}(B) = B(4) + p^2B(1) = 0$, so $\langle x\rangle$ would be a summand of $B$. Hence no such $x$ is possible, whence $p^2B(0,2) = 0$. Therefore $F(B) = B(4) + p^2B(0,2) = 0$, so the lemma says that any direct sum of copies of the simply presented valuated group of type $(0\text{-}1,2)\text{-}3$ in $B$ is a summand of $B$. Thus we may assume that $B$ has no such subgroup.

If $x$ is an element of order $p^3$ in $B$, then the value sequence of $x$ must be $(0,1,3)$. If $p^2x \in pB(2)$, then $B$ would contain a valuated subgroup of type $(0\text{-}1,2)\text{-}3$. If $p^2x \notin pB(2)$, then we can find a complementary summand $K$ of $\langle x\rangle$ containing $pA(2)$. It is easily verified that the decomposition $B = \langle x\rangle \oplus K$ respects values. But $B$ was assumed to have no cyclic summands. Thus $p^2B = 0$, so $B$ is a direct sum of cyclics. □

**COROLLARY 4.3.** If $B$ is a valuated group with $B(3) = 0$, then $B$ is a direct sum of cyclic valuated groups.

## 5. A Categorical Equivalence.

In order to construct an infinite indecomposable $p^3$-bounded valuated group of length 7, we introduce the following category.

**DEFINITION.** Let $K$ be a field. The *objects* of the category CAT are vector spaces $V$ over $K$ together with four distinguished subspaces $V_1$, $V_2$, $V_3$ and $V_4$ which are pairwise complementary summands. A *morphism* from $V$ to $W$ in CAT is a linear transformation $f$ such that $f(V_i) \subset W_i$ for each $i$.

**LEMMA 5.1.** Let $V$ be an object of CAT. Let $i$, $j$ and $k$ be distinct elements of $\{1, 2, 3, 4\}$. Let $\alpha_{ij}$ be the projection onto $V_i$

with kernel $V_j$. Then

a) $\alpha_{ij} = 1 - \alpha_{ji}$

b) $\alpha_{ij}\alpha_{ik} = \alpha_{ik}$

c) $\alpha_{ji}\alpha_{ki} = \alpha_{ji}$

d) $\alpha_{14}\alpha_{23}$ and $1 - \alpha_{14}\alpha_{23}$ are automorphisms of $V_1$.

**PROOF.** Part (a) is clear. Part (b) holds because $\alpha_{ik}$ maps into $V_i$ and $\alpha_{ij}$ is the identity on $V_i$. To show (c) we note that $\alpha_{ji}\alpha_{ki}$ clearly kills $V_i$; we must show that it fixes $V_j$. Let $c = a + b$ be in $V_j$ with $a \in V_i$ and $b \in V_k$. Then $\alpha_{ki}c = b = c - a$, and $\alpha_{ji}b = c$. To establish (d) we claim the following equations hold:

i) $(\alpha_{14}\alpha_{23})(\alpha_{13}\alpha_{24}) = \alpha_{14}$

ii) $(\alpha_{13}\alpha_{24})(\alpha_{14}\alpha_{23}) = \alpha_{13}$

iii) $(\alpha_{12}\alpha_{34})(1 - \alpha_{14}\alpha_{23}) = \alpha_{12}$

iv) $(1 - \alpha_{14}\alpha_{23})(\alpha_{12}\alpha_{34}) = \alpha_{14}$

As each map on the right hand side of these equations is equal to the identity on $V_1$, this will suffice. The equations are more or less routine consequences of (a), (b) and (c). The least routine is (iv) which we prove: $(1 - \alpha_{14}\alpha_{23})(\alpha_{12}\alpha_{34}) = \alpha_{12}\alpha_{34} - \alpha_{14}\alpha_{23}\alpha_{12}\alpha_{34} = \alpha_{12}\alpha_{34} - \alpha_{14}(1 - \alpha_{32})\alpha_{12}\alpha_{34} = [\alpha_{12}\alpha_{34} - \alpha_{14}\alpha_{12}\alpha_{34}] + \alpha_{14}\alpha_{32}\alpha_{12}\alpha_{34}$. The term in square brackets is 0; the other term is equal to $\alpha_{14}\alpha_{32}\alpha_{34} = \alpha_{14}\alpha_{34} = \alpha_{14}$. □

**THEOREM 5.2.** Let $V$ and $W$ be in CAT and $\lambda : V \to W$ a map in CAT. Then $\lambda\alpha_{ij} = \alpha_{ij}\lambda$. Hence $\lambda$ induces a linear transformation $f: V_1 \to W_1$ such that $f\alpha_{14}\alpha_{23} = \alpha_{14}\alpha_{23}f$. Conversely, given any linear transformation $f: V_1 \to W_1$ such that $f\alpha_{14}\alpha_{23} = \alpha_{14}\alpha_{23}$, then there is a unique map $\lambda : V \to W$ in CAT inducing $f$.

**PROOF.** The first part of the theorem is clear. Suppose we are given f. If f is the restriction of a CAT-map $\lambda$, then for a in $V_1$ and b in $V_2$ we have

$$\lambda(a + b) = \lambda(a) + \lambda(b) = f(a) + \alpha_{24}\lambda(b)$$

$$\text{and} \quad \alpha_{24}\lambda(b) = \alpha_{24}\alpha_{14}\lambda(b) = \alpha_{24}\lambda\alpha_{14}b = \alpha_{24}f\alpha_{14}b$$

so $\lambda(a + b) = f(a) + \alpha_{24}f\alpha_{14}b$. Conversely, suppose we define $\lambda$ by this last formula. Clearly $\lambda$ is a linear transformation that takes $V_i$ to $W_i$ for $i = 0, 1, 2$. Suppose $a + b \in V_3$. Then $b = -\alpha_{23}a$ so $\alpha_{24}f\alpha_{14}b = -\alpha_{24}f\alpha_{14}\alpha_{23}a = -\alpha_{24}\alpha_{14}\alpha_{23}f(a) = -\alpha_{24}\alpha_{23}f(a) = -\alpha_{23}f(a)$ so $\lambda(a + b) = f(a) - \alpha_{23}f(a) = \alpha_{32}f(a) \in W_3$. If $a + b \in V_4$, then $a = -\alpha_{14}b$ so $\lambda(a + b) = f(a) - \alpha_{24}f(a) = \alpha_{42}f(a) \in W_4$. □

**COROLLARY 5.3.** Let R be $K[x, 1/x, 1/(1-x)]$. Then the functor F from CAT to the category of R-modules, defined by $F(V) = V_1$, is a categorical equivalence.

**PROOF.** We endow $V_1$ with an R-module structure by setting $xa = \alpha_{14}\alpha_{23}a$ for a in $V_1$. Theorem 5.2 says that the functor F induces an isomorphism from Hom(V,W) to Hom(F(V),F(W)). If remains to show that every R-module is isomorphic to F(V) for some V in CAT. Let M be an R-module. Then we put a CAT structure on $V = M \oplus M$ by setting

$$V_1 = \langle (a,0) : a \in M \rangle$$

$$V_2 = \langle (0,a) : a \in M \rangle$$

$$V_3 = \langle (a,a) : a \in M \rangle$$

$$V_4 = \langle (xa,a) : a \in M \rangle$$

Then $\alpha_{14}\alpha_{23}(a,0) = \alpha_{14}(0,-a) = (xa,0)$. As x is invertible in R, we have $V = V_2 \oplus V_4$; as $1 - x$ is invertible in R, we have $V = V_3 \oplus$

$V_4$; as $1 - x$ is invertible in $R$, we have $V = V_3 \oplus V_4$. □

## 6. Infinite Indecomposable Valuated Groups.

**EXAMPLE 6.1.** Indecomposable $p^3$-bounded valuated groups of length 7.

Let $A'$ be the valuated group with two generators $x$ and $y$ subject to the conditions:

$$p^3x = p^3y = 0$$

$$vx = (0, 3, 5)$$

$$vy = (1, 2, 5)$$

$$v(p^2x + p^2y) = 6$$

Let $B'$ be the valuated group with one generator $z$ subject to the conditions: $p^2z = 0$; $vz = (4,5)$.

For a fixed cardinal $m$ let $A$ be a direct sum of $m$ copies of $A'$, and let $B$ be the direct sum of $m$ copies of $B'$. Let $V$ be a subspace of $A(5)$ that is a simultaneous complementary summand of $A(6)$, $p^2A(0,3)$ and $p^2A(1)$. By Corollary 5.3 and the fact that there exist indecomposable $R$-modules of arbitrarily large finite cardinality and arbitrary infinite cardinality, we can choose $V$ so that the subring of the endomorphism ring of $A(5)$ that respects $p^2A(0,3)$, $p^2A(1)$, $A(6)$, and $V$ has no nontrivial idempotents.

Identify $V$ with $B(5)$ and let $G$ be the amalgamated sum of $A$ and $B$ over $V$. Then $A$ is embedded in $G$ and $A(5) = G(5)$. Moreover

i) $G(6) = A(6)$

ii) $p^2G(0,3) = p^2A(0,3)$

iii) $p^2G(1) = p^2A(1)$

iv) $pG(4) = V$

v) $pG \cap G[p] = G(5) = G(3)[p]$

vi) $G[p] = G(2)[p] \subset G(3) + pG(1)$

We shall show that $G$ is indecomposable. Suppose $G = H \oplus K$. Then $G(5) = H(5) \oplus K(5)$, and the decomposition respects $G(6)$, $p^2G(0,3)$, $p^2A(1)$ and $V = pG(4)$. Hence one of the summands, say $K(5)$, is $0$. As $G(3)[p] = A(5) \subset H$ we have $G(3) \subset H$. Now $K(2) = K(2)[p] \subset K(3) + pK(1) = pK(1)$, but $pG \cap G[p] = G(5) \subset H$, so $K(2) = 0$. Hence $K = 0$ because $K[p] = K(2)[p]$. □

EXAMPLE 6.2. An easy infinite indecomposable valuated p-group of length $\omega$.

Let $x_n$ generate a cyclic group of order $p^{2n}$, and let $B$ be the direct sum of these cyclic groups. Let $H$ be the subgroup of $G$ generated by the elements $y_n = px_n - x_{n-1}$ for $n = 1, 2, \cdots$. We shall show that if the valuated group $H = A \oplus B$, then either $A = H$ or $B = H$. Note that the underlying group of $H$ is a direct sum of cyclics generated by the $y_n$, that the order of $y_n$ is $p^{2n-1}$, and that $p^{2n-2}y_n = p^{2n-2}x_n$ form a basis for $H[p]$ as a valuated vector space.

We may assume that the component of $y_1$ in $A[p]$ has value $1$. Let $\pi$ denote the projection of $H$ on $A$. We shall show that $v\pi p^{2n}y_{n+1} = 2n + 1$ by induction on $n$. Thus $A[p]$ contains elements of all odd values, and so equals $H[p]$; whence $A = H$. Suppose $v\pi p^{2n-2}y_n = 2n - 1$. Then

$$\begin{aligned} \pi p^{2n-1}y_{n+1} &= \pi p^{2n-2}y_n - \pi p^{2n-2}(y_n - py_{n+1}) \\ &= \pi p^{2n-2}y_n + Pp^{2n}x_{n+1} \end{aligned}$$

has height at least $2n - 1$ and value $2n - 1$ and is in $H[p^2]$. Thus if we write $\pi p^{2n-1}y_{n+1} = \Sigma a_iy_i$, then $a_i = 0$ for $i \leq n$ by height arguments alone. Moreover since $p^2 a_i y_i = 0$ we have $a_i$ divisible by $p^{2i-3}$ so ht $a_i y_i \geq 2i - 3 \geq 2n + 1$ if $i > n + 1$. Thus $v a_{n+1} y_{n+1} = 2n - 1$ and $\pi p^{2n-1} y_{n+1} = a_{n+1} y_{n+1} + p^{2n+1} w$, where $a_{n+1}$ is $up^{2n-1}$ for some unit $u$. Hence $\pi p^{2n} y_{n+1} = p a_{n+1} y_{n+1} + p^{2n+2} w = up^{2n+1} x_{n+1} + p^{2n+2} w$ has value $2n + 1$. □

## REFERENCES


[ALB] Albrecht, U. F., Valuated p-Groups and a Theorem of Szele, (preprint).

[AHW] Arnold, D., R. Hunter, and E. Walker, Valuated Groups, *Symposia Mathematica* XXII(1979) 77-84.

[BHW] Beers, D., R. Hunter, and E. Walker, Finite Valuated p-Groups, *Lecture Notes in Mathematics* 1006(1983) 471-507.

[BUT] Butler, M. C. R., On the Structure of Modules over Certain Augmented Algebras, *Proc. Lon. Math. Soc.* 21(1970) 277-295.

[BZ] Berman, S. D., and Z. P. Zilinskaja, On Simultaneous Direct Decompositions of a Finitely Generated Abelian Group and a Subgroup, *Soviet Math Dokl.* 14(1973) 833-837.

[FCHS] Fuchs, L., *Infinite Abelian Groups*, New York, 1970.

[HW] Hunter, R. and E. Walker, Valuated p-Groups, *Lecture Notes in Mathematics* 874(1981) 350-373.

[HRW1] Hunter, R., F. Richman and E. A. Walker, Finite direct sums of Cyclic Valuated p-Groups, *Pacific. J. Math.* **69**(1977) 97-104.

[HRW2] Hunter, R., F.Richman and E. A. Walker, Simply Presented Valuated Abelian p-Groups, *J. of Algebra*, 49(1977) 125-133.

[RW] Richman, F., and E. A. Walker, Valuated Groups, *J. of Algebra*, 56(1979) 145-167.

# A COMBINATORIAL THEOREM AND ENDOMORPHISM RINGS OF ABELIAN GROUPS II

Saharon Shelah (*)

University of Jerusalem

## §0 Introduction

This paper was originally part of [Sh 8] . It was separated for technical reasons and partly extendend, particularly in §§5,6. However we do not require knowledge of the first part.

Let us first deal with the combinatorics. In [Sh 2], [Sh 5] we pointed out that combinatorial proofs from [Sh 1], chap.VIII, should be useful for proving the existence of many non-isomorphic structures as rigid indecomposable systems. We applied this in [Sh 3] for separable

(*) The author would like to thank the United States Israel Binational Science Foundation for partially supporting this research.

p-groups illustrating the impossibility of a characterization of such groups by reasonable invariants. In [Sh 2] we built a rigid Boolean algebra in every $\lambda > \aleph_0$ ; see also [Sh 7] for more results and details. The main idea of the following proof is taken from [Sh 1], chap.VIII, Th. 2.6. We will continue with the combinatorics of [Sh 4], which has been utilized by Dugas and Göbel in [DG 1], [DG 2] and by Göbel and Shelah in [GS]. The nicest feature of these proofs was the fact that they were carried out in ZFC. Their main drawbacks were:

(i) The algebraic objects had strong limit singular cardinal numbers of not small cofinality.

(ii) The combinatorics was not separated from the proof; so analogous proofs have to repeat it.

(iii) The combinatorics contained things specific for modules, so that it is not immediately applicable to other structures.

The combinatorics in this paper is designed to overcome these drawbacks without using extra axioms of set theory. In section 1 we deal with the combinatorics for $\lambda$ with uncountable cofinality. This is accompanied with explanations for the case of the endomorphism rings of separable (abelian) p-groups. This is, in fact, repetitions of [Sh 8]. In section 2 we deal with the combinatorics for $\lambda$ with cofinality $\aleph_0$ and end with conclusions for all $\lambda$. In section 6 we point out some improvements.

Let us turn to abelian group theory. The existence of indecomposable and even endo-rigid groups was stressed in Fuchs [Fu] ; see there for previous history. Fuchs [Fu], with some help of Corner, proved the existence of indecomposable torsion-free abelian groups in every cardinal less than the first strongly inaccessible cardinal. Later Fuchs replaced the bound by the first measurable cardinal and Shelah [Sh 3] proved the existence of such groups in every cardinal. Eklof and Mekler [EM] proved, assuming V=L and $\lambda$ regular, not weakly compact, the existence of strongly $\lambda$-free indecomposable groups of power $\lambda$. They used Jensen's work on L, more specifically the diamond on non reflecting (=sparse) stationary subsets S of $\{\delta<\lambda : \mathrm{cf}\,\delta = \aleph_0\}$. The main algebraic fact they used was as follows.

(*) If $G = H^1 \oplus H^2$ and $G = \bigcup G_n$, $G_n \subseteq G_{n+1}$, with $G_n$ and $G_{n+1}/G_n$ free (abelian) groups, then for some group $G'$ extending G, $G'/G_n$ is free for each n; but the decomposition of G does not "extend" to one of $G'$ or even one of $G' \oplus G''$ ($G''$ free).

Dugas improved [EM], replacing indecomposable by endo-rigid. Hence his algebraic tool was like (*), replacing $H^1 \oplus H^2$ by an endomorphism of G. Then Shelah [Sh 6] proved the existence of strongly $\lambda$-free endo-rigid abelian groups of power $\lambda$ for $\lambda = \aleph_1$ under the hypothesis $2^{\aleph_0} < 2^{\aleph_1}$ or more generally for $\lambda$ satisfying $\nabla_\lambda$. The set theory used rested on Devlin and Shelah [DS]. Note that $\exists\lambda\ \nabla_\lambda$ is not probable in ZFC. The main algebraic fact needed was as follows.

(**) If $G = \bigcup G_n$ with $G_n$ and $G_{n+1}/G_n$ free, $a,b \in G$ and $b \notin a\mathbb{Z}$, then there are groups $H^1, H^2$ extending G with $H^i/G_n$ free, and there are no endomorphisms $h_i$ of $H^i$ such that $h_i(G) \subseteq G$, $h_1 \restriction G = h_2 \restriction G$ and $h_i(a) = b$.

Earlier Corner [C] dealt with a stronger problem, asking which rings can be represented as End(G). He proved that every reduced torsion-free countable ring R is representable. Dugas and Göbel [DG], following [Sh 6], used $\nabla_\lambda$ and removed Corner's countability restriction and added the very natural condition that "the p-adic integers cannot be embedded into the additive group of R". The combinatorics was as in [Sh 6], but the groups G in (**) were replaced by R-modules. The algebra rests on the notion <u>cotorsion-free</u>. Later Dugas and Göbel [DG 2] proved this result in ZFC, but as they use the method of [Sh 4] they obtain R-modules of strong limit $\lambda$ with cf $\lambda > |R|$. More details on the history can be found in [Fu], [DG], [DG 1], [DG 2]. The following papers are now based on the combinatorial result developed in the following sections: Corner and Göbel [CG], Göbel and Shelah [GS 1], [GS 2] and [Sh 10].

In section 5 we will apply the combinatorial proposition to obtain the following

<u>0.1.</u> <u>Theorem</u>: Let R be a ring whose additive group $R^+$ is cotorsion-free, i.e. $R^+$ is reduced and has no subgroups isomorphic to $\mathbb{Z}/p\mathbb{Z}$ or to the p-adic integers. For $\lambda = \lambda^{\aleph_0} > |R|$ there is an abelian group G of cardinality $\lambda$ whose endomorphism ring is isomorphic to R and as an R-module it is $\aleph_1$-free.

We can relax the demands on $R^+$ and may require that G extends a suitable group $G_0$ such that R is realized by End(G) modulo a suitable ideal of "small" endomorphisms.

Let us turn to p-groups. Here we merely complete [Sh 8]; see the history there. We refer to Pierce [P] and Fuchs [Fu] for End(G) of a separable p-group G, small endomorphisms and $E_s(G)$, cf. also [Sh 4] and see Dugas and Göbel [DG 1] for the representation of all suitable rings R as $End(G)/E_s(G)$ in the case of strong limits $\lambda$ with cf $\lambda > |R|$. In [Sh 8] we proved the following theorem for $\lambda$ of cofinality $> \aleph_0$ and in section 3 we will complete this for any $\lambda \geq |R|^{\aleph_0}$.

<u>0.2. Theorem</u>: Let R be a ring whose additive group is the completion of a direct sum of copies of the p-adic integers. If $\lambda^{\aleph_0} \geq |R|$ then there exists a separable p-group G with basic subgroup of cardinality $\lambda$ and $R \cong End(G)/E_s(G)$. As usually we get $End(G) = E_s(G) \oplus R$.

In section 4 we will show the necessity of the cardinality restriction. If $2^{\aleph_0} \leq \lambda < \lambda^{\aleph_0}$ and G has essential power $\lambda$, then $End(G)/E_s(G)$ has power $2^\lambda$. For $\lambda < 2^{\aleph_0}$ the problem is independent.

We should like to fulheartedly thank Rüdiger Göbel and Luigi Salce for taking care of the typing of this paper.

<u>Notation</u>: Groups are always abelian. The letters G,H and sometimes K,L,M are reserved for abelian groups, or modules. Let R be a ring and $R^+$ its additive group. We fix h for homomorphism and f,g for general functions. Let $\mathbb{Z}$ denote the integers, $\mathbb{Q}$ the rationals, $I_p$ the p-adic numbers, where p is a prime, $\mathbb{Z}_n = \mathbb{Z}/n\mathbb{Z}$ and $\mathbb{Z}_{p^\infty}$ the quasi-cyclic divisible p-group. We use $\lambda, \mu, \kappa$ for infinite cardinals, n,m,k,l (and

sometimes i,j) for integers and $i,j,\alpha,\beta,\gamma,\delta,\xi,\zeta$ for ordinals ( $\delta$ usually limit). Further $\omega$ is the first infinite ordinal. Let $A \subseteq^* B$ denote that A−B is finite.

## §1 The combinatorial principle

1.1 Context: Let $\lambda > \kappa$ be fixed infinite cardinals. We shall deal with the case $\mathrm{cf}\,\lambda > \aleph_0$, $\lambda^{\aleph_0} = \lambda^{\kappa}$ and usually $\kappa = \aleph_0$. Let L be a set of function symbols, each with $\leq \kappa$ places, of power $\leq \lambda$. Let $\mathfrak{M}$ be the L-algebra freely generated by $\underline{\underline{T}} \stackrel{\text{def}}{=} {}^{\kappa >}\lambda = \{\eta : \eta$ a sequence of length $< \kappa$ of ordinals $< \lambda\}$. We could replace $\underline{\underline{T}}$ by a set of urelements and let $\mathfrak{M}$ be the family of sets hereditarily of cardinality $\leq \kappa$ built from those urelements. For $\eta \in \underline{\underline{T}} \cup {}^{\omega}\lambda$ let $\mathrm{orco}(\eta) = \{\eta(n): n < \omega\}$, for a sequence $\bar{\eta} = \langle \eta_i : i < \beta \rangle$ let $\mathrm{orco}(\bar{\eta}) = \bigcup_{i<\beta} \mathrm{orco}(\eta_i)$, for $a = \tau(\bar{\eta}) \in \mathfrak{M}$ let $\mathrm{orco}(a) = \mathrm{orco}(\bar{\eta})$, and $\mathrm{orco}(\langle a_i : i < \beta \rangle) = \bigcup_{i<\beta} \mathrm{orco}(a_i)$, and similarly for a set. Now $\underline{\underline{T}}$ is naturally a tree and we consider the members of ${}^{\omega}\lambda$ as its branches.

1.2. Explanation: We shall explain here how this is used for the construction of a separable reduced p-group with a predetermined ring of endomorphisms modulo the small endomorphisms.

So let R be a ring with $R^+$ the p-adic completion of a direct sum of copies of the p-adic integers $I_p^+$. Let B be $\bigoplus_{\eta \in \underline{\underline{T}}} Rx_\eta$ with $p^{l(\eta)+1}x_\eta = 0$, i.e. B is an R-module freely generated by $x_\eta (\eta \in \underline{\underline{T}})$ except that $p^{l(\eta)+1}x_\eta = 0$. Let $\hat{B}$ be the torsion-completion of B; we can represent its elements as $\sum_{\eta \in \underline{\underline{T}}} p^{l(\eta) \dot{-} m} a_\eta x_\eta$, where $m < \omega$, each $a_\eta$ belongs to R, $\{\eta : p^{l(\eta) \dot{-} m} a_\eta x_\eta \neq 0,\ l(\eta) \leq k\}$ is finite for each k, and $n \dot{-} m = \max\{0, n-m\}$ with the natural equality and addition.

Identify $x_\eta$ with $\eta$ and the sum above with appropiate members of $\mathfrak{M}$, hence $\hat{B}$ is a subset of $\mathfrak{M}$. Note that each $y \in \hat{B}$ is a countable sum of $p^{l(\eta) \dot{-} m} a_\eta x_\eta$, hence it depends on only countably many members of $\underline{T}$.

Our desired group G will be an R-submodule of $\hat{B}$ containing B. So there is a natural embedding of R into End(G) and we identify $a \in R$ with the endomorphism $x \mapsto ax$. As we want $R = \mathrm{End}(G)/E_s(G)$, we will need for every endomorphism h of G some $a \in R$ such that h-a is small. Remember that h is small iff h maps $\hat{B}$ into B. We shall try to "kill" the other endomorphisms by the right choice of G.

1.3. <u>Definition</u>

1) Let $L_n$ be fixed vocabularies (=signatures), $|L_n| \leq \kappa$, $L_n \subseteq L_{n+1}$, (with each predicate function symbol finitary for simplicity), $P_n \in L_{n+1} - L_n$ monadic predicates.

2) Let $J_n$ be the family of sets (or sequences) of the form $\{\langle f_l, N_l \rangle : l \leq n\}$ satisfying

a) $f_l : {}^{l\geq}\kappa \to \underline{T}$ is a tree embedding, i.e.

(i) $f_l$ is length preserving, i.e. $f_l(\eta)$ has the same length as $\eta \in {}^{l\geq}\kappa$;

(ii) $f_l$ is order preserving, i.e. for $\eta, \nu \in {}^{l\geq}\kappa$, $\eta < \nu$ iff $f_l(\eta) < f_l(\nu)$.

b) $f_{l+1}$ extends $f_l$ (when $l+1 \leq n$).

c) $N_l$ is an $L'_l$-model of power $\leq\kappa$, $N_l \subseteq \mathfrak{M}$, where $L'_l \subseteq L_l$.

d) $L'_{l+1} \cap L_l = L'_l$ and $N_{l+1} \restriction L'_l$ extends $N_l$.

e) If $P_m \in L'_{m+1}$ and $m<l\leq n$ , then $P_m^{N_l} = |N_m|$.

f) Rang $(f_l) - \bigcup_{m<l}$ Rang $(f_m)$ is included in $|N_l| - \bigcup_{m<l} |N_m|$.

3) Let $J_\omega$ be the family of pairs $(f,N)$ sets (or sequences $\{(f_l,N_l) : l<\omega\}$ such that $\{(f_l,N_l) : l<n\}$ belongs to $J_n$ for $n<\omega$.

4) Let $J'_\omega$ be the family of $(f,N)$ such that for some $\{(f_l,N_l) : l<\omega\}$ $f = \bigcup_{l<\omega} f_l$ , $N = \bigcup N_n$ (i.e. $|N| = \bigcup_{n<\omega} |N_n|$, $L(N) = \bigcup_n L(N_n)$ and $N \restriction L(N_n) = \bigcup_{n\leq m<\omega} N_m \restriction L(N_n)$).

5) For any $(f,N) \in J'_\omega$ let $(f_n,N_n)$ be as above (it is easy to show that $(f_n,N_n)$ is uniquely determined, - notice d), e) in 2)).

6) Let $J'_n = \{(f_n,N_n)$: for some $(f_l,N_l)$ $(l<n)$, $\{(f_l,N_l) : l\leq n\} \in J_n\}$, and we adopt the conventions of 4).

7) Usually we identify $J_i$ and $J'_i$ (for $i\leq\omega$).

8) A branch of Rang(f) or of f (for f as in 3)) is just $\eta \in {}^\omega\lambda$ such that for every $n<\omega$, $\eta\restriction n \in$ Rang(f).

## 1.4. Explanation of our strategy

We will obtain $W = \{(f^\alpha,N^\alpha) : \alpha<\alpha^*\}$, so that every branch $\eta$ of $f^\alpha$ converges to some $\zeta(\alpha)$, $\zeta(\alpha) \in \mathrm{cf}\,\lambda$ non-decreasing. We have a free object generated by $\underline{\underline{T}}$ (B in our case) and by induction on $\alpha$ we define elements $a_\alpha$ and structures $B_\alpha$ (p-groups in our case) increasing continuously such that $B_{\alpha+1}$ extends $B_\alpha$ and $a_\alpha \in B_\alpha$. As usual $B_{\alpha+1}$ is "generated" by $B_\alpha$ and $a_\alpha$, and $a_\alpha$ is in the completion of $B_0$. Every element will "depend" on few $(\leq\kappa)$ members of $\underline{\underline{T}}$, and $a_\alpha$ is specially chosen: The set $Y_\alpha \subseteq \underline{\underline{T}}$ on which $a_\alpha$ "depends" is $Y_\alpha^0 \cup Y_\alpha^1$ where $Y_\alpha^0$ is bounded below $\zeta(\alpha)$ (i.e. $Y_\alpha^0 \subseteq {}^{\omega>}\zeta$ for some $\zeta<\zeta(\alpha)$) and $Y_\alpha^1$ is a branch of $f^{\alpha}$ (or something similar). See more in 1.8.

1.5. Definition of the game:

We define, for $W \in J_{\omega}$ a game $Gm(W) = Gm_{\lambda,\kappa}(W)$, which lasts $\omega$ moves:

In the n-th move player I chooses $f_n$, a tree-embedding of ${}^{n\geq}\kappa$ into ${}^{n\geq}\lambda$, extending $\bigcup_{l<n} f_l$ such that $\mathrm{Rang}(f_n) - \bigcup_{l<n} \mathrm{Rang}(f_l)$ is disjoint to $\bigcup_{l<n} |N_l|$. Then player II chooses $N_n$ such that $\{(f_l, N_l): l \leq n\} \in J_n$. At the end player I wins if $(\bigcup_{n<\omega} f_n, \bigcup_{n<\omega} N_n) \in W$.

1.6. Remark: We shall be interested in W such that player I wins the game, but W is thin. Sometimes we need a strengthening of the second player in two respects: he can force (in the n-th move) $\mathrm{Rang}(f_{n+1}) - \mathrm{Rang}(f_n)$ to be outside a "small" set, and in the zero move he can determine an arbitrary initial segment of the play.

1.7. Definition: We define, for $W \subseteq J_{\omega}$, a game $G'm(W)$ which lasts $\omega$ moves (but in the context of §2 we make a small change). In the zero move player I chooses $f_0$, a tree embedding of ${}^{0\geq}\kappa$ into ${}^{0\leq}\lambda$ and player II chooses $k<\omega$ and $\{(f_l, N_l): l \leq k\} \in J_k$. In the n-th move (n>0) player I chooses $f_{k+n}$, a tree embedding of ${}^{(k+n)\geq}\kappa$ into ${}^{(k+n)\geq}\lambda$, with $\mathrm{Rang}\,(f_{k+n}) - \bigcup_{l<k+n} \mathrm{Rang}\,(f_l)$ disjoint to $\bigcup_{l<k+n} N_l \cup \bigcup_{l<n} X_l$ and player II chooses $N_{k+n}$ such that $\{(f_l, N_l): l \leq k+n\} \in J_{k+n}$ and $X_n \subseteq T$, $|X_n| < \lambda$.

1.8. Remark: What do we want from W? Adding an element for each (f,N) we want to "kill" every undesirable endomorphism. For this W has to encounter every possible endomorphism, and this will follow from "W a barrier". For this $W = J_\omega$ is good enough, but we also want W to be thin enough so that various demands will have small interactions. For this serves disjointness and some further restrictions.

1.9. Definition:

1) We call $W \subseteq J_\omega$ a strong barrier if player I wins in $\underline{\underline{G}}m(W)$ and even $\underline{\underline{G}}'m(W)$; which just means he has a winning strategy.

2) We call W a barrier, if player II does not win in $\underline{\underline{G}}m(W)$ and even does not win in $\underline{\underline{G}}'m(W)$.

3) We call W disjoint if for any distinct $(f^l, N^l) \in W$ $(l=1,2)$ $f^1$ and $f^2$ have no common branch.

1.10. Explanation: What is the aim of W being a barrier or disjoint? Suppose h will be an undesirable endomorphism of G. If W is a barrier, for some $(f,N) \in W$ $N \restriction L_0 = (|N|, h \restriction |N|)$ and $(|N|, h \restriction |N|)$ is a "good approximation" of h. This is true as otherwise we can describe a winning strategy for II in $\underline{\underline{G}}m(W)$. If for each such $(f^*,N^*)$ there is no $y \in C$ satisfying the equations $h(a_\alpha)$ should satisfy, then $h \restriction |N|$ cannot extend to an endomorphism of G. This follows already for $W = J_\omega$. But we want a tight control over the elements in G, this is done using disjointness, (1.4) and more. In order to derive the existence of a strong disjoint barrier, we first define a strategy for player I and only then define W. We note

1.11. Observation:

1) If $\lambda^{\kappa}=\lambda$ then there is a one-to-one function cd from $J_{\omega}$ onto $\lambda$.

2) If $\lambda^{\kappa}=\lambda^{\aleph_0}$, then there are functions $cd_n$ from $J_n$ into $\lambda$ such that

a) if $m<n$ we can compute $cd_m(\langle (f_l,N_l): l<m\rangle)$ from $cd_n(\langle (f_l,N_l): l<n\rangle)$,

b) if $\langle (f_l^0,N_l^0): l<\omega\rangle \neq \langle (f_l^1,N_l^1): l<\omega\rangle$ are from $J_{\omega}$, then for every large enough n $cd_n(\langle (f_l^0,N_l^0): l<n\rangle) \neq cd_n(\langle (f_l^1,N_l^1): l<n\rangle)$,

c) if $\lambda^{\kappa}=\lambda$ then $cd_n$ is one-to-one.

3) There is a function $pr:\lambda \rightarrow \lambda$ which is onto and for every $\alpha<\lambda$ there are $\lambda$ many $\beta<\lambda$ with $pr(\beta)=\alpha$.

Remark: We shall use the functions $cd_n$ only when $cf\,\lambda>\aleph_0$.

Proof: We should only note that $|J_{\alpha}| = \lambda^{\kappa}$ for $\alpha\leq\omega$.

1.12. Lemma: If $\lambda^{\kappa}=\lambda^{\aleph_0}$, then there is a strong disjoint barrier W.

Proof: First we define the winning strategy for player I and later W. In the strategy we code the play. Suppose $\lambda^{\aleph_0}=\lambda$ hence $\lambda^{\kappa}=\lambda$ for the moment. For n=0 player I has no choice. Let n>0 and $\langle (f_l,N_l): l<n\rangle$ be the play so far. Then player I defines his move $f_n$, a tree embedding from ${}^{n\geq}\kappa$ into $\underline{\underline{T}}$ such that it extends $\bigcup_{l<n} f_l$. For $\eta\in {}^{n}\kappa$ let $f_n(\eta)$ = $f_{n-1}(\eta\restriction(n-1))^\frown\langle\gamma_\eta\rangle$ such that

(i) $f_n(\eta)\notin \bigcup_{l<n} X_l$;

(ii) $pr(\gamma_\eta) = cd_n(\langle (f_l,N_l): l<n\rangle)$;

(iii) $\eta \neq \nu \in {}^n\kappa$ implies $\gamma_\eta \neq \gamma_\nu$.

This is possible as by 1.11 (3), condition (ii) has $\lambda$ many solutions, whereas $|X_l| < \lambda$ for $l<n$ and for (iii), define $\gamma_\eta$ by induction on n for some well-ordering on ${}^n\kappa$; so $\leq \kappa$ many ordinals are excluded. Now let $W = \{<\bigcup_l f_l, \bigcup_l N_l> : <(f_l, N_l) : l<\omega>$ is a play of $\underline{\underline{G}}'_m$ in which player I uses the strategy defined above}. Trivially W is a barrier. Why is it disjoint? If $\eta$ is a branch of f for $(f,N) \in W$, then by (ii) above we can reconstruct the play from $\eta$.

<u>1.13. The existence lemma:</u>

1) Suppose $\lambda^{\aleph_0} = \lambda^\kappa$, cf $\lambda > \kappa$ and $C^* \subseteq \lambda$ closed unbounded. Then there is $W = \{(f^\alpha, N^\alpha) : \alpha < \alpha^*\} \subseteq J_\omega$ and a function $\zeta : \alpha^* \to C^*$ such that

a) W is a strong disjoint barrier.

b) For $\alpha < \beta < \alpha^*$, $\zeta(\alpha) \leq \zeta(\beta)$.

c) $cf(\zeta(\alpha)) = \aleph_0$ for $\alpha < \alpha^*$.

d) Every branch of $Rang(f^\alpha)$ is an increasing sequence converging to $\zeta(\alpha)$.

e) For every $n<\omega$ for some $\xi < \zeta(\alpha)$, $orco(N^\alpha_n) \leq \xi$.

f) If $\alpha + \kappa^{\aleph_0} \leq \beta < \alpha^*$ and $\eta$ is a branch of $Rang(f^\beta)$, then $\eta \restriction k \notin N^\alpha$ for some $k<\omega$.

g) If $\lambda = \lambda^\kappa$ we can demand: if $\eta$ is a branch of $Rang(f^\alpha)$ and $\eta \restriction k \in N^\beta$ for all $k<\omega$ (where $\alpha, \beta < \alpha^*$), then $N^\alpha \subseteq N^\beta$.

2) We can demand also:

h) For every stationary $S \subseteq \{\delta < \lambda \mid cf\,\delta = \aleph_0\}$, $\{(f^\alpha, N^\alpha) : \alpha < \alpha^*, \zeta(\alpha) \in S\}$ is a disjoint barrier.

1.14. Remark: By (e) the ordinal $\zeta(\alpha)$ is the "infinity" of $N^\alpha$ and by (d) the branches of Rang($f^\alpha$) converge to infinity. Conditions (f) and (g) strenghten disjointness.

Proof: 1) Again we first define the strategy of player I, using (1.11) and the functions $cd_n (n<\omega)$ from there. For n=0 player I has a unique choice. So suppose n>0 and $\langle(f_l, N_l): l<n\rangle$ is the play so far. We have to define $f_n$ extending $f_{n-1}$. Let $\gamma_\eta < \lambda$ for $\eta \in {}^n\kappa$, be such that (i)-(iv) below hold and then let $f_n(\eta) = f_{n-1}(\eta \restriction (n-1)^\frown\langle\gamma_\eta\rangle)$ for $\eta \in {}^n\kappa$.

The requirements are

(i) $f_n(\eta) \notin \bigcup_{l<n} X_l$;

(ii) $pr(\gamma_n) = cd_n(\langle(f_l, N_l): l<n\rangle)$;

(iii) if $\eta \neq \nu \in {}^n\kappa$, then $\gamma_\eta \neq \gamma_\nu$;

(iv) $\gamma > \sup(\text{orco}|N_{n-1}|)$, moreover there is a member of C* in the interval.

Note that $\sup(\text{orco}|N_{n-1}|) < \lambda$ as we have assumed cf $\lambda > \kappa$ and $\text{orco}|N_{n-1}|$ is a set of power $\leq\kappa$ (as $\|N_n\| \leq \kappa$, definition of $\mathfrak{M}$ and of orco (see 1.1)). The requirement (ii) has $\lambda$ solutions, (i), (iv) exclude less than $\lambda$ of them, and (iii) requires that we have ${}^n\kappa$ distinct ordinals satisfying (i), (ii), (iv). So we can carry on the definition, and then let W be as in the proof of 1.2. Clearly W is a strong disjoint barrier and (a) holds.

We define a function $\zeta$ from W to $\lambda$ : $\zeta((f,N)) = \sup(\text{orco}(\text{Rang}(f)))$. By (iv) above, $\zeta((f,N)) \in C^*$, and for every branch $\eta$ of Rang(f),

$\sup(\mathrm{orco}(\eta)) = \zeta((f,N))$. Now we define by induction on $i$ $(<|W|^+)$ for each ordinal $\zeta \in C^*$ a set $W_i^\zeta$, $W_i^\zeta \subseteq W^\zeta \overset{\mathrm{def}}{=} \{(f,N): (f,N) \in W, \zeta((f,N))=\zeta\}$ such that:

($\alpha$) $W_0^\zeta = \emptyset$;

($\beta$) $W_i^\zeta$ is increasing continuous (in i);

($\gamma$) $W_{i+1}^\zeta - W_i^\zeta$ has cardinality $\leq \kappa^{\aleph_0}$;

($\delta$) If $(f,N) \in W_{i+1}^\zeta$, $(f',N') \in W^\zeta$, $\eta$ a branch of $\mathrm{Rang}(f')$ and $\{\eta \restriction k: k<\omega\} \subseteq N$, then $(f',N') \in W_{i+1}^\zeta$;

($\varepsilon$) If $W \neq W_i^\zeta$, then $W_{i+1}^\zeta \neq W_i^\zeta$.

This is straightforward; only (iv) requires the following observation: $|N|$ contains at most $\kappa^{\aleph_0}$ branches of $\underline{\underline{T}}$. By ($\varepsilon$), $W^\zeta = \bigcup_i W_i^\zeta$.

To finish the proof of 1.13 (1), choose for every $\zeta$, i a well-ordering $<^*_{\zeta,i}$ of $W_i^\zeta$ of order type $\leq \kappa^{\aleph_0}$. Define a well-ordering $<^*$ of W such that $\zeta((f^0,N^0)) < \zeta((f^1,N^1))$ implies $(f^0,N^0) <^* (f^1,N^1)$ and $(f^0,N^0) \in W_i^\zeta$, $(f^1,N^1) \in W_j^\zeta$, $i<j$ implies $(f^0,N^0) <^* (f^1,N^1)$ and if $(f^0,N^0), (f^1,N^1) \in W_i^\zeta$, $(f^0,N^0) <^* (f^1,N^1)$ if and only if $(f^0,N^0) <^*_{\zeta,i} (f^1,N^1)$. Now let $\{(f^\alpha,N^\alpha): \alpha<\alpha^*\}$ be a list of the members of W such that for $\alpha<\beta<\alpha^*$, $(f^\alpha,N^\alpha) <^* (f^\beta,N^\beta)$, and let $\zeta(\alpha) \overset{\mathrm{def}}{=} \zeta((f^\alpha,N^\alpha))$. Now we have already cheeked (a), and (b) is trivially satisfied. We have already observed that a branch $\eta$ of $\mathrm{Rang}(f^\alpha)$ converges to $\zeta(\alpha) = \zeta((f^\alpha,N^\alpha))$, hence (c) and (d) hold. The demand (iv) above guarantees (e), and the condition ($\delta$) (and the choice of $<^*_{\zeta,i}$ to have order type $\leq \kappa^{\aleph_0}$) ensures (f). For (g) use (1.11)(2)(c). In case 2 the same construction works. It can also easily be proved by taking elementary submodels $M_\eta$ of $(H(\lambda),\in)$ which contains all relevant information, $\lambda$ large enough, $M_n \in M_{n+1}$, $\sup(\lambda \cap \bigcup_n M_n) \in S$.

1.15 Remark: We may also want to build $2^{(\lambda^{\aleph_0})}$ objects of power $\lambda^{\aleph_0}$, each one like G, with no homomorphisms from one to the other, except the necessary ones. This can be done alternatively as follows.

1) Together with G we also build G′ extending G and elements $a_i \in G'$ $(i < \lambda^{\aleph_0})$ and let for $A \subseteq \lambda^{\aleph_0}$, $G_A = \langle B \cup \{a_i : i \in A\}\rangle$. We then try to guarantee that $A \nsubseteq B$ implies that there are only necessary homomorphisms from $G_A$ to $G_B$. This clearly suffices.

2) For each $A \subseteq {}^{\omega}\lambda$ we build $G^A$. We use W not only to approximate endomorphisms of $G^A$, but also look for $N^\alpha$ which is a submodel of $(G^A, G^B, h)$ where $A \neq B \subseteq {}^{\omega}\lambda$, h a homomorphism from $G^A$ to $G^B$. For $(N^\alpha, f^\alpha)$ we try to add an element y to $G^A$ and omit the corresponding type from $G^B$ which prevents h to map y into $G^B$. Note that $N^\alpha$ "knows" $A \cap N^\alpha$, $B \cap N^\alpha$, but A,B themselves.

## §2 The combinatorial principle for $\lambda$ of cofinality $\aleph_0$

If we want to get a p-group (or similar algebraic objects) of density character $\lambda$, the combinatorics of §1 does not help us. Here we shall deal with this case. We also formulate a conclusion which holds for every $\lambda$, $\lambda^{\aleph_0}=\lambda^{\kappa}$ thus enables us to give a uniform proof.

2.1. Context: As we want to deal not only with the main case, $\lambda>\aleph_0=\mathrm{cf}\lambda$, but also with $\lambda=\aleph_0$ we will have two possibilities

(1) $\lambda=\aleph_0=\lambda_n$ and $\lambda^*_n=n!$ for each n.

(2) $\lambda>\aleph_0$, $\mathrm{cf}\,\lambda=\aleph_0$, $\kappa<\lambda$, $\lambda=\sum_{n<\omega}\lambda_n$, $\lambda_n=\lambda*$ regular and $\kappa<\lambda_n<\lambda_{n+1}<\lambda$.

In both cases let D be a non-principal ultrafilter on $\omega$, $\underline{\underline{T}}=\bigcup_{n<\omega}\prod_{m<n}\lambda_m$ and let L, $\mathfrak{M}$ be as in 1.1.

In the definition of $\underline{\underline{G}}'m(W)$ we make a change and demand $|X_l|<\lambda^*_{k+l-1}$ and stipulate $\lambda^*_{0-1}=1$.

2.2. Definition: For $f,g\in{}^{\omega}\mathrm{Ord}$ (i.e. a function from $\omega$ to the class of ordinals) let $f<_D g$ iff $\{n: f(n)<g(n)\}\in D$ (and similarly $\leq_D$).

2.3. Claim: There are a regular cardinal $\mu$, $\lambda\leq\mu\leq\lambda^{\aleph_0}$ and functions $g_\xi$ in $\prod_{n<\omega}\lambda_n$ for $\xi<\mu$ such that:

(a) for $\xi<\gamma<\mu$, $g_\xi<_D g_\gamma$;

(b) for every $g\in\prod_{n<\omega}\lambda_n$ for some $\xi$, $g<_D g_\xi$;

(c) $g_\xi(n)$ is divisible by $\lambda^*_{n-1}$.

Proof: It is well known that the ultraproduct $\prod_{n<\omega}(\lambda_n,<)/D$ is a linear order, and let $\langle g_\xi : \xi<\mu\rangle$ be an increasing unbounded sequence ($\mu$ regular). Now $\mu$ is at most the power of the ultraproduct which is $\lambda^{\aleph_0}$, and as D is non-principal easily $\mu>\lambda$. Taking care of (c) is easy.

2.4. Remark: If $\lambda>2^{\aleph_0}$ we can choose D the filter of co-bounded subsets of $\omega$, and (2.3) still holds; see [Sh 9].

2.5. Remark: The $g_\xi$ are needed to slice $\prod_{n<\omega}\lambda_n/D$ similar to the range of the function $\zeta$ in Th. 1.13.

2.6. Notation: We identify any set $a \subseteq \mathfrak{M}$ with the function $\chi_a \in {}^{\omega}\mathrm{Ord}$, $\chi_a(n) = \sup(\lambda_n \cap \mathrm{orco}(a))$ using $<_D$.

2.7. Observation: 1) If $\lambda^{\aleph_0}=\lambda^*$ there are functions $cd_n$ from $J_n$ to $\lambda^*_{n-1}$ such that:

(a) If $m<n$ then we can compute $cd_m(\langle(f_l,N_l): l<m\rangle)$ from $cd_n(\langle(f_l,N_l): l<n\rangle)$.

(b) If $\langle(f^0_l,N^0_l): l<\omega\rangle \neq \langle(f^1_l,N^1_l): l<\omega\rangle$ are in $J_\omega$, then $cd_n(\langle(f^0_l,N^0_l): l<\omega\rangle \neq cd_n(\langle(f^1_l,N^1_l): l<\omega\rangle)$ for every large enough n.

(2) There are functions $pr'_n$ from $\lambda_n$ to $\lambda^*_{n-1}$ such that for every $\alpha<\lambda_n$ divisible by $\lambda^*_{n-1}$, $\gamma<\lambda^*_{n-1}$ there are $\lambda_n$ ordinals $\beta$ satisfying $\alpha<\beta<\alpha+\lambda^*_n$, $pr'_n(\beta)=\gamma$.

2.8. The existence theorem: Suppose $\lambda^{\aleph_0}=\lambda^{\kappa}$.

1) Then there are $W=\{(f^{\alpha},N^{\alpha}): \alpha<\alpha^*\}\subseteq J_{\omega}$ and a function $\zeta:\alpha^*\longrightarrow\mu$ such that:

a) W is a disjoint barrier.

b) For $\alpha<\beta<\mu$, $\zeta(\alpha)\leq\zeta(\beta)$.

c) $cf(\zeta(\alpha))=\aleph_0$ for every $\alpha<\alpha^*$.

d) For every branch $\eta$ of $Rang(f^{\alpha})$, $\eta<_D g_{\zeta(\alpha)}$ but for every $\xi<\zeta(\alpha)$ $g_{\xi}<_D\eta$.

e) For every $n<\omega$ for some $\xi<\zeta(\alpha)$, $orco(|N^{\alpha}_{n}|)<_D g_{\xi}$.

f) If $\zeta(\alpha)=\zeta(\beta)$, $\alpha+\kappa^{\aleph_0}\leq\beta$, $\eta$ a branch of $Rang(f^{\beta})$, then for some $k$, $\eta\restriction k\notin N_{\alpha}$.

2) For every stationary set $S\subseteq\{\delta<\mu: cf\,\delta=\aleph_0\}$ $\{(f^{\alpha},N^{\alpha}):\alpha<\alpha^*, \zeta(\alpha)\in S\}$ is a disjoint barrier.

Proof: We first define for every ordinal $\zeta<\mu$ of cofinality $\aleph_0$ a subset $W^{\zeta}$ of $J_{\zeta}$. $W^{\zeta}$ is the set of $(f,N)\in J_{\omega}$ satisfying:

(i) for $\eta\in{}^{n+1}\kappa$, $pr'_{n}(f_{n}(\eta)(n)=cd_{n}(\langle(f_{1},N_{1}): 1\leq n\rangle)$.

(ii) conditions (d) and (e) of 2.8. (1) hold with $\zeta$ taking the place of $\zeta(\alpha)$.

Now we let $W\overset{def}{=}\bigcup_{\zeta<\mu}W^{\zeta}$. The choice of the list $\{(f^{\alpha},N^{\alpha}):\alpha<\alpha^*\}$ of W and the function is just as in the proof of 1.13, except for the proof of one half of (a): W is a barrier, and to this the rest of the proof is dedicated. For notational simplicity we shall deal with the game $\underline{\underline{Gm}}(W)$ only. Suppose St* is a winning strategy for player II in $\underline{\underline{Gm}}(W)$. Let $\vartheta$ be a large enough regular cardinal. We can choose elementary submodels $M_n$ of $(\underline{H}(\lambda), \in)$, such that St*, $\mathfrak{M}$, $J_\omega$, $\langle g^*_\xi : \xi < \mu \rangle$ belong to each $M_n$, $\{i : i \leq \kappa\} \subseteq |M_n|$, $M_n \in M_{n+1}$ and $\|M_n\| = \kappa$. Let $\zeta(n) = \sup(|M_n| \cap \mu)$ and $\zeta = \bigcup_{n<\omega} \zeta(n)$. As $M_n \in M_{n+1}$ (and $\|M_n\| = \kappa < \lambda < \mu$, $\mu$ regular) clearly $\zeta(n) \in M_{n+1}$, hence $\zeta(n) < \zeta(n+1)$. Also the function $f^n$, $\mathrm{Dom}(f^n) = \omega$, $f^n(k) = \sup(|M_n| \cap \lambda_k)$ belongs to $M_{n+1}$; and as by 2.3 for some $\xi$, $f^n <_D g_\xi$, there is such $\xi \in M_{n+1}$, hence (as $\xi < \zeta(n+1)$, $g_\xi <_D g_{\zeta(n+1)}$) clearly $f^n <_D g_{\zeta(n+1)}$.

Now we shall define a play $\langle (f_l, N_l) : l < \omega \rangle$ of the game $\underline{\underline{G}}\mathrm{m}$. We shall define $f_n, N_n$ by induction on n so that

(*) $\langle (f_m, N_m) : m < n \rangle$ form an initial segment of a play of $\underline{\underline{G}}\mathrm{m}(W)$ in which player II uses the strategy St*, and it belongs to $M_n$.

For n=0 player I has a unique choice for $f_0$, and clearly $f_0 \in M_0$. As St* $\in M_0$ clearly $N_0 \in M_0$.

So suppose $\langle (f_m, N_m) : m \leq n \rangle$ satisfies (*). Let $k^0_n = \mathrm{Max}\{l : l \leq n+1$ <u>and</u> $l=0$ or $g_{\zeta(0)}(n), \dots, g_{\zeta(l)}(n) < g_\zeta\}$ ($k^0_n$ is well-defined as the set is finite and non-empty). We shall define $f_{n+1}$ such that for $\eta \in {}^{n+1}\kappa$, $f_{n+1}(\eta) = f_n(\eta \restriction n)^\frown \langle \gamma_\eta \rangle$, where $g_{\zeta(k_n)} < \gamma_\eta < g_{\zeta(k_n)} + \lambda^*_{n-1} < \lambda_n$, (i) above holds and $\gamma_\eta \neq \gamma_\nu$ if $\eta \neq \nu$. By 2.7. this is possible. Moreover we can choose $f_{n+1} \in M_{n+2}$ as $M_{n+1} \in M_{n+2}$ (and $\langle (f_l, N_l) : l \leq n \rangle$, $f_{n+1}$ belongs to $M_{n+2}$) also $N_{n+1}$ belongs to $M_{n+2}$.

So $\langle (f_l, N_l) : l < \omega \rangle \in J_\omega$ is the result of a play of $\underline{\underline{G}}\mathrm{m}(W)$ in which player II uses his strategy St*. However we shall show that he loses the play, i.e. $\langle (f_l, N_l) : l < \omega \rangle \in W$, thus getting the desired contradiction.

In fact $\langle (f_l, N_l) : l < \omega \rangle \in W^\zeta$; the least trivial part is why condition (d) holds. Now $\eta <_D g_{\zeta(\alpha)}$ as for each branch $\eta$ of $\mathrm{Rang}(\bigcup f_m)$, for every n, $\eta(n) < g_{\zeta(k(m))}(n) + \lambda^*_{m-1}$; now if $k(n)>0$ $g_{\zeta(k(m))}(n) + \lambda^*_{m-1} \leq g_\zeta(n)$ (see 2.3) and $\{n: \eta(n) < g_\zeta(n)\} \in D$ as required.

On the other hand for each $m<\omega$, $A_m = \{n<\omega: n>m$ and $g_{\zeta(m)}(n) < g_\zeta(n)\} \in D$, hence $\bigcap_{m\leq 1} A_m \in D$, and for each $n \in \bigcap_{m\leq 1} A_m$, $g_{\zeta(1)}(n) \leq g_{\zeta(k_n)}(n)$ but $\eta(n) > g_{\zeta(k_n)}(n)$, hence $\{n: g_{\zeta(1)} \leq \eta(n)\} \supseteq \bigcap_{m\leq 1} A_m$, hence $g_{\zeta(1)} <_D \eta$.

2) The proof is similar except that we can demand $\zeta \in S$.

2.9. Remark: 1) In 2.8 (1)(d) we can demand $(\forall n)\, \eta(n) < g_{\zeta(\alpha)}(n)$ as w.l.o.g. $g_\zeta(n) > 0$ for every $\zeta$ and n, and in the proof when $k_m=0$ use 0 instead $g_{\zeta(k^0_m)}(n)$.

2) In the proof we could have chosen an infinite $A^* \subseteq \omega$, $A^* \notin D$, and restrict (i) to $n \in A^*$. In this case we can demand only $|X_1| < \lambda_{k+1-1}$ in the present variant of the definition of $\underline{\underline{\mathrm{Gm}}}'(W)$.

In fact we can conclude from 1.13, (2)-(8) an assertion, which is the one we shall use in §3, thus getting a uniform proof of Th. 3.5 for all $\lambda$. So here we are in the context common to §1 and §2.

2.10. Conclusion: Suppose $\lambda^\kappa = \lambda^{\aleph_0}$, $\lambda > \kappa$. Then there is $W = \{(f^\alpha, N^\alpha): \alpha < \alpha^*\}$ such that:

(a) W is a disjoint barrier (for $\lambda, \kappa$).

(b) For every $\alpha \leq \beta < \alpha^*$ and branch $\eta$ of $\mathrm{Rang}(f^\beta)$ and $n<\omega$, for every large enough k, $\eta \upharpoonright k \notin N^\alpha_n$.

(c) If $\alpha + \kappa^{\aleph_0} \leq \beta < \alpha^*$ and $\eta$ is a branch of $\mathrm{Rang}(f^\beta)$, then for every large enough k, $\eta \restriction k \notin N^{\alpha}$.

Proof: If (1.13) or (2.8) apply, this is immediate. The remaining case is $\aleph_0 < \mathrm{cf}\,\lambda \leq \kappa$ (but the main case is anyhow $\kappa = \aleph_0$). For them note

2.11. Fact: 1) If $\kappa < \lambda^* \leq \lambda$, $(\lambda^*)^{\aleph_0} = \lambda^{\kappa}$, we can repeat 1.13 (and everything else in §1) letting $\underset{=}{T}$, $\mathfrak{M}$ be defined using $\lambda$, by letting player II choose embeddings into $\bigcup_{n<\omega} {}^{n}(\lambda^*)$ (with the obvious changes).
2) The same holds for 2.8.

§3 On separable p-groups with predetermined endomorphism ring

We prove here that for suitable R, $R \cong End(G)/E_s(G)$ for some G of density character $\lambda$, $|G| = \lambda^{\aleph_0}$.

It might be possible to predetermine $\dim p^n G[p] = \lambda'_n \leq \lambda$ with $\lambda = \lim\sup \lambda'_n$. Replace B in (3.5) by B′ using $p^{k(n)+1} x^n_\eta = 0$ for $p^{n+1} x^n_\eta = 0$ with some sequence k(n) such that $\lambda_n \leq \lambda'_{k(n)}$. Extend B′ to obtain the right G; but we have not checked the details.

3.1. Definition: A separable p-groups G is an abelian p-group such that every element belongs to a finite direct summand. We will deal with separable groups in §§3 and 4.

3.2. Definition: A map h from G into $G_0$ is called small if for every m and for every large enough n $h(p^n G[p^m]) = 0$ (where $p^n G[p^m] = \{p^n x : x \in G, p^{n+m} x = 0\}$).

3.3. Definition: For an abelian group G let End(G) be the ring of endomorphisms of G and let $E_s(G)$ be the set of all small endomorphisms of G.

Trivially

3.4. Lemma: $E_s(G)$ is an ideal of End(G).

3.5. Theorem: Let R be a ring such that $R^+$ is the p-adic completion of a free p-adic module. Suppose $\lambda \geq |R|$.

1) There is a separable p-group G with End(G) isomorphic to $R \oplus E_s(G)$ and $|G| = \lambda^{\aleph_0}$ and G has a basic subgroup of power $\lambda$.

2) There are groups $G_i$ $(i<2^{(\lambda^{\aleph_0})})$ as in (1) such that homomorphisms from $G_i$ to $G_j$, $i \neq j$ are small.

3.5.A. Remark: We can replace $\lambda \geq |R|$ by $\lambda^{\aleph_0} > |R|$ (or even $\forall n\ \lambda^{\aleph_0} > |R^+/p^n R^+|$), $\lambda > \aleph_0$, without change in the proof. If $\lambda^{\aleph_0} = |R^+/p^n R^+|$ we get (1) and with more care, also (2).

Proof: By [Sh 5] we can restrict ourselves to the case cf $\lambda = \aleph_0$. We can choose regular $\lambda_n < \lambda$, $\aleph_0 < \lambda_n < \lambda_{n+1}$ such that $\lambda = \sum_{n<\omega} \lambda_n$. Let $\kappa = \aleph_0$. We shall use freely the notation of §2 in general and of (2.8) in particular.

Stage A: Let B be the R-module freely generated by $\{x_\eta : \eta \in \underline{\underline{T}}\}$ with $p^{l(\eta)+1} x_\eta = 0$. So every $b \in B$ is of the form $\sum_{\eta \in \underline{\underline{T}}} r_\eta x_\eta (r_\eta \in R)$ where $\{\eta : r_\eta \neq 0\}$ is finite. Let H be the torsion-completion of B so that any $b \in H$ is a formal infinite sum $\sum_{\eta \in \underline{\underline{T}}} r_\eta p^{l(\eta) \dot{-} m} x_\eta$ such that for every l, $\{\eta : \eta \in \underline{\underline{T}},\ l(\eta) \leq l$ and $r_\eta p^{l(\eta) \dot{-} m} x_\eta \neq 0\}$ is finite. This implies that $\underline{d}(b) \overset{def}{=} \{\eta \in \underline{\underline{T}} : r_\eta p^{l(\eta) \dot{-} m} x_\eta \neq 0\}$ is countable. We have $\sum_\eta r^1_\eta p^{l(\eta) \dot{-} m(1)} x_\eta = \sum_\eta r^2_\eta p^{l(\eta) \dot{-} m(2)} x_\eta$ iff for every $\eta$, $r^1_\eta p^{l(\eta) \dot{-} m(1)} - r^2_\eta p^{l(\eta) - m(2)}$ is divisible by $p^{l(\eta)+1}$, and we can define H as the set of those sums with the obvious addition (see [Fu]).

Note that $r_\eta$ is not uniquely determined by b, but $r_\eta p^{l(\eta)\dot{-}m} x_\eta$ is. Note that H extends B and is torsion-complete. This means that $\sum_{n<\omega} y_n$ exists if $(\exists l)(\forall n)$ $p^l y_n = 0$ and for every l for every large enough n, $y_n$ is divisible by $p^l$. In fact, if $y_n = \sum_{\eta \in T} a^n_\eta x_\eta$, $\sum_{n<\omega} y_n = \sum_\eta (\sum_n a^n_\eta) x_\eta$ and $\sum_n a^n_\eta$ exists by the choice of R.

Note that

(A)(1) $\underline{d}(y+z) \subseteq \underline{d}(y) \cup \underline{d}(z)$.

(A)(2) $\underline{d}(\sum_{n<\omega} y_n) \subseteq \bigcup_{n<\omega} \underline{d}(y_n)$ when $\sum_{n<\omega} y_n$ is well-defined.

(A)(3) if $y \in H$ is divisible by $p^n$, then $l(\eta) \geq n-1$ for all $\eta \in \underline{d}(y)$.

(A)(4) if $y = \sum_{\eta \in T} r_\eta x_\eta$, $r_\eta x_\eta$ is uniquely determined by $\eta$ and we say "$x_\eta$ appears in y as $r_\eta x_\eta$"; really $r_\eta + p^{l(\eta)+1} R$ is uniquely determined.

We shall build an R-module G, such that $B \subseteq G \subseteq H$. Let $G^+$ denote the additive group of G, and homomorphisms will be $\mathbb{Z}$-homomorphisms.

Stage B: Recall from [Fu]:

Fact: If $B \subseteq G \subseteq H$, then every endomorphism h of $G^+$ extends uniquely to an endomorphism of $H^+$. If h is small, the range of the extension is in $G^+$.

Stage C: The construction

We now define the R-module G. First identify members of H with members of $\mathfrak{M}$. If $A \subseteq H$, let SG(A) be the R-submodule of H generated by A. We define by induction on $\alpha < \alpha^*$ the following:

(1) The truth value of $\alpha \in J_0$, $\alpha \in J_1$, $\alpha \in J_2$ so that exactly one holds.

(2) For $\alpha \in J_0 \cup J_1$ we define elements $a_{\alpha,l}$ $(l<\omega)$, $b_\alpha$ of H.

(3) For $\alpha \in J_0 \cup J$ we fix a branch $\nu_\alpha$ of Rang$(f^\alpha)$ such that

(4) $a_{\alpha,m} = a^0_{\alpha,m} + a^1_{\alpha,m}$ (both in H) where $a^1_{\alpha,m} = \sum_{k\geq m} p^{k-m} x_{\nu_\alpha \restriction k}$ so that $pa^i_{\alpha,l+1} - a^i_{\alpha,l} \in B$.

(5) $a^0_{\alpha,l} \in N^\alpha_0$

(6) $\nu_\alpha \neq \nu_\beta$ for $\beta<\alpha$

For $J \subseteq J_1 \cap \alpha$ let $G^\alpha = SG(B \cup \{a_{\beta,l} : \beta<\alpha, \beta \in J_0 \cup J\})$ and let $G^{\alpha^*}_J = G_J$

(7) $b_\beta \notin G^\alpha_J$ for $\beta<\alpha$, when $\beta \in J_0$; and $a_{\beta,l} \in G^\alpha_J$ iff $\beta \in (J_0 \cup J) \cap \alpha$ when $\beta<\alpha$.

(8) If $N^\alpha = (|N^\alpha|, L, h, \ldots)$, L a subgroup of $G_{J\cap\alpha} \cap |N^\alpha|$ for some $J \subseteq J_1 \cap \alpha$, h an endomorphism of L, and we can find $\nu_\alpha$, $a^i_{\alpha,l}$ $(i=0,1;\ l<\omega)$, $b_\alpha$, $l(\alpha)<\omega$, satisfying (2)-(7) for $\alpha+1$ (stipulating $\alpha \in J_0$) such that for every endomorphism h′ of H extending h, $h'(a_{\alpha,l(\alpha)}) = b_\alpha$, then $\alpha \in J_0$

(9) If the hypothesis of (8) fails, but we can find $a_\alpha$, $a_{\alpha,l}$, $\nu_\alpha$ such that conditions (2)-(7) hold, then $\alpha \in J_1$ ; otherwise $\alpha \in J_2$.

<u>Remark</u>: Really $J_2 = \emptyset$

<u>Stage D</u>: <u>Claim</u>:

1) Every element x of G (where $J \subseteq J_1$) can be represented as $\sum_{l=1}^{k} r_l a_{\alpha_l,m} + b$, where $b \in B$ and $\alpha_k<\ldots<\alpha_1$.

2) If k>0, $\zeta(\alpha_1) = \zeta(\alpha_k)$, then Rang$(\nu_{\alpha_1}) \subseteq^* \underline{d}(x)$ ; moreover there is $m^*<\omega$ such that $\underline{d}(x) \cap \{\rho \in \underline{\underline{T}} : \nu_{\alpha_1} \restriction m^* \leq \rho\}$ is equal to

$\{\nu_{\alpha_1} \upharpoonright i : m^* \le i < \omega\}$, and if $x = \sum_{\eta \in \underline{T}} r_\eta x_\eta$, then $\eta = \nu_{\alpha_1} \upharpoonright l$, $l \ge m$ implies $r_\eta x_\eta = r_1 p^{l(\eta)-m} x_\eta$.

3) The representation in 1) is totally determined by m, hence k, $\langle \alpha_1, \ldots, \alpha_k \rangle$ depend on x only, and if $\sum_{l=1}^{k} r'_l a_{\alpha_1, m(1)} + b'$ is another representation, then $r_1 p^{m(2)-m} a_{\alpha_1, m(2)} = r'_1 p^{m(2)-m(1)} a_{\alpha_1, m(1)}$ for every large enough m(2).

Remark: The claim explains the peculiar choice of the $a_\alpha$ : by having special domains for them we have specific severe restraints of the domain of any $x \in G$.

Proof: 1) By the definition of G we can represent x as $\sum_{l=1}^{k} r_l a_{\alpha_l, m_l} + b$, $b \in B$, $\alpha_1 < \alpha^*$. Of course w.l.o.g. $\alpha_k < \ldots < \alpha_1$. If k=0 we finish, otherwise let $m = \mathrm{Max}\{m_1, \ldots m_k\}$. As $a_{\alpha,l} - p a_{\alpha,l+1} \in B$ we can easily transform this to $\sum_{l=1}^{k} r_l a_{\alpha_l, m} + b'$.

2) and 3) Easy.

Stage E: Claim: If h is an endomorphism of H mapping B into $G_{J_1}$ such that for no $r \in R$ with h−r a small endomorphism of $G_J$, then for some $a^*_l \in H$, $p a^*_{l+1} - a^*_l \in B$ and for some $l_1$ $h(a^*_{l_1}) \notin SG(G_{J_1} \cup \{a^*_l : l < \omega\})$.

Proof: The proof is by cases.

Case I: There is $l(*) < \omega$ such that for every $n < \omega$ there are $r \in R$, $\eta \in \underline{T}$ satisfying $l(\eta) \ge n$ and $\underline{d}(h(p^{l(\eta)-l(*)} r x_\eta)) \nsubseteq \{\eta\}$. In this case we can easily choose by induction on $i<\omega$ $r_i \in R$, $\nu_i$, $\eta_i \in \underline{T}$ and $n_i<\omega$ such that:

(i) $\nu_i \in \underline{d}(h(p^{l(\eta)-l(*)} r_i x_{\eta_i}))$

(ii) $l(\eta_i) > n_i > \mathrm{Max}\{i+l(\eta_i)+l(*)+l(\nu_j): j<i\}$

(iii) $\nu_i \neq \eta_i$.

Now for every function $s \in {}^{\omega}\mathbb{Z}$ and $k<\omega$ we define $a_k^s = \sum_{k\leq i<\omega} s(i)p^{l(\eta_i)-l(*)-k} r_i x_{\eta_i}$.

We shall prove that for some s, $\langle a_k^s: k<\omega\rangle$ satisfies the requirements on $\langle a_k^*: k<\omega\rangle$ in the claim. Note that $a_k^s - pa_{k+1}^s \in B$ for every k (and $a_k^s \in H$).

We now define by induction on n, $K_n$ such that

(i) $K_n$ is a countable subset of $G_J$,

(ii) $K_n \subseteq K_{n+1}$

(iii) for $i<\omega$, $r_i x_{\eta_i} \in K_0$

(iv) if $x \in K_n$ then $h(x) \in K_{n+1}$

(v) if $x,y \in K_n$, then $x+y \in K_{n+1}$, $x-y \in K_{n+1}$

(vi) If $\rho \in \underline{d}(y)$, $y,z \in K_n$, $\alpha<\alpha^*$, $r \in R$, $\rho < \nu_\alpha$ and $\underline{d}(z-ra_{\alpha,m}) \cap \{\eta \in \underline{T}: \rho<\eta\}=\emptyset$, then $ra_{\alpha,m} \in K_{n+1}$ (note that $ra_{\alpha,m}$ is uniquely determined). Let $K=\bigcup_{n<\omega} K_n$ and $W^* = \bigcup\{\underline{d}(y): y \in K\}$; so clearly $W^*$ is countable.

We want to prove that $h(a_0^s) \notin SG(G_J \cup \{a_i^s : i<\omega\})$ for some s. We suppose this does not hold for a given s and shall get restrictions on s, so that this will guide us in choosing an appropiate s.

As $a_i^s - pa_i^s \in G_J$ for some $j_s$, and $r^s \in R$, $h(a_0^s) - r^s a_{j_s}^s \in G_J$. Hence for every $y \in K \cap G_J$ $h(a_0^s) - r^s a_{j_s}^s + y \in G_J$, hence applying stage D(1)

(*) $h(a_0^s) - r^s a_{j_s}^s + y = r_1 a_{\alpha_1,m} + r_2 a_{\alpha_2,m} + \ldots + r_k a_{\alpha_k,m} + b$, where

$\alpha_k < \ldots < \alpha_1 < \alpha^*$ , $b \in B$.

Of course $r_l, \alpha_l$ , m and b depend on s,y. Note that by stage D(2)(3) s and y determine k, $\alpha_l (l=1,..,k)$ uniquely, and essentially $r_l$ . Let $\underline{S}^0_y = \{s \in {}^\omega\mathbb{Z}$ : for y we get $\alpha_1$ minimal}, $S^1_y$ = $\{s \in {}^\omega\mathbb{Z}: h(a^s_0) = r^s a^s_{j_s} + r^s y \in B\}$. Our argument will rest on the computation of the domain. As for the left hand side

$$\underline{d}(h(a^s_0) - r^s a^s_{j_s} + y) \subseteq \underline{d}(h(a^s_0)) \cup \underline{d}(r^s a_{j_s}) \cup \underline{d}(y) \subseteq$$
$$\underline{d}(h(\sum_{i<\omega} s(i) p^{l(\eta_i) \dot{-} l(*)} r_i x_{\eta_i})) \cup \underline{d}(r^s a_{j_s}) \cup W^* \subseteq$$
$$\underline{d}(\sum_{i<\omega} s(i) p^{l(\eta_i) - l(*)} h(r_i x_{\eta_i})) \cup \{\eta_i : i<\omega\} \cup W^* \subseteq$$
$$\underline{d}(h(r_i x_{\eta_i})) \cup W^* \subseteq \bigcup_{i<\omega} \underline{d}(h(r_i x_{\eta_i})) \cup W^* \subseteq W^*.$$

Now we apply stage D(1) on the right hand side of (*). So there is $\rho \in \underline{\underline{T}}$ in the domain of the right hand side (hence of W*) such that:

$(**)_1$ $\underline{d}(r_1 a_{\alpha_1, m} + \ldots + r_k a_{\alpha_k, m} + b) \cap \{\eta \in \underline{\underline{T}}: \rho \leq \eta\}$ is a branch (except the first $l(\rho)$ elements) (in fact $\{\nu_{\alpha_1} \restriction i: l(\rho) \leq i < \omega\}$);

$(**)_2$ for some $r \in R$ every $\nu \in \{\eta \in \underline{\underline{T}}: \rho \leq \eta\}$ appears in $r_1 a_{\alpha_1, m} + \ldots + r_k a_{\alpha_k, m} + b$ as $p^{i-m} r x_\nu$ or $0 x_\nu$ .

We can substitute the left hand side of (*) and get $(**)'_1$, $(**)'_2$.

This is quite a strong restriction. From s (and remembering (i),(ii),(iii) above) we know much on $\underline{d}(h(a^s_0) - r_s a^s_{j_s})$, and so get a contradiction.

Now ${}^\omega\mathbb{Z}$ is a topological space having the Baire property. As W* is countable and the $\rho$ above is necessarily in W*, it suffices to prove

$(+)_1$ for every $\rho \in W$ , $y \in K$ the set of $s \in \underline{S}^0_y$ for which $(**)'_1$, $(**)'_2$ hold is meagre (= of the first category);

$(+)_2$ for every $y \in K$ the set $\underline{S}^1_y$ is meagre.

So let $\rho \in W$ , $u \in \bigcup_n {}^n\mathbb{Z}$, $n<\omega$ and we should find a function $t \in \bigcup_n {}^n\mathbb{Z}$ extending u such that no $s \in {}^\omega\mathbb{Z}$ extends t.

If $y + \sum_{i \in \mathrm{Dom}\, s} s(i)p^{l(\eta_i)-l(*)} h(r_i x_{\eta_i})$ does not satisfy $(**)_1'$ or $(**)_2'$, then there is $k<\omega$ such that this is examplified even if we restrict ourselves to $\eta \in \underline{\underline{T}}$ of length $\leq k$. However $i>k+l(*)$ implies $l(\eta_i) \geq i > k+l(*)$ which implies that $p^{l(\eta_i)-l(*)} h(r_i x_{\eta_i})$ is divisible by $p^{k+1}$ (in H). Hence every $\eta \in \underline{d}(p^{l(\eta_i)-l(*)} h(r_i x_{\eta_i}) \pm r_i x_{\eta_i})$ has lenght $>k$. So if $n' = \mathrm{Max}\{n, k+l(*)\}$ and $t(i)=0$ whenever $n \leq i < n'$, t extends s, then t is as required. This argument is a little inaccurate, because $\sum_{i \in \mathrm{Dom}\, s} r_s r_i x_{\eta_i}$ is represented in the left hand side of (*), but as this involves finitely many members of $\underline{\underline{T}}$ it can be correct trivially.

So we can assume that $y + \sum_{i \in \mathrm{Dom}\, s} s(i)p^{l(\eta_i)-l(*)} h(r_i x_{\eta_i})$ satisfies $(**)_1'$, $(**)_2'$ if we ignore $\{\eta_i : i \in \mathrm{Dom}\, s\}$. Moreover w.l.o.g. this holds for every $t \in \bigcup_{n<\omega} {}^n\mathbb{Z}$ extending s.

Now if we can find $s_a, s_b, s_c \in \underline{S}_y^0$ as exemplified by distinct $\alpha_{1,a}$, $\alpha_{2,b}$, $\alpha_{3,c}$, such that $s_a$, $s_b$, $s_c$ extend s and $|\{\alpha_{1,a}, \alpha_{2,b}, \alpha_{3,c}\}| \geq 3$ let $s^* \in {}^\omega\mathbb{Z}$ be defined by $s^*(n) = s_a(n) - s_b(n) + s_c(n)$ and then $t = s^* \restriction n$, n large enough, extends s as required. Otherwise only say $\alpha_{1,a}$, $\alpha_{2,b}$ appear and the contradiction is even easier.

So it remains to prove $(+)_2$, i.e. suppose some $\underline{S}_y^1$ is not meager and get a contradiction. As easily $\underline{d}(b) \subseteq W^*$, clearly $b \in K$, so w.l.o.g. $b=0$. Looking at "how $x_{\nu_i}$ may appear" this is trivial.

Let $h(rx_\eta) = a_{r,\eta} x_\eta + h_1(rx_\eta)$ $(a_{r,\eta} \in R)$. Note that $h_1(rx_\eta)$ is not necessarily an endomorphism but a function from $G_J$ to $G_J$ which is small.

<u>Case II</u>: For some $l(*)$ for every $m>l(*)$ there are $\eta^a, \eta^b \in \underline{T}$, $l(\eta^a)$, $l(\eta^b)>m$ and $r^a, r^b \in R$ such that $p^{l(\eta^a)-l(*)} a_{r^a, \eta^a} x_{\eta^a} + p^{l(\eta^b)-l(*)} a_{r^b, \eta^b} x_{\eta^b}$ is not an R-multiple of $r^a x_{\eta^a} + r^b x_{\eta^b}$.

We can easily assume $\eta^a \neq \eta^b$ (as we can try a third candidate). The proof is like case I (using $r_i^a x_{\eta_i^a} + r_i^b x_{\eta_i^b}$ instead of $r_i x_{\eta_i}$)).

<u>Case III</u>: not case I nor case II. In this case easily for some $r \in R$, $h-r$ is small.

<u>Stage F</u>: <u>Claim</u>: If $h: G_\emptyset \longrightarrow G_{J_1}$ is a homomorphism, then for some $r \in R$, $h-r$ is small.

<u>Proof</u>: Suppose there is no such r. We can extend h uniquely to an endomorphism $\hat{h}$ of H. Let $\hat{h}(\sum_{\eta \in \underline{T}} r_\eta x_\eta) = \sum_{\eta \in \underline{T}} h(r_\eta x_\eta)$ and $l(*)$, $a_l^* \in B$ $(l<\omega)$ be from stage E.

Now we can define a strategy for player II in the play $\underline{Gm}(\{(f^\alpha, N^\alpha): \alpha<\alpha^*\})$ (see §2). He plays so that $|N_m| \cap H$ is closed under h, $a_1^* \in N_0$, $h^{N_m} = h \restriction |N_m|$, $L^{H_m} = |N_m| \cap G_{J_1}$. An $(f^\alpha, N^\alpha)$ is a barrier for some $\alpha$, $(f^\alpha, N^\alpha)$ is the result of such a play. Let $l(\alpha) = l(*)$ and $\nu$ be a branch of $\mathrm{Rang}(f^\alpha)$ not in $\{\nu_\beta : \beta<\alpha\}$. We want to show that in (8) of stage C there are $a^0_{\alpha,l}$ $(l<\omega)$, $b_\alpha$ as required there ($a^1_{\alpha,m}$ is already defined from $\nu$). Ignoring requirement (7) for a moment this clearly suffices, as then $\hat{h}(a_{\alpha,l(\alpha)})$ necessarily belongs to $G_{J_1}$, but it is $b_\alpha$ and $b_\alpha \notin G_{J_1}$.

First try $a^0_{\alpha,m}=0$, then the only thing that can go wrong is

$h(a^0_{\alpha,1(*)}) \in SG(G^{\alpha}_{J_1\cap\alpha} \cup \{a^0_{\alpha,m} : m<\omega\})$, i.e. for some m and $r_1 \in R$

(1) $h(a^1_{\alpha,1(*)}) - r_1 a^1_{\alpha,m} \in G^{\alpha}_{J_1\cap\alpha}$.

If this fails try $a^0_{\alpha,m} = a^*_m$ and again the only thing that can go wrong is $h(a^1_{\alpha,1(*)} + a^0_{\alpha,1(*)}) \in SG(G^{\alpha}_{J_1\cap\alpha} \cup \{a^0_{\alpha,m} : m<\omega\})$, so for some $r_2 \in R$

(2) $h(a^1_{\alpha,1(*)} + a^0_{\alpha,1(*)}) - r_2(a^1_{\alpha,1(*)} + a^0_{\alpha,1(*)}) \in G^{\alpha}_{J_1\cap\alpha}$.

Subtracting (2) from (1) we get

(3) $h(a^0_{\alpha,1(*)}) - r_2 a^0_{\alpha,1(*)} + (r_1-r_2)a^1_{\alpha,1(*)} \in G^{\alpha}_{J_1\cap\alpha}$.

As $a^0_{\alpha,1}$, $h(a^0_{\alpha,1}) \in N^{\alpha}_0$, computation of the domain (using stage D) leads to $\nu=\nu_\beta$ for some $\beta<\alpha$; contradiction.

We shall denote the successful try by $a^{0,\nu}_{\alpha,1}$, $a^{1,\nu}_{\alpha,1}$. As $\nu_\beta(\beta<\alpha)$ can be a branch of Rang($f^\alpha$) only if $\beta<\alpha<\beta+2^{\aleph_0}$ (see (f) of Th. 2.8.) all except $<2^{\aleph_0}$ branches of Rang($f^\alpha$) will do. We still have to deal with requirement (7) from stage C. We deal with it for each $\beta$.

If $\beta+2^{\aleph_0}\leq\alpha$, comparison of domains leads to a contradiction.

As $|\{\beta: \beta<\alpha<\beta+2^{\aleph_0}\}|<2^{\aleph_0}$, it suffices to prove that for each such $\beta$, $b_\beta \in SG(G^{\alpha}_{J_1\cap\alpha} \cup \{a^{0,\nu}_{\alpha,1} + a^{1,\nu}_{\alpha,1} : 1<\omega\}$ for at most one $\nu$ (as then all branches of Rang($f^\alpha$) except $<2^{\aleph_0}$ will do).

So suppose $\nu^1\neq\nu^2$ are branches of Rang($f^\alpha$) (but $\notin\{\nu_\gamma:\gamma<\alpha\}$) and for i=1,2:

$$b_\beta \in SG(G^{\alpha}_{J_1\cap\alpha} \cup \{a^{0,\nu^i}_{\alpha,1} + a^{1,\nu^i}_{\alpha,1} : 1<\omega\}).$$

So (for i=1,2) for some m: $b_\beta - r^i(a^{0,\nu^i}_{\alpha,m} + a^1_{\alpha,m}) \in G^{\alpha}_{J_1\cap\alpha}$.

Subtracting we get

$$(r^1 a^{0,\nu^1}_{\alpha,m} - r^2 a^{0,\nu^2}_{\alpha,m}) + (r^1 a^{0,\nu^1}_{\alpha,m} - r^2 a^{0,\nu^2}_{\alpha,m}) \in G^{\alpha}_{J_1\cap\alpha}.$$

Computing domains we get a final contradiction.

Stage G: Proof of the Theorem

It is easy to show that $|J_1| = \lambda^{\aleph_0}$. Now by stage F each $G_J (J \subseteq J_1)$ has no "undesirable" endomorphisms. Let $\{J^\xi : \xi < 2^{\aleph_0}\}$ be a family of subsets of $J_1$ with $|J^\xi - J^\zeta| = \lambda^{\aleph_0}$ for $\xi \neq \zeta$. So it suffices to prove for $\xi \neq \zeta$ that every homomorphism h from $G_{J^\xi}$ to $G_{J^\zeta}$ is small. But by stage F h-r is small for some $r \in R$ (and clearly r is unique). Let $\alpha \in J^\xi - J^\zeta$, and consider $h(a_{\alpha,l(\alpha)})$; for every m>l $a_{\alpha,l(\alpha)} - p^{m-l(\alpha)} a_{\alpha,m} \in B$, and we know that $(h-r)(p^{m-l(\alpha)} a_{\alpha,m}) = 0$ for every large enough $m<\omega$. So $(h-r)(a_{\alpha,l(\alpha)}) = (h-r)(y_\alpha)$ for some $y_\alpha \in B$. As $|J^\xi - J^\zeta| > \lambda + |R| = |B|$, for some $\alpha \neq \beta \in J^\xi - J^\zeta$ $y_\alpha = y_\beta$. So $(h-r)(a_{\alpha,l(\alpha)} - a_{\beta,l(\beta)}) = 0$, hence $h(a_{\alpha,l(\alpha)} - a_{\beta,l(\beta)}) = r(a_{\alpha,l(\alpha)} - a_{\beta,l(\beta)})$. But by stage D $r(a_{\alpha,l(\alpha)} - a_{\beta,l(\beta)}) \in G_{J^\zeta}$ implies r=0, so h is small.

## §4 The necessity of $|G|^{\aleph_0} = |G|$ in the groups we have constructed

In the previous section (and in [Sh 8] we have constructed an abelian p-group G with a prescribed ring $R = \mathrm{End}(G)/E_s(G)$. For an arbitrary $\lambda$ we build such a group of power $(|R| + \lambda^{\aleph_0})$. The restriction $|G| \geq |R|$ is obvious. The stronger restriction $|G| \geq |R|^{\aleph_0}$ is also quite necessary as we shall show in (4.2).

4.1. Context: Let G be an abelian reduced p-group. So there are $\lambda_n, x_i^n$ $(i < \lambda_n)$ such that $\{x_i^n : i < \lambda_n, n < \omega\}$ generate an abelian group freely except that $p^{n+1} x_i^n = 0$, $B \subseteq G$, B is dense in G. So renaming we can assume that G is contained in $\hat{B}$, the torsion-completion of B.

So for every $x \in S$, $x = \sum_{(n,i) \in S} a_i^{n,x} x_i^n$, where $S_x \subseteq \{(n,i): n < \omega, i < \lambda_n\}$ is countable, $a_i^{n,x} \in \mathbb{Z}$, and for every n $\{i : (n,i) \in S_x\}$ is finite.

It is known that $G = G^0 + G^1$ (direct sum), $G^1$ is bounded (i.e. $(\exists n)(\forall x \in G^1)\ p^n x = 0)$ and for $G^0$ the cardinals $\lambda_n$ satisfies $(\forall n)(\exists^\infty m)$ $\lambda_n \leq \lambda_m$. It is known that ess pow(G) = Min$\{\lambda$: for every n large enough, $\lambda_n \leq \lambda\}$.

4.2. Theorem: If G is an abelian separable reduced p-group of ess pow(G) $= \lambda > 2^{\aleph_0}$, then $\mathrm{End}(G)/E_s(G)$ has power $2^\lambda$.

Remark: The proof will give much more explicit information.

Proof: It is easy to show that for $G = G^0 + G^1$, $G^1$ bounded, the rings $End(G)/E_s(G)$ and $End(G^0)/E_s(G^0)$ are isomorphic and ess pow(G) = ess pow($G^0$).

So we can assume G as in (4.1), and for every n there are infinitely many m with $\lambda_n < \lambda_m$. Let $\lambda = \sum_{m<\omega} \lambda_m$.

We know that every endomorphism of G is determined by its restriction to B. Now the number of functions from B to G is $\leq |G|^{|B|} \leq (\lambda^{\aleph_0})^\lambda = 2^\lambda$. So $End(G)/E_s(G)$ has power $\leq 2^\lambda$.

4.3. Fact: Suppose $A_m \subseteq \lambda_m$, $A = \bigcup_m \{n\} \times A_n$, and let $G_A = \bigoplus_{(n,i)\in A} \mathbb{Z} x_i^n$. A sufficient condition for $G_A$ to be a direct summand of G is:

(*) For every $x \in G$, $S_x \cap A$ is finite (on $S_x$ see 4.1)

Proof of the fact: We shall define a projection h from G onto $G_A$: $h(x) = h(\sum_{(n,i)\in S_x} a_i^{n,x} x_i^n) = \sum_{(n,i)\in A} a_i^{n,x} x_i^n$. This is well-defined (and the result is in $G_A$) by (*); and the checking is easy.

4.4. Fact: There are $A_n \subseteq \lambda_m$, $\sum |A_m| = \lambda$ such that $\bigcup_{n<\omega} \{n\} \times A_m$ satisfies (*).

This suffices for 4.2. We have already proved $|End(G)/E_s(G)| \leq 2^\lambda$. Clearly by 4.3 $|End(G)/E_s(G)| \geq |End(G_A)/E_s(G_A)|$ and clearly $|End(G_A)/E_s(G_A)| \geq 2^\lambda$.

Proof of 4.4.: Choose $n(k) < \omega$ for $k < \omega$ such that $n(k) < n(k+1)$, $\lambda_{n(k)} \leq \lambda_{n(k+1)}$, and $\lambda = \sum_k \lambda_{n(k)}$. Now we can choose by induction on k, for every $\eta \in \prod_{m\le k} \lambda_{n(m)}$ subsets $A_\eta$ of $\lambda_{n(k)}$ such that:

(i) $|A_\eta| = \lambda_{n(l(\eta))}$

(ii) for $\eta \ne \nu \in \prod_{m\le k} \lambda_{n(m)}$, $A_\eta \cap A_\nu = \emptyset$.

This is easily done. Now for every $\eta \in \prod_{k<\omega} \lambda_{n(k)}$ let $A_l^\eta$ be $A_{\eta\restriction(k+1)}$ if $l=n(k)$ and $\emptyset$ otherwise. Clearly $A_l^\eta \subseteq \lambda_l$, $\sum_l |A_l^\eta| = \sum_k \lambda_{n(k)} = \lambda$, hence it suffices to prove that for some $\eta \in \prod_{k<\omega} \lambda_{n(k)}$, $A^\eta = \bigcup_l \{l\} \times A_l^\eta$ satisfies (*).

As the number of $\eta \in \prod_k \lambda_{n(k)} = \lambda^{\aleph_0}$ is $> |G| + 2^{\aleph_0}$, it suffices to prove:

(**) for every $x \in G$ the number of $\eta \in \prod_k \lambda_{n(k)}$ for which $S_x \cap A^\eta$ is infinite, is $\le 2^{\aleph_0}$.

This is easy: for suppose $\eta_i \in \prod_k \lambda_{n(k)}$ are distinct, for $i<(2^{\aleph_0})^+$, the number of possible $S_x \cap A^{\eta_i}$ is $\le 2^{\aleph_0}$ (= the number of subsets of $S_x$). Hence for some $i \ne j$ $S_x \cap A^{\eta_i} = S_x \cap A^{\eta_j}$ is infinite but $S_x \cap (\{l\} \times \lambda_l)$ is finite for each $l$. Hence for no $n$ is $A^{\eta_i} \cap A^{\eta_j} \subseteq \bigcup_{l\le n} (\{l\} \times \lambda_l)$. But for some $n$ $\eta_i(n) \ne \eta_j(n)$, hence easily $A^{\eta_i} \cap A^{\eta_j} \subseteq \bigcup_{l<n} \{l\} \times \lambda_l$; contradiction.

<u>4.5. Lemma</u>: 1) Assume MA + $2^{\aleph_0} > \lambda$. If $G$ is a separable (abelian) p-group, $\lambda$ = ess pow(G), then $\mathrm{End}(G)/E_s(G)$ has power $2^\lambda$.

2) Assume V (=the universe of set theory) is a generic extension of V′ by adding $\lambda$ many Cohen reals, $\lambda > \aleph_0$. Assume R is a ring, $R^+$ the completion of a direct sum of copies of $\underline{I}_p^+$, $|R^+/p^n R^+| = \lambda$. Then in V there is a separable p-group G, $|G| = \lambda$, $\mathrm{End}(G) \cong E_s(G) + R$.

Proof: We define a forcing notion P: $P = \{(A,B): B,A \subseteq \{x_i^n: i<\lambda_n, n<\omega\}$, $B \cap A = \emptyset$, A finite and for some finite $Y \subseteq G$, $B = \bigcup_{y \in Y} d(y) - \omega\}$; order natural. P satisfies the countable chain condition: if $(A_\alpha, B_\alpha) \in P$ for $\alpha < \omega_1$ w.l.o.g. for some n(*) $A_\alpha \subseteq \{x_i^n: i<\lambda_n, n<n(*)\}$. Note that $A_\alpha \cap \{x_i^n: i<\lambda, n<n(*)\}$ is finite for each $\alpha$, the rest is by the $\Delta$-system lemma.

So by MA we can easily get $A_n \subseteq \lambda_n$, $|A_n| = \lambda_n$ satisfying (*) of Fact 4.3, and we finish as in the proof of 4.2.

2) Left to the reader (provided that he knows what Cohen reals are).

4.6. Remark: We may wonder when we can have $|R| > \lambda^{\aleph_0}$. Now

(*) there are left ideals $I_\alpha$ $(\alpha<\lambda)$ of R such that

(i) $\bigcap_{\alpha<\lambda} I_\alpha = \{0\}$

(ii) if $\langle x_\alpha : \alpha < \lambda \rangle$ is a sequence of members of R satisfying $(x_\alpha + I_\alpha) \cap (x_\beta + I_\beta) \neq \emptyset$ for every $\alpha, \beta < \lambda$, then $\bigcap_{\alpha<\lambda} (x_\alpha + I_\alpha) \neq \emptyset$.

This seems necessary (if $R = \mathrm{End}(G)/E_S(G)$, $B \subseteq G$ is basic, $B = \{x_i: i<\lambda\}$, let $I_i = \{r \in R: r\, x_i = 0\}$) and sufficient (proof as in (1)) with B being the R-module freely generated by $\{x_\eta^i: i<\lambda, \eta \in \underline{T}\}$ except $p^{l(\eta)+1} x_\eta^i = 0$, $r x_\eta^i = 0$ $(r \in I_i)$.

## §5 Abelian groups with predetermined ring of endomorphisms

What can we say about End(G)=R ?. Clearly R is a ring with identity 1 and G is an R-module. If $n\cdot 1=0$, then $n\cdot G=0$ and G is a direct sum of cyclic groups. So we may discard this case. Trivially G has a divisible sub-group $D\neq 0$ iff $R^+$ has a divisible subgroup $\neq 0$. In this case every homomorphism h of G into D extends to an endomorphism of G, so we cannot control End(G). If G is uncountable then End(G) has cardinality $2^{|G|}$.

Hence we assume that G is reduced, i.e. G has no nonzero divisible subgroup. Define the $\mathbb{Z}$-adic metric d on it:

$$d(x,y) = \mathrm{Min}\{2^{n} : n! \text{ divides } x-y\}.$$

We can now define the completion $G^c$ of G.

5.1. Theorem: Suppose R is a ring with 1, characteristic 0 such that $R^+$ is reduced. Suppose also $G_0$ an R-module with $G_0^+$ reduced. Suppose further that $\lambda^{\aleph_0}\geq |G_0|+|R|$, $\mathrm{cf}\,\lambda > \aleph_0$.

1) There is an R-module K extending $G_0$ such that:

(a) K has cardinality $\lambda^{\aleph_0}$.

(b) $K/G_0$ is an $\aleph_1$-free R-module.

(c) If $h\in \mathrm{End}(K)$, we find $r\in R$ such that $h-r=h'$ is inessential, i.e. in this context $\mathrm{Rang}(h')\subseteq SG(G_0\cup A\cup G[q])$ for some finite $A\subseteq G$, $q\in\mathbb{Z}$.

2) If $R^+$, $G_0$ are cotorsion-free (i.e. in addition $R^+$, $G_0$ have no subgroups isomorphic to the additive groups of p-adic integers),

then h-r=0 in (c).

Proof:

Stage A: Let B be the free R-module generated by $G_0 \cup \{x_\eta : \eta \in \underline{\underline{T}}\}$ except the equations which hold in $G_0$. Let H be the completion of B; so every $y \in B$ has the form $y = \sum_{\eta \in \underline{\underline{T}}} r_\eta x_\eta + \sum_{n<\omega} g_n$, $g_n \in G_0$, $r_\eta \in R^c$ for all but finitely many $\eta \in \underline{\underline{T}}$ and $n<\omega$, (n!) divides $r_\eta$ and (n!) divides $g_n$ (in $G_0$). Note that $\underline{d}(y) = \{\eta \in \underline{\underline{T}} : r_\eta \neq 0\}$ is countable but $\underline{d}(y) \cap \prod_{n<m} \lambda_n$ may be infinite. However (A)(1),(2),(3),(4) (from (3.5)) still hold (replacing $p^n$ by n!).

We can define $\underline{d}_n(y) = \{\eta \in \underline{\underline{T}} : r_\eta$ is divisible by n} and use it similarly.

Stage B: As in (3.5).

Stage C: The construction

We identify H with a subset of $\mathfrak{M}$. We define by induction on $\alpha<\alpha^*$:

1) The truth value of $\alpha \in J_0$.

2) For $\alpha \in J_0$ we define $a_{\alpha,l} \in H$ for $l<\omega$.

3) For $\alpha \in J$, a branch $\nu_\alpha$ of Range($f^\alpha$) such that, for $\alpha \in J_0$

4) $a_{\alpha,m} = a^0_{\alpha,m} + a^1_{\alpha,m}$ (both in H), $a^1_{\alpha,m} = \sum_{k<\omega} \left( \prod_{m\leq i\leq k} i! \right) x_{\nu_\alpha \restriction k}$ and (l!) $a^i_{\alpha,l+1} - a^i_{\alpha,l} \in B$

5) $a^0_{\alpha,m} \in N^\alpha_0$

6) $\nu_\alpha \neq \nu_\beta$ for $\beta<\alpha$

Let $G^\alpha = SG(B \cup \{a_{\beta,m} : \beta<\alpha,\ m<\omega\})$.

7) $b_\beta \notin G^\alpha$ for $\beta<\alpha$.

8) If $N^\alpha = (|N^\alpha|, L, h, \ldots)$, L a subgroup of $G^\alpha$, h an endomorphism of L, and we cannot find $\nu^\alpha$, $a^i_{\alpha,m}$ $(i=0,1;\ m<\omega)$, $b_\alpha$, $l(\alpha)<\omega$ satisfying (2)-(7) for $\alpha+1$ (stipulating $\alpha \in J_0$) such that for every endomorphism h′ of a group G′, $G^\alpha \subseteq G' \subseteq H$, extending h, $h'(a_{\alpha,l(\alpha)}) = b_\alpha$, then $\alpha \in J_0$ and for every endomorphism h′ of a group G′, $G_\alpha \subseteq G' \subseteq H$, extending h, $h'(a_{\alpha,l(\alpha)}) = b_\alpha$.

<u>Stage D</u>: As in (3.5). We can note there that $G^\alpha \cap G_0^c = G_0$.

<u>Stage E</u>: <u>Claim</u>: If h is an endomorphism of $G=G^{\alpha^*}$ such that for no $r \in R$ h-r is inessential, then for some $a^*_l \in H$, $la^*_{l+1} - a^*_l \in B$ and $h(a^*_l) \notin SG(G \cup \{a^*_l : l<\omega\})$.

<u>Proof</u>: It is done by cases:

<u>Case I</u>: For every finite $W \subseteq \underline{T}$ and $l<\omega$ $(l>0)$ for some $\eta, \nu \in \underline{T} - W$ and $r \in R$, $\eta \neq \nu$ and $\nu$ appears in $h(lrx_\eta)$. This is handled like case I of stage E in (3.5) (but here, in order to guarantee that $\nu$ appears in the infinite sum, we split the sum into two parts - in the first it appears, but the coefficient is not divisible by some k, in the second it appears with coefficient divisible by k).

So for some finite W* and $l^* \in \{1,2,3,\ldots\}$ for every $r \in R$, $\eta \in \underline{T} - W$, for some $a_{r,\eta} \in R$ $h(l^*rx_\eta) = a_{r,\eta}\ x_\eta \in SG(G_0 \cup \{x_\eta : \eta \in W^*\})$.

Case II: Not case I, but for every finite $W \subseteq \underline{\underline{T}}$ and $k<\omega$ for some $\eta^a, \eta^b \in \underline{\underline{T}} - W \cup W^*$, $r^a \in R$, $r^b \in R$ the following holds: $1r^a \neq 1r^b$. As in (3.5).

Case III: Neither case I nor case II.

Let $W^{**}$, $1^{**}$ exemplify the failure of case II (and w.l.o.g. $W^* \subseteq W^{**}$). Clearly there is $r^* \in R^c$, such that for every $\eta \in \underline{\underline{T}} - W^* \cup W^{**}$ and $r \in R$, $1^{**}1^*a_{r,\eta} = 1^{**}r^*$. So for $\eta \in \underline{\underline{T}} - W^{**}$ $h(1^{**}1^*rx\ ) - 1^{**}r^*rx_\eta \in SG(G_0 \cup \{x_\eta : \eta \in W^*\})$, so choosing $r=1$ ($\eta \in \underline{\underline{T}} - W^{**}$) we see that $1^*$ divides $r^*$ and the result is in $R$, so let $r^*=1^*r^{**}$. We can conclude that for some $\eta \in \underline{\underline{T}} - W^* \cup W^{**}$, $1^{**}1^*(h(rx_\eta)-r^{**}(rx_\eta)) \in SG(G_0 \cup \{x_\eta : \eta \in W^*\})$. So $(h-r^{**})(rx_\eta) \in SG(G_0 \cup \{x_\eta : \eta \in W^*\} \cup G[1^{**}1^*])$.

We still have to prove that $h'(SG(G_0 \cup \{x_\eta : \eta \in W^*\} \subseteq SG(G_0 \cup A \cup G(k))$ for some finite $A$ and $k$, where $h'=h-r^{**}$. We can choose by induction on $n$ finite subsets $W_n \subseteq \underline{\underline{T}}$, $U_n \subseteq \alpha^*$ and $g_n \in SG(G_0 \cup \{x_\eta : \eta \in W^*\})$ such that $W^{**} \subseteq W_n \subseteq W_{n+1}$, $U_n \subseteq U_{n+1}$ and

$$h((n!)g_n) \notin SG(G_0 \cup \{x_\eta : \eta \in W_n\} \cup \{a_{\alpha,1} : \alpha \in U_n, 1<|U_n|\}).$$

We then get a contradiction as in case I of stage E in (3.5).

Stage F: Claim: For every endomorphism $h$ of $G$, for some $r \in R$, $h-r$ is inessential.

As in (3.5).

Stage G: $K/G_0$ is an $\aleph_1$-free $R$-module. Easy.

Proof of theorem 5.1:

1) As in (3.5).

2) So suppose h is inessential and nonzero. By stage G there is a finitely generated R-submodule L of K, $L \cong \bigoplus_{i=1}^{n} R^{+}$, $L \cap G_0 = \{0\}$, and $n < \omega$, $\mathrm{Rang}(n\,h) \subseteq L + G_0$, but as G is torsion-free, we can disregard n. Using projection w.l.o.g. $\mathrm{Rang}(h) \subseteq G_0$ or $\mathrm{Rang}(h) \subseteq L'$, $L' \cong R^{+}$. Also Rang(h) is complete. (Otherwise we can prove the conclusion of stage E (if $\sum \prod_{i=1}^{n} (i!)h(a_n) \notin G$, $a_n \in G$, choose $\eta \notin \{\eta_i : i\}$, $\eta$ an $\omega$-branch of $\underline{T}$, and try $a_n^* = \sum_{m \geq n} \prod_{i=n}^{m} i!(h(a_n) + x_{\eta \restriction m})$ and also try $a_n^* = \sum_{m \geq n} \prod_{i=1}^{m} i!\,1 x_{\eta \restriction m})$ and then get a contradiction as in stage F).

## §6 Revisiting the combinatorics

The combinatorics in §1 and §2 can be strengthened and modified in various ways, which may be useful in other contexts.

6.1. Claim: In the context of section 1 or 2 suppose $\chi$ is a cardinal satisfying $\lambda^{\chi} = \lambda$, $\chi \geq \kappa^{\aleph_0} + \kappa^+$ (e.g. $\chi = (\kappa^{\aleph_0})^+$).

Then we can prove Th. 1.13 or 2.8 when weaken (b) and strengthen (f) to

(b′) W is a disjoint barrier (not necessarily strong).

(f′) If $\alpha<\beta<\alpha^*$, $\eta$ is a branch of Rang($f^{\beta}$), then for some k, $\eta \upharpoonright k \notin N^{\alpha}$.

Proof: Let $W^* = \{(f^{\alpha}_*, N^{\alpha}_*) : \alpha<\alpha^*\}$, $\zeta^*$ be what we get applying (1.13) (or (2.8)) for $\lambda, \chi$ (instead of $(\lambda,\kappa)$). We consider only those $\alpha$'s for which $N^{\alpha}_*$ codes a tree $\{(f^{\alpha,\eta}, N^{\alpha,\eta}) : \eta \in {}^{\omega>}\chi\}$ such that:

(i) $N^{\alpha}_{*,n}$ codes $\{(f^{\alpha,\eta}, N^{\alpha,\eta}) : \eta \in {}^{n\geq}\chi\}$ and includes each $|N^{\alpha,\eta}|$.

(ii) $\langle (f^{\alpha,\eta\upharpoonright l}, N^{\alpha,\eta\upharpoonright l}) : l<l(\eta)\rangle \in J_n$ for $\eta \in {}^{n}\chi$.

(iii) Rang($f^{\alpha,\eta}$) $\subseteq$ Rang($f^{\alpha}$).

(iv) if $\eta, \nu \in {}^{(n+1)}\chi$, $\eta(1) \neq \nu(1)$, then Rang($f^{\alpha,\eta}$) $\cap$ Rang($f^{\alpha,\nu}$) $\subseteq$ Rang($f^{\alpha,\eta\upharpoonright 1}$).

We now define by induction on $\alpha$ ($\alpha$ as above) a member $\rho_\alpha$ of ${}^{\omega}\chi$ such that

(*) if $\beta<\alpha<\beta+\chi^{\aleph_0}$, then for every $\omega$-branch $\eta$ of Rang($\bigcup_{k<\omega} f^{\alpha,\rho_\alpha\upharpoonright k}$), for some $m<\omega$, $\eta(0) \notin \bigcup_{k<\omega} N^{\beta,\rho_\beta\upharpoonright k}$.

Why is this possible? For a given $\alpha$ there are less than $\aleph_0$ possible $\beta$'s, and for each $\beta$ the number of "unsuccessful" $\rho \in {}^{\omega}\chi$ is $\le |\bigcup_{k<\omega} N^{\beta,\rho_\beta \restriction k}| = \kappa$. So the number of unsuccessful $\rho \in {}^{\omega}\chi$ is $\le \kappa|\{\beta : \beta < \alpha < \beta + \chi^{\aleph_0}\}|$. As $\chi \ge \kappa^{\aleph_0} + \kappa^{+}$ we finish the proof of (*).

Now $W = \{ (\bigcup_{k<\omega} f^{\alpha,\rho_\alpha \restriction k}, \bigcup_{k<\omega} N^{\alpha,\rho_k \restriction k}) : \alpha < \alpha^*$, $\alpha$ as above$\}$ is the required barrier (using the same function $\zeta^*$).

We use only some $\alpha < \alpha^*$, but this is a minor point.

Why is W a barrier? Suppose player II has a winning strategy of $\underset{=}{Gm}_{\lambda,\kappa}(W)$ and we can easily describe one for $\underset{=}{Gm}'_{\lambda,\kappa}(W^*)$.

<u>6.2.</u> <u>Remark</u>: 1) In (6.1), we may sometimes weaken the demand on $\chi$ to $\chi \ge (\kappa^{+})^{\omega}$. We need that ${}^{\omega\ge}\chi$ is not $\bigcup\{T_i : i < \chi^{\aleph_0}\}$, each $T_i$ is closed under initial segments and $|T_i \cap {}^{\omega>}\chi| \le \kappa$.

<u>6.3.</u> <u>Concluding Remark</u>: 1) We can let player II determine $f_n$ for, say, odd n, with no significant change.

2) We can make the games last $\vartheta$ moves ($\vartheta \ne \omega$), which gives no significant change.

3) For strong limit singular $\lambda$, we can use the theorem from Rubin, Shelah [R Sh] to get similar theorems, weakening a little the "barrier" condition. (E.g. if $cf\lambda = \aleph_0$, we know that for any model with universe of power $\lambda$ and $\chi < \lambda$ operations, there is a $\Delta$-system tree of sub-models. Now we can list all such trees $\{T_i : i < \lambda^{\aleph_0} = 2^{\lambda}\}$ and choose by induction a branch from each, so that they are as disjoint as possible. For $cf\lambda > \aleph_0$, $\lambda = \sum_{i<cf\lambda} \lambda_i$, $\lambda_i$ increasing continuous,

$2^{\lambda_i} < \lambda_{i+1} < \lambda$, we deal with each $\lambda_j$, $cf\delta = \aleph_0$ ; see also [Sh 7]

4) Note that some of the combinatorics of [Sh 4] is not used here: if we have enough elements, for some large subset of them their domain behaves as a $\Delta$-system, with the same coefficient of the common parts, so the difference of any two has domain disjoint to the "heart", so we can make it to be disjoint to a predescribed set.

6.4. Definition: We define the game $\underline{\underline{Gm}}''(W)$ as $\underline{\underline{Gm}}'(W)$, but ($\vartheta$ a regular cardinal, $\underline{\underline{T}} = {}^{\vartheta>}\lambda$)

(i) the game lasts $\vartheta$ moves.

(ii) $\mathrm{Dom}(f_\alpha)$ is any subset of ${}^{\vartheta>}\kappa$, closed under initial segments but with no $\vartheta$-branch.

(iii) in odd stages $\alpha$ player I chooses $f_\alpha$ .

6.5. Theorem: Suppose $\vartheta \leq \kappa$, $\lambda^{<\vartheta} = \lambda^\kappa$. Then for some $W=\{(f^\alpha, N^\alpha): \alpha<\alpha^*\}$ and function $\zeta$

1) If $cf\lambda > \vartheta > \aleph_0$, then

(a) W is disjoint, and in $\underline{\underline{Gm}}''(W)$ player II has a winning strategy.

(b) For $\alpha<\beta<\alpha^*$, $\zeta(\alpha) \leq \zeta(\beta)$.

(c) $cf(\zeta(\alpha)) = \vartheta$.

(d) Every branch $\eta$ of $f^\alpha$ satisfies: $(\forall i<\vartheta)$ $\eta(i)<\zeta(\alpha)$ and $\zeta(\alpha) = \bigcup_{i<\vartheta}\eta(i)$.

(e) for every $i<\vartheta$ for some $\xi<\zeta(\alpha)$, $\mathrm{orco}(N_i^\alpha) \leq \xi$.

(f) If $\alpha+\kappa^\vartheta \leq \beta < \alpha^*$ and $\eta$ is a $\vartheta$-branch of $\mathrm{Rang}(f^\beta)$, then $\eta \restriction i \notin N^\alpha$ for some $i<\vartheta$.

(g) If $\lambda=\lambda^{\kappa}$ we can demand: if $\eta$ is a $\vartheta$-branch of Rang($f^{\alpha}$) and $\eta \restriction i \in N^{\beta}$ for every $i<\vartheta$ (where $\alpha,\beta<\alpha^*$), then $N^{\alpha} \subseteq N^{\beta}$.

2) As (1) is the parallel to (1.13)(1), so parallel to (1.13)(2), (2.8),(2.9),(6.1) holds.

<u>On the proof</u>: The point is that, if $\eta$ is a $\vartheta$-branch, for it to "code the play" it is enough that for a closed unbounded set of $i<\vartheta$, $\eta(i)$ code appropiate information on the first i moves. (When $\lambda<\lambda^{<\vartheta}$, remember we can split $\vartheta$ to $\vartheta$ disjoint stationary sets).

<u>6.6. Remark</u>: So clearly we could have divided the choices of the $f_{\alpha}$'s between player I and II differently, as long as for each $\vartheta$-branch $\eta$ of $\bigcup_{i<\vartheta} f_i$, {i: player II chooses $\eta(i)$} belong to D, for some fixed filter D on $\vartheta$.

<u>6.7. Theorem</u>: In (6.1) we can strengthen it by replacing (b) by

(b)" W is disjoint and player II has no winning strategy in $\underline{\underline{\text{Gm}}}$"(W).

<u>Point of the Proof</u>: Unlike (6.5) we do not have a filter on $\omega$, but we can try for each $\eta$ all infinite subsets of $\omega$ as "the set of choices of player II".

## References


[CG] A.L.S. Corner and R. Göbel, Prescribing endomorphism algebras - a uniform treatment, Proc. London Math. Soc. (to appear).

[C] A.L.S. Corner, Every countable reduced torsion free ring is an endomorphism ring, Proc. London Math. Soc. (3) 13 (1963), 687-710.

[DS] K. Devlin and S. Shelah, A weak form of the diamond which follows from $2^{\aleph_0}<2^{\aleph_1}$, Israel J. Math. 29 (1978), 239-247.

[DG] M. Dugas and R. Göbel, Every cotorsion free ring is an endomorphism ring, Proc. London Math. Soc. (3) 45 (1982), 319-336.

[DG 1] ----------------------, On endomorphism rings of primary abelian groups, Math. Ann. 261 (1982), 359-385.

[DG 2] ----------------------, Every cotorsion free algebra is an endomorphism algebra, Math. Z. 181 (1982), 451-470.

[Fu] L. Fuchs, Infinite abelian groups, Academic Press, New York, vol.I 1970, vol.II 1973.

[EM] P.C. Eklof and A.M. Mekler, On constructing indecomposable groups in L, J. Algebra 49 (1977), 96-103.

[GS] R. Göbel and S. Shelah, Semi-rigid classes of torsion-free abelian groups, J. Algebra (to appear).

[GS 1] ------------------------, Modules over arbitrary domains I, Math. Z. (to appear).

[GS 2] ------------------------, Modules over arbitrary domains II, submitted to Fund. Math.

[P] P.S. Pierce, Homomorphisms of primary abelian group, pp. 214-310 in: Topics in abelian groups, Chicago, Scott and Foresman 1963, ed. J.M. Irwin and E.A. Walker.

[R Sh] M. Rubin and S. Shelah, Combinatorial problems on trees, partitions, $\Delta$-systems and large free subsets, Annals of Pure and Applied Logic.

[Sh 1] S. Shelah, Classification Theory, North Holland Publ. Co Amsterdam, 1978.

[Sh 2] ---------, Why there are many non-isomorphic models for unsuperstable theories, Proc. of the International Congress of Math. Vancouver (1974), 553-557.

[Sh 3] ---------, Infinite abelian groups, Whitehead problem and some constructions, Israel J. Math. 18 (1974), 243-256.

[Sh 4] ---------, Existence of rigid like families of abelian p-groups, Model Theory and Algebra, ed. D. H. Saraceno and V. B. Weisfening, Lecture Notes in Math. Springer Verlag Vol. 498, Berlin 1975, 384-402.

[Sh 5] ---------, The lazy model theorist guide to stability, Logique et Analyse 71-72 (1975), 241-308.

[Sh 6] ---------, On endo-rigid strongly $\aleph_1$-free abelian groups in , Israel J. Math. 40 (1981), 291-295.

[Sh 7] ---------, Construction of many complicated uncountable structures and Boolean algebras, Israel J. Math. 45 (1983), 100-146.

[Sh 8] ---------, A combinatorial principle and endomorphism rings of p-groups, Proceedings of the 1980/1 Jerusalem Model Theory Year, Israel J. Math (to appear).

[Sh 9] ---------, On powers of singular cardinals, Notre Dame J. of

Formal Logic.

[Sh 10] ---------, Existence of endo-rigid Boolean algebras, (to appear in a Springer Lecture Notes′ volume).

# Almost Σ-cyclic Abelian p-groups in L

Manfred Dugas and Rüdiger Göbel

**Universität Essen**

## §1 Introduction

We will deal with primary abelian groups in the universe V = L. Our main result will fill in a missing theorem on endomorphism rings. Assuming ZFC only, we have the two parallel results on torsion-free respectively primary abelian groups; see [DG 2,3,4] and [CG].

THEOREM A: For a ring R the following are equivalent:

(1) $R^+$ is cotorsion-free.

(2) There exists a cotorsion-free abelian group G with End G $\cong$ R.

Here $R^+$ denotes the additive group of the ring R. An abelian group G is cotorsion-free iff G does not contain a cotorsion subgroup different from 0. This is equivalent to say that G is torsion-free and $\mathbb{Q}^+ \not\subseteq G$, $J_p \not\subseteq G$, where $\mathbb{Q}$ denotes the field of rational numbers and $J_p$ the additive group of the ring of p-adic numbers. This result can be extended to derive arbitrarily large rigid systems and semi-rigid classes (which are not sets); see [DG 4], [CG], [GS]. In the case of p-groups, we have the

THEOREM A*: For a ring R the following are equivalent:

(1) $R^+$ satisfies the Pierce-condition, i.e. $R^+$ is the p-adic completion of a free $J_p$-module.

(2) There exists a p-group G with End G = R ⊕ $E_s(G)$, where $E_s(G)$ denotes the ideal of small endomorphisms of End G.

Recall that a homomorphism $\sigma$ from G is small if for any n there exists an m such that $p^m G[p^n]\sigma = 0$.

Like Theorem A, the second theorem has parallel extensions towards rigid systems in the appropriate sense; see [DG 2,3], [CG]. Both results have many applications by choosing suitable rings R. Assuming now V = L, Theorem A has been sharpened in [DG 1]. For this we say that an abelian group is Σ-*cyclic* if it is a direct sum of cyclic groups. A group is κ-*cyclic* for some cardinal κ, if any subgroup of cardinal < κ is Σ-cyclic. Furthermore H is *strongly* κ-*cyclic* if any subgroup $H_o \subseteq H$ of cardinal $|H_o| < \kappa$ is contained in some subgroup $H_1 \subseteq H$ of cardinal $|H_1| < \kappa$ with $H/H_1$ κ-cyclic. In the torsion-free case we replace "cyclic" respectively "Σ-cyclic" by "free". This definition extends to R-modules in the obvious way. Then we have the following

THEOREM B (ZFC + V = L): The following conditions on a ring R are equivalent:

(1) $R^+$ is cotorsion-free.

(2) There exists a strongly $\kappa$-free R-module G with End G $\cong$ R (for any regular not weakly compact cardinal $\kappa > |R|$).

If $R^+$ is free the derived abelian groups G are strongly $\kappa$-free. Hence all "known pathologies" take place very close to free groups, see also [C 1,2] for suitable rings. Naturally the question arises whether the situation is similar in the case of p-groups, and the answer will be given in our Theorem B*. In order to derive the existence of rigid classes which are not sets, we recall Richman's [R] notion of thin groups, which is parallel to slender groups. A p-group G is thin if every homomorphism of a torsion-complete p-group into G is small in the sense of R.S.Pierce; see L. Fuchs [F, Vol.I, p. 195 and Vol.II, p.24, Exercise 7]. In the case of separable p-groups (which we will consider only) this is equivalent to say that G contains no unbounded torsion-complete subgroup; see C. Megibben [M]. It is easy to see that the class of thin groups is closed under extensions, direct sums and subgroups. In particular, all countable reduced groups and all $\kappa$-cyclic p-groups are thin. Then we have the following

THEOREM B*(ZFC + V = L): The following conditions on a ring R are equivalent:

(1) $R^+$ satisfies the Pierce-condition.

(2) There exists a strongly-$\kappa$-cyclic p-group G with End G $\cong$ R $\oplus$ $E_s(G)$.

(3) For each regular, not weakly compact cardinal $\kappa > \kappa_0$ (with $\kappa_0$ from a dense subgroup $\bigoplus_{\kappa_0} J_p$ of $R^+$) there exist strongly $\kappa$-cyclic p-groups $A_\kappa^\alpha$ ($\alpha < 2^\kappa$) such that

   (a) End G $\cong$ R $\oplus$ $E_s(A_\kappa^\alpha)$.

   (b) If $\varphi : A_\kappa^\alpha \to A_{\kappa'}^{\alpha'}$ is a homomorphism and $(\alpha,\kappa) \neq (\alpha',\kappa')$, then $\varphi$ is small.

   (c) If G is any separable, thin p-group of cardinal $|G| < \kappa$, then $\varphi : A_\alpha^\kappa \to G$ is small for each homomorphism $\varphi$ and $\alpha < 2^\kappa$.

Theorem B* anwers Problem 56 of L. Fuchs [F, Vol.II,p.55] for all regular cardinals in L: For which cardinals $\kappa$ does a p-group G of cardinality $\kappa$ exist, which is not $\Sigma$-cyclic but $\kappa$-cyclic. Choose $R = J_p$ and apply the theorem for $\kappa$ not weakly compact. It is well-known, that such G does not ex-exist for weakly compact cardinals. The existence of such p-groups G for $|G| = \aleph_n$ ($n \in \omega$) in ZFC was shown by R. Nunke [N].

We also want to point out that there is a similar result in L on $\aleph_1$-separable p-groups of cardinal $\aleph_1$, realizing rings R with Pierce-condition and $\kappa_0 \leq \aleph_1$ modulo suitable "inessential" endomorphisms; see P. Eklof and A. Mekler [EM]. For other results concerning these groups, see also M. Huber [H].

## §2 The Step-Lemmas

The algebraic requirements of theorems of type A*,B* are always built into step-lemmas which are needed to apply the combinatorial machinery of predictions in L. In order to derive this ring realization, we will need (2.1), and Theorem B*(3) will follow from a second step-lemma (2.2). The symbol $X \sqsubset Y$ will denote that X is a direct summand of Y, $\omega$ is the first infinite ordinal and $R(p^\infty)$ for a prime p and a ring R will denote a natural p-divisible abelian group $\bigcup_{n\in\omega} p^{-n}R$.

Step-Lemma 2.1: Let R be a ring with Pierce-condition for p and $I_{nk}$ be a set with at least two elements for all $n,k \in \omega$ and

(+) $A = \oplus\{ (n,\alpha,k)R : n,k \in \omega,\ \alpha \in I_{nk}\}$

where $\operatorname{ann}_R(n,\alpha,k) = p^kR$. Let

$A_t = \oplus\{(n,\alpha,k)R : n,k \in \omega,\ n \leq t,\ \alpha \in I_{nk}\}$ for $t \in \omega$. If $\varphi \in \operatorname{End}_{\mathbb{Z}} A = \operatorname{End} A$ and $\varphi \notin R + E_s(A)$, then there exists a Σ-cyclic R-module A' extending A such that

(i) $A_t \sqsubset A'$ for almost all $t \in \omega$.

(ii) $\varphi$ does not extend to an endomorphism of A'.

(iii) $A'/A \cong R(p^\infty)$.

Proof: If $a \in A$, then [a] denotes the finite support of a with respect to the direct sum decomposition (+). Passing to the p-adic (torsion) completion $\bar{A} \subset \hat{A}$, we have at most countable supports [a] for any $a \in \hat{A}$. For all $m \in \omega$ let

$S_m = \{p^{k-m}(n,\alpha,k) : n,k \in \omega,\ \alpha \in I_{nk},\ k \geq m\}$,

$T_m = \{(n,\alpha,k) : n,k \in \omega,\ \alpha \in I_{nk},\ k \geq m,\ p^{k-m}\varphi(n,\alpha,k) \notin (n,\alpha,k)R\}$ and $T_m^* = \{k \in \omega : \exists (n,\alpha)$ such that $(n,\alpha,k) \in T_m\}$. Moreover let $H_k^m = \{n \in \omega : \exists\ \alpha \in I_{nk}$ such that $(n,\alpha,k) \in T_m\}$. The sets $T_m$, $T_m^*$ and $H_k^m$ measure the "activity" of the given map $\varphi$ on the "socles", as $S_o = \emptyset$, $S_m \subseteq A[p^m]$ for $m > 0$. We will see that the condition $\varphi \notin R \oplus E_s(A)$ forces $\varphi$ to be active enough to ensure the existence of A'. This will follow by cases.

Case 1: There exists $m \in \omega$ such that $T_m^*$ is infinite.

We fix this m and consider first a subcase

(a) $\{k : H_k^m$ infinite$\}$ is infinite or there is a cofinite subset $\Delta \subseteq \omega$ with $H_k^m$ finite for all $k \in \omega$ but $\bigcup_{k \in \Delta} H_k^m$ is infinite.

Now we can choose strictly increasing sequences $(n_l)_{l\in\omega}$, $(k_l)_{l\in\omega}$ and $\alpha_l \in I_{n_l k_l}$ such that

(*) $\qquad (n_l,\alpha_l,k_l) \in T_m$ for all $l \in \omega$.

Since $\varphi \in$ End A, all supports $[\varphi(n_l,\alpha_l,k_l)]$ are finite. Passing to subsequences we may assume that

$W_l = [\varphi(n_l,\alpha_l,k_l)p^{k_l-m}] \cup [(n_l,\alpha_l,k_l)]$ are pairwise disjoint sets $(l \in \omega)$.

Take $x = \sum_{l=0}^{\infty} p^{k_l-m}(n_l,\alpha_l,k_l) \in \bar{A}$ in the torsion-completion $\bar{A}$ of A. Then we define a "divisibility chain"

$$x_d = \sum_{k_l-m \geq d} p^{k_l-m-d}(n_l,\alpha_l,k_l) \text{ where } x_o = x,$$

and let A' be the R-submodule of $\bar{A}$ generated by

$A \cup \{x_d : d \in \omega\}$, i.e. $A' = A + \sum_{d \geq 0} x_d R$. Let

$B_x = \{(n,\alpha,k) : (n,\alpha,k) \neq (n_l,\alpha_l,k_l)$ for all $l \in \omega\}$.

Visibly $A' = \bigoplus_{b \in B_x} bR \oplus \bigoplus_{d \in \omega} x_d R$ is a free R-module and

$A'/A \cong R(p^\infty)$. Suppose $\varphi$ extends to an endomorphism $\varphi'$ of $A'$.

Then we can find $r \in R$ and $d \in \omega$ such that $\varphi'(x) - x_d r \in A$.

By continuity we get

$$\sum_{l=0}^{\infty} p^{k_l - m}\varphi(n_l,\alpha_l,k_l) - \sum_{k_l - m \geq d} p^{k_l-m-d}(n_l,\alpha_l,k_l)r \in A.$$

Since the sets $W_l$ $(l \in \omega)$ are pairwise disjoint, and elements in A have finite support, we derive

$$p^{k_l-m}\varphi(n_l,\alpha_l,k_l) - p^{k_l-m-d}(n_l,\alpha_l,k_l)r = 0$$

for almost all $l \in \omega$, which contradicts (*).

Therefore (ii) and (iii) follow. In order to verify (i) we will determine a complement of $A_t$. We easily find $d_o \in \omega$ such that $[x_d] \cap \bigcup_{a \in A_t} [a] = \emptyset$ for all $d \geq d_o$. Therefore

$$A' = A_t \oplus \bigoplus_{\alpha \geq d_o} x_d R \oplus \bigoplus\{(n,\alpha,k)R : (n,\alpha,d) \in B_x, n > t\}.$$

Now consider the complementary subcase, i.e.

(b) The set $\Delta = \{k \in \omega : H_k^m \text{ finite}\}$ is cofinite in $\omega$ and $\bigcup_{k \in \omega} H_k^m$ is finite as well.

We can choose $k_o \in \omega$ such that $H_k^m$ is finite for all $k \geq k_o$ and since $\bigcup_{k \geq k_o} H_k^m$ is finite, we may pick an upper bound $n_o^* \in \omega$ of this set. Therefore

(**) $\qquad (n,\beta,k) \notin T_m$ for all $\beta \in I_{nk}$, $n \geq n_o^*$, $k \geq k_o$.

Since $T_m^*$ is infinite by hypothesis, we find some $n_1^* \in \omega$ and a strictly increasing sequence $(k_l)_{l\in\omega}$ such that

(***) $(n_l^*, \alpha_l, k_l) \in T_m$ for some $\alpha_l \in I_{n_l^* k_l}$.

Passing to a subsequence $(n_l)_{l\in\omega}$ of $\omega$, we may assume

$n_o > n_o^* + n_1^*$, $\beta_l \in I_{n_l k_l}$ and the sets

$W_l = [p^{k_l - m}\varphi(n_l^*, \alpha_l, k_l)] \cup \{(n_l, \beta_l, k_l)\}$ $(l \in \omega)$

are pairwise disjoint.

Let $x = \sum_{l=0}^{\infty} p^{k_l - m}((n_l^*, \alpha_l, k_l) + (n_l, \beta_l, k_l))$ and define a divisibility chain $x_d$ and A' as in (a). Using (**) and (***), any extension $\varphi'$ of $\varphi$ has infinite support on $\varphi'(x) - x_d r$ for all $r \in R$, $d \in \omega$. Therefore $\varphi'(x) \notin A'$ and similarly to (a) conditions (i), (ii) and (iii) hold. In the remaining

Case 2: We consider $T_m^*$ to be finite for all $m \in \omega$.

For each $m \in \omega$, we can choose some upper bound $k(m) \in \omega$ of $T_m^* \subseteq \omega$. We derive

(α) $p^{k-m}\varphi(n,\alpha,k) = (n,\alpha,k)p^{k-m} r_{n\alpha k,m}$ for suitable

$r_{n\alpha k,m} \in R$ and all $(n,\alpha)$.

First we want to deal with a subcase

(a) There exists $m \in \omega$ such that any $r \in R$ gives rise to an infinite subset $\Delta \subseteq \omega$ and certain $(n,\alpha,k)$ for $k \in \Delta$ with $r \not\equiv r_{n\alpha k,m} \mod p^m R$.

Fixing m, we can choose a strictly increasing sequence $(k_l)_{l\in\omega}$ with $k_o > k(m)$ and $(n_l, \alpha_l)$ such that $r_{n_l \alpha_l k_l, m} + p^m R$ is not constant after finitely many steps.

Suppose that the sequence $n_l$ $(l \in \omega)$ can be choosen strictly increasing. Then let $x = \sum_{l=0}^{\infty} p^{k_l - m}(n_l, \alpha_l, k_l)$ and define $x_d$

and A' as in case 1. If $\varphi'$ extends $\varphi$, there exist $r \in R$ and $d \in \omega$ such that $\varphi'(x) - x_d r \in A$. Now it is easy to compute $\varphi'(x) = \sum_{l=0}^{\infty} p^{k_l - m}(n_l,\alpha_l,k_l) r_{n_l \alpha_l k_l, m}$ which implies $r \equiv r_{n_l \alpha_l k_l, m} \bmod p^m R$ for almost all $l \in \omega$. This contradicts (a). So we find $n = n_l$ for almost all $l \in \omega$. Then we consider any strictly increasing sequence $(n_l^*)_{l\in\omega}$ with $n_o^* > n$ and let

$$x = \sum_{l=0}^{\infty} p^{k_l - m}((n_l^*,\alpha_l^*,k_l) + (n,\alpha_l,k_l)).$$

This again leads to an R-module A' with (i),(ii) and (iii). Finally we deal with the remaining subcase

(b) For all $m \in \omega$ there exist $r(m) \in R$ and $k(m) \in \omega$ such that $r(m) \equiv r_{n,\alpha,k,m} \bmod p^m R$ for all $n,\alpha$ and $k \geq k(m)$.

This and ($\alpha$) imply

($\beta$) $p^{k-m}\varphi(n,\alpha,k) = r(m)p^{k-m}(n,\alpha,k)$ for all $k \geq k(m)$.

Obviously $p^{k-m-1}\varphi(n,\alpha,k) = r(m+1)p^{k-m-1}(n,\alpha,k)$ for $k \geq k(m+1)$ and hence

$$p^{k-m}(r(m) - r(m+1))(n,\alpha,k) = 0 \text{ for } k \geq \max\{k(m),k(m+1)\}.$$

We conclude $r(m) \equiv r(m+1) \bmod p^m R$, i.e. $r(m)$ is a p-adic Cauchy sequence. Since R is complete, we find a limit $r \in R$ such that $r(m) \equiv r \bmod p^m R$ for all $m \in \omega$, and (++) turns into

($\gamma$) $p^{k-m}\varphi(n,\alpha,k) = p^{k-m}\, r(n,\alpha,k)$ for $k \geq k(m)$.

If $B^* = \langle (n,\alpha,k) : n,k \in \omega,\ \alpha \in I_{nk} \rangle$, then $(\varphi - r)p^{k(m)}B^*[p^m] = 0$ by ($\gamma$) and $\varphi - r$ is small on $B^*$. We want to show that $\varphi - r$ is small on A; suppose $\varphi - r$ is not small. Finally we consider an arbitrary element $(n,\alpha,k)s$ with

$s \in R \setminus pR$. Since small "does not depend on finitely many elements", we find sequences $s_l \in R \setminus pR$ and $k_l < k_{l+1} \in \omega$ such that

($\delta$) $p^{k_l-m}\varphi(s_l(m_l,\alpha_l,k_l)) \neq p^{k_l-m} r\, s_l(n_l,\alpha_l,k_l)$.

Since $|I_{n_l k_l}| \geq 2$, we can pick $\beta_l \in I_{n_l k_l} \setminus \{\alpha_l\}$ and change the R-basis of A substituting $(n_l,\alpha_l,k_l)$ by $(n_l,\alpha_l,k_l) + (n_l,\alpha_l,k_l)s_l$. Now we repeat our cases so far and derive suitable extensions as in case (1) and (2a). In the new case (2b) we find another limit $r^* \in R$ such that

($\gamma^*$) $p^{k_l-m}\, \varphi((n_l,\alpha_l,k_l) + s_l(n_l,\alpha_l,k_l))$

$$= p^{k_l-m} r^*((n_l,\beta_l,k_l) + s_l(n_l,\alpha_l,k_l))$$

holds for all $k_l \geq k^*(m)$ and some $k^*(m) \geq k(m)$.

Hence $p^{k_l-m}\varphi((n_l,\alpha_l,k_l)) = p^{k_l-m} r^*(n_l,\alpha_l,k_l)$ for all $k_l \geq k^*(m)$, which implies $r = r^*$. We derive $p^{k_l-m}\varphi(s_l(n_l,\alpha_l,k_l)) = p^{k_l-m} r s_l(n_l,\alpha_l,k_l)$ contradicting ($\delta$). Therefore $p^{k-m}\varphi(s(n,\alpha,k)) = p^{k-m} rs(n,\alpha,k))$ for all $k \geq k(m)$ and $\varphi - r$ is small on A. □

<u>Step-Lemma 2.2</u>: Let R be a ring with Pierce-condition for p, $I_{nk}$ be sets of cardinal $\geq 2$ for all $n,k \in \omega$ and

(+) $\qquad A = \oplus \{(n,\alpha,k)R : n,k \in \omega,\ \alpha \in I_{nk}\}$

where $\mathrm{ann}_R(n,\alpha,k) = p^k R$. Let $A_t = \oplus\{(n,\alpha,k)R : n,k \in \omega,\ n \leq t,\ \alpha \in I_{nk}\}$ for $t \in \omega$. If G is a separable, thin p-group and $\varphi : A \to G$ a non-small homomorphism then there exists a $\Sigma$-cyclic R-module A' extending A such that

(i) $A_t \sqsubset A'$ for almost all $t \in \omega$

(ii) $\varphi$ does not extend to a homomorphism of $A'$ into $G$

(iii) $A'/A \cong R(p^\infty)$.

<u>Proof</u>. Since $\varphi$ is not small, we can find a strictly increasing sequence $(k_l)_{l\in\omega}$ and some $m \in \omega$ with

(*) $\varphi(p^{k_l-m}(n_l,\alpha_l,k_l)r_l) \neq 0$ $(l \in \omega)$

for suitable $n_l \in \omega$, $\alpha_l \in I_{n_l k_l}$ and $r_l \in R$.

First we will see that w.l.o.g.

(**) $r_l = 1$ $(l \in \omega)$.

Take any $\beta_l \in I_{n_l k_l} \setminus \{\alpha_l\}$ for all $l \in \omega$. If $\varphi(p^{k_l-m}(n_l,\beta_l,k_l)) \neq 0$ for infinitely many $l \in \omega$, then (**) holds. Therefore suppose $\varphi(p^{k_l-m}(n_l,\beta_l,k_l)) = 0$ for all $l \in \omega$. Then we substitute $(n_l,\beta_l,k_l)$ by $(n_l,\beta_l,k_l) + (n_l,\alpha_l,k_l)\cdot r_l$ in the R-basis of A. This obviously does not change $A_t$ and (**) follows with respect to the new basis.

Next we want to show that w.l.o.g.

(***) The sequence $(n_l)_{l\in\omega}$ can be choosen to be strictly increasing.

If $(n_l)_{l\in\omega}$ is unbounded, (***) follows passing to a subsequence. Hence we can find a bound $n_o^* \in \omega$ with $n_l \leq n_o^*$ for all $l \in \omega$. Now we choose another strictly increasing sequence $(n_l^*)_{l\in\omega}$ with the same $n_o^*$ and $\alpha_l^* \in I_{n_l^* k_l}$. If $\varphi(p^{k_l-m}(n_l^*,\alpha_l^*,k_l)) \neq 0$ for infinitely many $l \in \omega$, then (***)

holds. Hence we may assume $\varphi(p^{k_l-m}(n_l^*,\alpha_l^*,k_l)) = 0$ for all $l \in \omega$. Again substitute $(n_l^*,\alpha_l^*,k_l)$ by $b_l = (n_l,\alpha_l,k_l) + (n_l^*,\alpha_l^*,k_l)$ in the R-basis of A. Since $p^{k_l-m}\varphi(b_l) \neq 0$ and $o(b_l) = p^{k_l}$, conditions (*), (**) and (***) hold. If $U = \bigoplus_{l\in\omega} (n_l,\alpha_l,k_l)\mathbb{Z}$, then $\varphi \restriction U : U \to G$ is not small by these conditions. Since G is separable and thin, the unique extension $\hat{\varphi} : \bar{U} \to \bar{G}$ of $\varphi \restriction U$ between the torsion-completions $\bar{U}$ and $\bar{G}$ does not map $\bar{U}$ into G, as follows from Megibben's observation mentioned in the introduction. Hence there exists some $x \in \bar{U}\setminus U$ such that $\hat{\varphi}(x) \notin G$. Let $x = \sum_{l=1}^{\infty} (n_l,\alpha_l,k_l)z_l$ and eliminate summands $(n_l,\alpha_l,k_l)z_l = 0$. As in the proof of Step-Lemma 2.1 we build a divisibility chain $x_d = \sum_{d\leq h_p(z_l)} (n_l,\alpha_l,k_l)z_l p^{-d}$ where $h_p(z)$ denotes the p-height of z in $\hat{\mathbb{Z}}$. If $A' = A + \sum_{d\in\omega} x_d R$, then (ii) and (iii) follow immediately. If d is large enough, then $[x_d] \cap [a] = \emptyset$ for all $a \in A_t$. This implies (i) as in (2.1).

## §3 Proof of Theorem B

In this section we work in classical statistics of the universe L. Hence we prepare the underlying sets in such a way that the outcomes of the ◊-machinery will be the right kind of abelian groups. The set theoretic preliminaries are standard and may be found in T. Jech [J]; see also [DG 1]. By hypothesis, our ring R satisfies the Pierce-condition;

hence $R^+ = \widehat{\bigoplus_{\kappa_0} J_p}$ is the p-adic completion of a free p-adic module of a certain rank $\kappa_0$ (which may be finite). Let $\kappa$ be the given regular, not weakly compact cardinal $> \kappa_0$ and observe that $R/p^nR \cong \bigoplus_{\kappa_0} \mathbb{Z}/p^n\mathbb{Z}$.

Choose any sparse, stationary set $E \subseteq \{\lambda < \kappa, \mathrm{cf}(\lambda) = \omega\}$. From Solovay's decomposition theorem we get a partition $E = E_e \dot{\cup} E_k \dot{\cup} \dot{\bigcup}_{\alpha<\kappa} E_\alpha$ into (pairwise disjoint) stationary sets $E_\alpha$ ($\alpha \in \kappa \cup \{e,k\}$). Let $H = \bigcup_{\alpha<\kappa} H_\alpha$ be any $\kappa$-filtration of some set $H$ of cardinal $\kappa$. In order to obtain a maximal rigid system, fix a set $\mathfrak{X}$ of incomparable subsets of $\kappa$ with $|\mathfrak{X}| = 2^\kappa$. We decorate $E$ with the diamonds $\diamondsuit_\kappa(E_\alpha)$ for $\alpha \in \kappa \cup \{e,k\}$ and derive Jensen-functions $\{\varphi_\alpha : H_\alpha \to H_\alpha : \alpha \in E_e\}$ guessing endomorphisms of $H$ and Jensen-sets $\{U_\alpha \subseteq H_\alpha : \alpha \in E_k\}$ guessing kernels of homomorphism into p-groups of cardinal $< \kappa$ and additional Jensen-sets for each $\gamma < \kappa$ of the form $\{(\varphi_\alpha, +_\alpha) \subseteq H_\alpha^5 : \alpha \in E_\gamma\}$ where $H_\alpha^5 = H_\alpha \times H_\alpha \times H_\alpha \times H_\alpha \times H_\alpha$.
The last sets of Jensen-set are supposed to guess the additive structure $+_\alpha$ on $H_\alpha$ (i.e. $+_\alpha \subseteq H^3$) and homomorphisms on these groups (i.e. $\varphi_\alpha \subseteq H^2$); observe that $2 + 3 = 5$. Compare T. Jech [J,p.226] for Jensen-functions and sets. If $X \in \mathfrak{X}$, we define by transfinite induction on $H_\alpha$ ($\alpha < \kappa$) an abelian group $H_\alpha^X$ and R-module structure and derive finally the required p-group $H^X = \bigcup_{\alpha<\kappa} H_\alpha^X$.
The induction depends on X and is carried out as follows.

(1) $H_0^X = 0$ and each $H_\alpha^X$ ($\alpha < \kappa$) is a Σ-cyclic R-module.

(2) If $\alpha$ is a limit, then $H_\alpha^X = \bigcup_{\beta<\alpha} H_\alpha^X$.

(3) If $\alpha < \beta < \kappa$ and $\alpha \notin \bigcup\{E_\tau : \tau \in X \cup \{e,k\}\}$, then $H_\alpha^X \sqsubset H_\beta^X$ as R-modules.

(4) If $H_\alpha^X$ has been defined, let

(4.0) $H_{\alpha+1}^X = H_\alpha^X \oplus \bigoplus_{\aleph_0} (\bigoplus_{n\in\omega} R/p^nR)$ except in the following cases:

(4.1) If $\alpha \in E_e$ and $\varphi_\alpha : H_\alpha^X \to H_\alpha^X$ is a homomorphism not in $R \oplus E_s(H_\alpha^X)$, then choose a sequence $\alpha_n \in \alpha \setminus E$ strictly increasing with $\sup_{n\in\omega} \alpha_n = \alpha$. By (3) we have $H_{\alpha_n}^X \sqsubset H_{\alpha_{n+1}}^X \sqsubset H_\alpha^X$ and we can apply our Step-Lemma 2.1. Hence we obtain an R-module $A' = H_{\alpha+1}^X$ extending $A = H_\alpha^X$ such that $\varphi_\alpha$ does not lift to an endomorphism of $H_{\alpha+1}^X$. The consequence (2.1)(i) takes care of the requirement (3) in the construction and (2.1)(iii) will be used later.

(4.2) If $\alpha \in E_k$ and $U_\alpha \subset H_\alpha^X$ is a subgroup, then let $\pi : H_\alpha^X \sqsubset H_\alpha^X/U_\alpha$ be the canonical projection. If $\pi$ is a non-small homomorphism into a separable and thin p-group $H_\alpha^X/U_\alpha$, then we apply similarly our Step-Lemma 2.2. Hence we obtain an extension $H_{\alpha+1}^X \supset H_\alpha^X$ and $H_{\alpha_n}^X \sqsubset H_{\alpha+1}^X$ such that $\pi$ does not lift to a homomorphism $\pi' : H_{\alpha+1}^X \to H_\alpha^X/U_\alpha$. Again (2.2)(i) takes care of (3) and (2.2)(iii) is used later.

(4.3) If $\alpha \in E_\gamma$ for some $\gamma \in X$, $(H_\alpha, +_\alpha)$ is a separable and thin p-group and $\varphi_\alpha : H_\alpha^X \to (H_\alpha, +_\alpha)$ is a non-small homomorphism, then we apply our Step-Lemma 2.2 once more.

Hence we obtain $H^X_{\alpha+1} \supset H^X_\alpha$ such that $H^X_{\alpha_n} \sqsubset H^X_{\alpha+1}$, $H^X_{\alpha+1}/H_\alpha \cong R(p^\infty)$ and $\varphi_\alpha$ does not extend to a homomorphism $\varphi'_\alpha : H^X_{\alpha+1} \to (H_\alpha, +_\alpha)$.

Now it is easy to check that $H^X$ is a strongly $\kappa$-cyclic p-group and an R-module.

In order to derive Theorem B*(3) we will need an easy

Lemma 3.1: Let $A = \bigcup_{\alpha<\kappa} A_\alpha$ be a $\kappa$-filtration of the p-group A and $\varphi : A \to H$ a homomorphism into the p-group H. If $\{\nu \in \kappa : \varphi \upharpoonright A_\nu$ is small$\}$ is unbounded in $\kappa$ and $cf(\kappa) > \omega$, then $\varphi$ is small.

Proof: Let $S = \{\nu \in \kappa : \varphi \upharpoonright A_\nu$ is small . If $\nu \in S$, let $n = n(\nu,m)$ be minimal such that $p^n A_\nu[p^m]\varphi = 0$. If the sets $\{n(\nu,m) : \nu \in \kappa\} = N_m$ are bounded by some $n=n(m) \in \omega$, then also $p^n A[p^m]\varphi = 0$ and $\varphi$ is small. Hence suppose that there exists an $m \in \omega$ such that $N_m$ is unbounded. We can choose $\nu_1 \in \kappa$ such that $n(\nu_1,m) < n(\nu_{1+1},m)$ is strictly increasing. Since S is unbounded in $\kappa$ and $cf(\kappa) > \omega$, we find $\nu \in S$ with $\nu_1 < \nu$ for all $1 \in \omega$. Since $\nu \in S$, there exists $k \in \omega$ such that $p^k A_\nu[p^m]\varphi = 0$. However $A_{\nu_1} \subseteq A_\nu$ implies $p^k A_{\nu_1}[p^m]\varphi = 0$. Therefore $n(\nu_1,m) \leq k$ for all $1 \in \kappa$ by minimality of $n(\nu_1,m)$. This contradicts our choice of $n(\nu_1,m)$. □

In order to derive Theorem B*, we only have to check (1) → (3). The converse (2) → (1) is a classical result of R.S. Pierce; see e.g. L. Fuchs [F; Vol.I, p. 197]; and (3) → (2) is trivial. Since $|\mathfrak{X}| = 2^\kappa$, it remains to prove for any $X \neq Y \in \mathfrak{X}$ :

(3.2) $\operatorname{End} H^X = R \oplus E_s(H^X)$

(3.3) If $\varphi : H^X \to H^Y$ is a homomorphism, then $\varphi$ is small.

(3.4) If $\varphi : H^X \to G$ is a homomorphism into a separable and thin p-group of cardinal $< \kappa$, then $\varphi$ is small.

(3.5) If $\kappa < \kappa'$ are regular, not weakly compact cardinals $> \kappa_0$ and $Y'$ is from the indexing set associated with the construction at the cardinal $\kappa'$, then any homomorphism $\varphi : H^X \to H^{Y'}$ is small.

Proof of (3.2): Since $H^X$ is an R-module, we have $R \subseteq \operatorname{End} H^X$ identifying R with scalar-multiplication. Therefore also $R + E_s(H^X) \subseteq \operatorname{End} H^X$. The group $H^X$ is a natural unbounded R-module; hence scalar-multiplications which are small must be 0. We derive $R \oplus E_s(H^X) \subseteq \operatorname{End} H^X$.

Suppose for contradiction that we find some $\varphi \in \operatorname{End} H^X \setminus R \oplus E_s(H^X_\alpha)$. If $C = \{\alpha \in \kappa : \varphi(H^X_\alpha) \subseteq H^X_\alpha\}$, then C is a cub in $\kappa$. Suppose that $C_0 = \{\alpha \in \kappa : \varphi \restriction H^X_\alpha \in R \oplus E_s(H^X_\alpha)\}$ is unbounded in $\kappa$. For all $\nu \in C_0$ there exists $r_\nu \in R$ such that $(\varphi - r_\nu) \restriction H^X_\nu$ is small. Take any $\nu < \mu \in C_0$. Then $(\varphi - r_\nu) \restriction H^X_\nu$ and $(\varphi - r_\mu) \restriction H^X_\mu$ are small. Hence $r_\nu - r_\mu$ is small on $H^X_\nu$. Since $H^X_\nu$ is an unbounded R-module and $r_\nu - r_\mu$ acts by scalar multiplication, we derive $r_\nu = r_\mu = r$. Therefore $(\varphi - r) \restriction H^X_\nu$ is small for all $\nu \in C_0$ and $C_0$ is unbounded. From (3.1) we derive that $\varphi - r$ is small, which contradicts $\varphi \notin R \oplus E_s(H^X)$. Hence $C_0$ must be bounded and $C^* = C \setminus C_0$ is still a cub in $\kappa$. From $\diamondsuit_\kappa(E_e)$ we have that $\{\nu \in E_e : \varphi_\nu = \varphi \upharpoonright H^X_\nu\}$ is a stationary set. Hence we find $\alpha \in E_e \cap C^*$ such that $\varphi_\nu = \varphi \upharpoonright H^X_\nu$. By (4.1) we see that cannot be lifted to $\varphi' : H^X_{\nu+1} \to H^X_{\nu+1}$. From (2.1)(iii) we have $H^X_{\nu+1}/H^X_\nu \cong R(p^\infty) \cong \bigoplus_{\kappa_0} \mathbb{Z}(p^\infty)$ and from the construction follows that $H^X/H^X_{\nu+1}$ is $\kappa$-cyclic. Therefore $H^X_{\nu+1}$ is the p-adic closure of $H^X_\nu$ in $H^X$. This implies $\varphi(H^X_{\nu+1}) \subseteq H^X_{\nu+1}$, i.e. $\varphi \upharpoonright H^X_\nu = \varphi_\nu$ lifts to $\varphi \upharpoonright H^X_{\nu+1} : H^X_{\nu+1} \to H^X_{\nu+1}$. From this contradiction it follows (3.2).

<u>Proof of (3.3)</u>: Suppose $\varphi : H^X \to H^Y$ is a non-small homomorphism. As in the previous proof we see from (3.1) that $C = \{\nu < \kappa : \varphi(H^X_\nu) \subseteq H^X_\nu,\ \varphi \upharpoonright H^X_\nu \text{ non small}\}$ is a cub in $\kappa$. Recall that $H^X_\nu = H_\nu = H^Y_\nu$ as <u>sets</u>. Let $+_Y : H^Y \times H^Y \to H^Y$ denote the addition on $H^Y$. Using $\diamondsuit_\kappa(E_\gamma)$ $(\gamma < \kappa)$ we see that

$$W_\gamma = \{\nu \in E_\gamma : +_Y \upharpoonright (H_\nu \times H_\nu) = +_\nu,\ \varphi \upharpoonright H_\nu = \varphi_\nu\}$$

is a stationary set in $\kappa$. The addition $+_\nu$ was defined as a Jensen-set in the construction. Since C is a cub we find $\alpha \in C \cap W_\gamma$ and (4.3) implies that $\varphi \upharpoonright H^X_\alpha : H^X_\alpha \to H^Y_\alpha$ cannot be extended to a map $\varphi' : H^X_{\alpha+1} \to H^Y_\alpha$. From (2.2)(iii) we derive that $H^X_{\alpha+1}$ is the p-adic closure of $H^X_\alpha$ in $H^X$. Since $\alpha \in E_\gamma$ and $\gamma \notin Y$, the group $H^Y_\alpha$ is necessarily closed in $H^Y$. We derive $\varphi(H^X_{\alpha+1}) \subseteq H^Y_\alpha$ which contradicts (4.3), and (3.3) follows.

<u>The proof of (3.4)</u> is similar to (3.3), but using (4.2) and (2.2)(iii).

Proof of (3.5): Let $\varphi$ be a homomorphism which satisfies the hypothesis of (3.5). Since $H^{Y'}$ is strongly $\kappa'$-cyclic and $|H^X| = \kappa < \kappa'$, we have that $\varphi(H^X)$ is $\Sigma$-cyclic. If $\varphi$ is not small, we find a projection $\pi$ from $\varphi(H^X)$ into a countable summand of $\varphi(H^X)$ such that $\varphi\pi$ is still not small. Since $\aleph_0 < \kappa$, this contradicts (3.4). □

## §4 References


C1 A.L.S. Corner, On endomorphism rings of primary abelian groups, Quart. J. Math. Oxford 20 (1969), 277-296.

C2 A.L.S. Corner, Additive categories and a theorem of W.G. Leavitt, Bull. Amer. Math. Soc. 75 (1969), 78-82

CG A.L.S. Corner, R. Göbel, Prescribing endomorphism algebras - a unified treatment, to appear in Proc. Lond. Math. Soc.

DG1 M. Dugas, R. Göbel, Every cotorsion-free ring is an endomorphism ring, Proc. Lond. Math. Soc. 45 (1982), 319-336.

DG2 M. Dugas, R. Göbel, On endomorphism rings of primary abelian groups, Math. Annalen 261 (1982), 359-385

DG3 M. Dugas, R. Göbel, Endomorphism algebras of torsion modules II, pp. 400-411, Abelian Group Theory, Proceedings, Honolulu 1982/83, Springer LNM 1006 (1983).

DG4 M. Dugas, R. Göbel, Every cotorsion-free algebra is an endomorphism algebra, Math. Zeitschr. 181 (1982), 451-470.

EM P. Eklof, A. Mekler, On endomorphism rings of $\aleph_1$-separable primary groups, pp. 320-339, Abelian Group Theory, Proceedings, Honolulu 1982/83, Springer LNM 1006 (1983).

F L. Fuchs, "Infinite abelian groups", Vol.I(1970), Vol. II(1973), Academic Press, New York.

GS R. Göbel, S. Shelah, On semi-rigid classes of torsion-free abelian groups, to appear in Journ. Alg. (1984).

H M. Huber, Methods of set theory and the abundandce of separable abelian p-groups, pp. 304-319, Abelian Group Theory, Proceedings, Honolulu 1982/83, Springer LNM 1006 (1983).

J T. Jech, Set Theory, Academic Press, New York 1978.

M C. Megibben, Large subgroups and small homomorphisms, Michigan Journ. 13 (1966), 153-160.

N R.J. Nunke, On the structure of Tor, Proc. Colloqu. Abelian groups, pp. 115-124, Budapest 1964 and Pac. J. Math. 22 (1967, 453-464.

R F. Richman, Thin abelian groups, Pacific J. Math. (1968), 599-606.

# ESSENTIALLY C-INDECOMPOSABLE $p^{\omega+n}$-PROJECTIVE p-GROUPS

DOYLE CUTLER
UNIVERSITY OF CALIFORNIA AT DAVIS

In Cutler and Missel[1], a class of examples of $p^{\omega+2}$-projective p-groups is constructed having the property

(1) every summand that is a direct sum of cyclic groups must be bounded.

In this paper we will generalize this construction. To describe the generalization, recall that in the construction, a $p^{\omega+1}$-projective p-group $G$ was used such that $G[p] = S[p] \oplus P$ where among other things $P$ was assumed to be isometric to the socle of a torsion complete p-group. We replace this with the assumption that, as a valuated vector space, any free summand of $P$ has finite support.* We will actually construct

*L. Fuchs suggested that this be generalized in this way.

$p^{\omega+n}$-projective p-groups having property (1) for all integers $n > 1$.

All groups in this paper will be p-primary abelian groups for a fixed but arbitrary prime p. If P is a subgroup of a p-group G and we consider P as a valuated group or valuated vector space without specifying the valuation we assume the valuation to be the restriction of the height function in G. We will use $\oplus'$ and $\cong'$ for valuated direct sum and isometry, respectively. A group G will be said to be essentially C-indecomposable if it satisfies property (1). Otherwise our notation and terminology will be that of Fuchs[2]. We will assume that the reader is familiar with Fuchs and Irwin[3], Fuchs[4], and Cutler and Missel[1]. For more information on valuated groups the reader is referred to Richman[5].

We will need the following technical lemma in the construction.

Lemma. Let G be a $p^{\omega+n}$-projective p-group such that $G[p^n] = S[p^n] \oplus P$ with S a pure subgroup of G, both S and G/P are direct sums of cyclic groups, and the restriction of the natural homomorphism $\varphi: G \to G/S[p^n]$ to P preserves heights in G. Suppose that P[p] as a valuated vector space has the property that any free summand must have finite support. Then if $Q \subseteq G[p^n]$ such that G/Q is a direct sum of cyclic groups, there exists a positive integer k such that $p^kG \cap P \subseteq Q$.

Proof. Note that $P[p^k]/P[p^{k-1}]$, $k \leq n$, (as a valuated vector space with valuation being heights computed in $G/P[p^{k-1}]$) is isomorphic to $p^{k-1}G \cap P[p]$ (as a subgroup of $p^{k-1}G$). One can see this as follows. Note that since the restriction of $\varphi$ to P preserves heights, $P[p^k]/P[p^{k-1}] \cong' (S[p^{k-1}] + P[p^k])/(S[p^{k-1}] + P[p^{k-1}])$. Also $G/G[p^{k-1}] \cong p^{k-1}G$ and

$(S[p^{k-1}] + P[p^k])/(S[p^{k-1}] + P[p^{k-1}])$ maps onto $p^{k-1}G \cap P[p]$ under this isomorphism proving the isometry. Therefore $P[p^k]/P[p^{k-1}]$ has only free summands of finite support.

Let $Q \subseteq G[p^n]$ such that $G/Q$ is a direct sum of cyclic groups. Then $R = P \cap Q$ also has this property by Fuchs[4]. Let $\eta$ be the natural homomorphism from $G$ onto $G/R$. Then $\eta|P$ is a height nondecreasing homomorphism from $P$ into a direct sum of cyclic groups. By Lemma 1 of Fuchs and Irwin[3], $P[p] = \ker(\eta|P[p]) \oplus T$ where $T$ is decomposable with finite values. Thus $T$ has finite support since the only summands of $P[p]$ with finite values have finite support. Hence there exists an integer $m$ such that $p^mG \cap P[p] \subseteq R[p]$. Assume that there exists an integer $r$ such that $p^rG \cap P[p^k] \subseteq R[p^k]$. Note that $G/(p^rG \cap R)$ is a direct sum of cyclic groups. Also $(p^rG \cap P[p^{k+1}])/(p^rG \cap P[p^k]) \cong' p^{r+k}G \cap P[p]$ (as a subgroup of $p^kG$). Let $\lambda\colon G/(p^rG \cap R[p^k]) \to G/(p^rG \cap R)$ be natural. Then $\lambda|(p^rG \cap P[p^{k+1}])/(p^rG \cap R[p^k])$ is a height nondecreasing homomorphism into a direct sum of cyclic groups. Hence, $N = (p^rG \cap P[p^{k+1}]) \div (p^rG \cap R[p^k]) = \ker(\lambda|(N)) \oplus' T$ where $T$ is decomposable with finite values. Thus $T$ must have finite support. Hence there exists an integer $t$ such that $p^tG \cap P[p^{k+1}] \subseteq R$. Thus there exists an integer $s$ such that $p^sG \cap P \subseteq R$.

Construction: We will now give our construction.

Let $G$ be a proper $p^{\omega+n}$-projective p-group $(n \geq 1)$ with $p^n$-socle $S[p^n] \oplus P$ such that

(2) both $G/P$ and $S$ are direct sums of cyclic groups,

(3) $S$ is a pure subgroup of $G$ with $pS[p] = S$,

(4) the natural homomorphism $\varphi: G \to G/S[p^n]$ preserves heights of elements in $P$, and

as valuated vector spaces

(5) $pS[p]$ is isometric to a basic subspace of $P[p]$ and

(6) $P[p]$ has the property that every free valuated vector space summand has finite support.

Note: We can replace (4) by $S[p^n] \oplus' P$ if we assume $p^{\omega}G = 0$.

Let $C$ be a pure subgroup of $G$ such that $C[p^n] = C \cap P$ and $C[p]$ is a basic subspace of $P[p]$. (This is possible since in fact the natural homomorphism $\nu: G \to G/S$ preserves heights of elements of $P$ and hence if $B$ is a basic subgroup of $G/S$ then $(\nu|P)^{-1}(B[p^n])$ supports such a $C$.) Thus $C \cong pS$, and $S \oplus C$ is a basic subgroup of $G$. Let $\psi$ be an isomorphism from $pS$ onto $C$. Define $L = \{x-\psi(x) | x \in S[p]\}$, $A = G/L$, and $Q = (P[p]+L)/L$. We will show that $A$ is the desired group.

Note that $A/Q \cong G/(P[p]+L) = G/G[p] \cong pG$. Let $T = \{y-\psi(y)\colon y \in pS\}$. Then $T$ is a pure subgroup of $G$, and $T[p] = L$. Thus $T/L$ is a pure subgroup of $A$, and

(7) $A[p] = T[p^2]/L \oplus Q$.

The proof of this is the same as that of (1) in Cutler and Missel[1]. Decompose $C = C_0 \oplus C'$ where $C_0$ is a maximal $p^n$-bounded summand of $C$. Decompose $P = C_0 \oplus C'[p^n] \oplus M$ where $M = \oplus_{z \in Z}\langle pz\rangle$ with $z \in G$. Let $N' = C[p^{n+1}] \oplus (\oplus_{z \in Z}\langle z\rangle)$ and $N = (N'+L)/L$. Note that $Q = N[p]$.

We will need the following facts, the proofs of which will be given below.

(8) $A/N$ is a direct sum of cyclic groups.

(9) $N/Q \cong' pG \cap P$ where the valuation on $N/Q$ is the height function with heights computed in $A/Q$ and the valuation on $pG \cap P$ is the height function in $pG$.

(10) If $R$ is a subgroup of $A[p^{n+1}]$ such that $A/R$ is a direct sum of cyclic groups then there exists an integer $m > 0$ such that $p^m A \cap N \subseteq R \cap N$.

From (8) it follows that $A$ is $p^{\omega+n+1}$-projective. From (10) it follows that $A$ is not $p^{\omega+n}$-projective. We will need (9) to prove (10).

We will now show that any summand of $A$ which is a direct sum of cyclic groups must be bounded. Assume that $A = E \oplus A'$ where $E$ is an unbounded direct sum of cyclic groups. Let $R \subseteq A'[p^{n+1}]$ such that $A'/R$ is a direct sum of cyclic groups. Decompose $S = \oplus_{i\in\omega} S_i$ where $S_i = \oplus_{y\in Y(i)}\langle y\rangle$ with $o(y) = p^{i+1}$. By (10), there exists an integer $m > 0$ such that $p^m A \cap N \subseteq R \cap N$. Let $x \in E[p]$ of height greater than $m$. Since $A[p] = (T[p^2]/L) \oplus' Q$ and $T/L$ is a pure subgroup of $A$,

$$x = (\Sigma_{i>n+2}\Sigma_{y\in Y(i)} k_y p^{i-2}(py - \psi(py)) + L) + q$$

where $0 \le k_y < p$ with $k_y = 0$ for almost all $y$, $q \in p^m A \cap Q$, and the height of $x$ is the minimum of the heights of $q$ and $j-2$ where $j$ is the smallest $i$ such that $y \in Y(i)$ and $k_y \neq 0$. Set

$$z = (\Sigma_{i>m+2}\Sigma_{y\in Y(i)} k_y p^{i-2}\psi(py) + L) - q$$

and note that the height of $x$ is smaller than the height of $x+z$. Since the $(k_y p^{i-2}\psi(py) + L)$'s and $q$ are in $p^m A \cap N$ we have $z \in A'$. Thus $x \in E$, $z \in A'$, and $h(x) < h(x+z)$ which contradicts our assumption that $A = E \oplus A'$ with $E$ an unbounded direct sum of cyclic groups. Hence $E$ must be bounded.

We will now prove (8), (9), and (10). To prove (8) note that $A/N = (G/L)/((N'+L)/L) \cong G/(N'+L) \cong (G/P)/(N'+L)/P$. Since $(N'+L)/P = (G/P)[p]$ and $G/P$ is a direct sum of cyclic groups, we have $A/N$ a direct sum of cyclic groups.

To prove (9) note that under the isomorphism $\beta: A/Q \to pG: (x+L)+Q \to px$ (we have this from $A/Q = (G/L)/(P[p]+L)/L \cong G/(P[p]+L) = G/G[p] \cong pG$), $\beta|(N/Q)$ is an isomorphism from $N/Q$ onto $pG \cap P$.

In order to prove (10) let $R$ be a subgroup of $A[p^{n+1}]$ such that $A/R$ is a direct sum of cyclic groups. Then $R \cap N$ also has this property. We will show that there is an integer $m$ such that $p^m A \cap N \subseteq R \cap N$. Let $\varphi$ be the natural homomorphism from $A$ onto $A/(R \cap N)$. Then $\varphi$ restricted to $Q$ is a height nondecreasing homomorphism from $Q$ into a direct sum of cyclic groups. Hence it follows from Lemma 1 of Fuchs and Irwin[3] that $Q = B \oplus' \ker(\varphi|Q)$ where $B$ is a free valuated vector space with finite values. Since $C[p]$ is a basic subspace of $P[p]$, $Q = (P[p]+L)/L$, and, modulo $L$, the valuation on $C[p]$ is increased by one, $Q$ is isometric to $P[p]$ where the valuation on $P[p]$ is one plus the height function in $G$. Hence $B$ must have finite support. Therefore there exists an integer $k$ such that $p^k A \cap Q \subseteq \ker(\varphi|Q)$.

Let $\alpha: A/(p^k A \cap Q) \to A/(R \cap N)$ be the natural homomorphism and consider the restriction $\alpha: p^k A/(p^k A \cap Q) \to p^k(A/(R \cap N)) = (p^k A+(R \cap N)) \div (R \cap N) \cong p^k/(p^k \cap (R \cap N))$. Note that

$$p^k A/(p^k A \cap Q) \cong (p^k A+Q)/Q = p^k(A/Q) \cong p^k(pG) = p^{k+1}G$$

and

$$(p^k A \cap N)/(p^k A \cap Q) \cong ((p^k A \cap N)+Q)/Q \cong p^{k+1}G \cap P;$$

the second two isomorphisms being restrictions of the first two, respectively. Now $p^{k+1}G[p^n] = p^{k+1}S[p^n] \oplus p^{k+1}G \cap P$. Hence our lemma applies and there is an integer $j$ such that

$$p^j(p^kA/(p^kA\cap Q)) \cap (p^kA\cap N)/(p^kA\cap Q) \subseteq (p^kA\cap (R\cap N))/(p^kA\cap Q).$$

Setting $m = j + k$ we have $p^mA\cap N \subseteq R\cap N$ as desired.

## REFERENCES


1. Cutler, D. and C. Missel, The Structure of C-decomposable $p^{\omega+n}$-projective abelian p-groups. Communications in Algebra, 12, 301, 1984.

2. Fuchs, L., Infinite Abelian Groups, Vol. 1 and 2, Academic Press, New York, 1970 and 1973.

3. Fuchs, L. and J. Irwin, On $p^{\omega+1}$-projective p-groups, Proc. London Math. Soc., 30, 459, 1975.

4. Fuchs, L., On $p^{\omega+n}$-projective abelian p-groups, Publ. Math. Debrecen, 23, 309, 1976.

5. Richman, Fred, A guide to valuated groups, in Lecture Notes in Mathematics, 616, Springer Verlag, 1977.

# STRAIGHT AND STRONGLY STRAIGHT PRIMARY MODULES OVER PRINCIPAL IDEAL DOMAINS

K. BENABDALLAH* and D. BOUABDILLAH
UNIVERSITÉ DE MONTRÉAL

## INTRODUCTION

Straight and strongly straight primary abelian groups were introduced by the first author and K. Honda in [2]. It is commonly believed that most results in abelian group theory carry over to modules over principal ideal domains with only minor adjustments. It turns out that the concept of straightness and strong straightness behave differently according to the principal ideal domains chosen. This phenomenon justifies the study of these concepts in the more general setting of primary modules over arbitrary principal ideal domains. We obtain the following characterization: A primary module is strongly straight if and only if every isometry between its socle and that of a straight module extends to an isomorphism. As a consequence of this characterization the class of strongly straight primary modules is seen to contain all torsion complete primary modules, all direct sums of cyclic primary modules and all divisible primary modules. We show also that every subsocle of a strongly straight primary

* Research under canadian CRSNG grant no A5591.

module M supports a pure submodule K which is straight and M/K is straight. In particular strongly straight primary modules belong to the elusive class of pure complete modules. Over the polynomial ring K[t] where K is a field, straight t-primary modules are strongly straight, however over $\mathbb{Z}$ the reduced Prüfer group is straight but not strongly straight. It follows that over $\mathbb{Z}$ the class of strongly straight primary modules is not closed under direct sums whereas over K[t] the t-primary strongly straight modules form a class closed under direct sums. We conclude this article with some open questions.

## 1. DEFINITIONS AND ELEMENTARY PROPERTIES

Throughout this section we let R be a principal ideal domain, p a prime element of R , M a p-primary R-module and S a system of representatives of R module pR containing 0 . All modules considered in this article are p-primary R-modules.

Notation: Let B be a subset of M and let n be a non-negative integer. We denote by B(n) the set of elements of order $p^{n+1}$ in B

$$B(n) = \{b \in B \mid O(b) = p^{n+1}\} .$$

DEFINITION 1.1. A subset B of M is said to be a *straight basis* of M if for every non-negative integer n , B(n) is a maximal independant subset of M(n) .

As in the case of $\mathbb{Z}$-modules it is easy to verify the following:

PROPOSITION 1.2. Let $B$ be a subset of $M$. The following properties are quivalent:

i) $B$ is a straight basis of $M$ ;

ii) for every non-negative integer $n$, $p^n(B(n))$ is a basis of $(p^nM)[p]$ ;

iii) for every non-negative integer $n$, $\{b+M[p^n] \mid b \in B(n)\}$ is a basis of $M[p^{n+1}]/M[p^n]$ .

PROPOSITION 1.3. Let $B$ be a straight basis of $M$. Then every element of $M$ can be written uniquely as a linear combination of elements of $B$ with coefficients taken in $S$. More precisely, if $0(x) = p^{n+1}$, $x \in M$, $n \geq 0$ then $x = f_o+\ldots+f_n$ where $f_n \neq 0$ and for each $i$, $f_i$ is a linear combination of elements of $B(i)$ with coefficients in $S$.

PROOF. We proceed by induction on the order of the elements of $M$. If $x \in M$, and $0(x) = p$ then $x \in M[p]$. Since $B(0)$ is a basis of $M[p]$ viewed as a vector space over the field $R/pR$, $x$ admits the desired representation. We assume that the result is true for all elements of $M$ of order less or equal to $p^n$ and consider $x \in M$, $0(x) = p^{n+1}$. Then $p^nx \in (p^nM)[p]$ and from proposition 1.2(ii),

$$p^nx = \Sigma\, a_ip^nb_i \quad , \quad a_i \in S \quad , \; b_i \in B(n) \; .$$

Thus: $0(x-\Sigma a_ib_i) \leq p^n$ and by induction: $x-\Sigma a_ib_i = f_o+\ldots+f_{n-1}$, where the $f_i$'s satisfy the condition described in the statement above. Let $f_n = \Sigma\, a_ib_i$, then: $x = f_o+\ldots+f_n$, as desired. The uniqueness follows from the fact that: $f_n \neq 0$ implies $0(f_n) = p^{n+1}$.

DEFINITION 1.4. For every non-negative integer $n$ let $C_n$ be a basis of $(p^n M)[p]$. The family $\{C_n\}_{n\geq 0}$ is said to be a *sequence of bases for the socle* of $M$.

From proposition 1.2 (ii), every straight basis $B$ of $M$ induces a sequence of bases for the socle of $M$, namely $\{p^n(B(n))\}_{n\geq 0}$.

DEFINITION 1.5. Let $B$ a straight basis of $M$. We say that $B$ is *S-normal* if for every integer $n \geq 0$, and for every $b \in B(n+1)$,

$$pb = \Sigma\, a_i b_{ni} \quad , \text{ where } b_{ni} \in B(n) .$$

A module $M$ is said to be an *S-straight module* if it contains an S-normal straight basis. $M$ is said to be *strongly S-straight* if every sequence of bases for the socle of $M$ is induced by an S-normal straight basis of $M$. In other words, $M$ is strongly S-straight if for given sequence $\{C_n\}_{n\geq 0}$ of bases for the socle of $M$, there exists $B$ an S-normal straight basis of $M$ such that for every $n \geq 0$, $C_n = p^n(B(n))$.

PROPOSITION 1.6. The direct sum of a family of S-straight modules is an S-straight module.

LEMMA 1.7. Cyclic modules are S-straight for any system of representatives $S$ of $R$ modulo $pR$.

PROOF. Let $M = Rx$ and let $0(x) = p^{n+1}$. For $a \in S$, $a \neq 0$, the set $\{y_i\}_{i=0}^{n}$ defined by: $y_o = p^n x$, $y_1 = ap^{n-1}x$, $y_2 = a^2 p^{n-2}x, \ldots, y_n = a^n x$

is an S-normal straight basis of $M$ .

LEMMA 1.8. Divisible modules are strongly S-straight for any system of representatives $S$ of $R$ modulo $pR$ containing $0$ .

PROOF. Let $M$ be a divisible $R$-module and $\{C_n\}_{n=0}$ a sequence of bases for the socle of $M$ . We construct $B$ inductively letting $B(0) = C_0$ . Assume that $B(i)$ have been constructed for $0 \leq i < n$ . Let $c \in C_n$ , then $c = \Sigma\, a_i c_{n-1i}$ , for $a_i \in S$ , and $c_{n-1i} \in C_{n-1}$ , let $b_{n-1i}$ be the already chosen element of $B(n-1)$ such that $p^{n-1} b_{n-1i} = c_{n-1i}$ and choose an element $b$ in $M$ such that $pb = \Sigma\, a_i b_{n-1i}$ . The set of elements chosen in this manner constitute $B(n)$ . It is clear that $B = \cup B(n)$ is an S-normal straight basis of $M$ which induces the sequence of bases $\{C_n\}_{n=0}$ .

PROPOSITION 1.9. Direct sums of p-primary cocyclic R-modules are S-straight for every set of representatives $S$ of $R$ modulo $pR$ containing $0$ .

We conclude this section with the following straight forward adaptation of theorem 3.5 of [2].

THEOREM 1.10. Let $V_0$ be an elementary p-primary R-module and let $\{V_n\}_{n\geq 1}$ be a descending chain of submodules of $V_0$ . For each $n \geq 0$ let $C_n$ be a basis of $V_n$ . Then there exists an S-straight p-primary R-module $M$ with an S-normal straight basis $B$ such that $p^n M[p] = V_n$ and $p^n(B(n)) = C_n$ for every $n \geq 0$ .

## 2. STRONGLY S-STRAIGHT PRIMARY MODULES

We maintain the convention of section 1. Namely, R is a principal ideal domain, p is a prime in R , M is a p-primary R-module and S a fixed system of representatives of R modulo pR containing 0 .

DEFINITION 2.1. Let M and N be two p-primary R-modules, we say that an isomorphism f between M[p] and N[p] is an *isometry on the socle* if for every $x \in M[p]$ the p-height of x in M is the same as the p-height of f(x) in N .

We have the following important characterization of strongly S-straight modules.

THEOREM 2.2. Let M be a primary R-modules, then M is strongly S-straight if and only if every isometry on the socle between M and any S-straight module N extends to an isomorphism between M and N .

PROOF. If M is strongly S-straight then an obvious adaptation of the proof of theorem 3.7 in [2]yields that M satisfies the second property. Conversely if M satisfies the second property, let $\{C_n\}_{n\geq 0}$ be a family of bases of the socle of M . From theorem 1.10 there exists an S-straight R-module M' and a straight S-normal basis B' in M' such that $p^n(B'(n)) = C_n$ and $(p^nM')[p] = (p^nM)[p]$ for every $n \geq 0$ . The identity map between M[p] and M'[p] is an isometry and as such it extends to an isomorphism f between M and M' . Without loss of generality we may assume that $f : M' \to M$ . Let $B = f(B')$ . Clearly B is an S-normal straight basis of $M$ and $p^n(B(n)) = p^n f(B'(n))$ $= f(p^n(B'(n)) = f(C_n) = C_n$ . Therefore $M$ is strongly S-straight.

Next we establish two useful and important properties of strongly S-straight primary modules.

PROPOSITION 2.3. Summands of strongly S-straight modules are strongly S-straight.

PROOF: Let $M$ be a strongly S-straight module and let $M = H \oplus K$ where $H$ and $K$ are submodules of $M$ . For every $n \geq 0$ let $H_n$ be a basis of $(p^nH)[p]$ and $K_n$ a basis of $(p^nK)[p]$ . Then $C_n = H_n \cup K_n$ is a basis of $(p^nM)[p] = (p^nH)[p] \oplus (p^nK)[p]$ . Let $B$ be an S-normal straight basis of $M$ such that $p^n(B(n)) = C_n$ for every $n \geq 0$ . Let $f : M \to H$ be the projection of $M$ over $H$ along $K$ and let $H^n = \{f(b) \mid b \in B(n)\}$ . We leave it to the reader to check that $\bigcup_{n\geq 0} H^n$ is an S-normal straight basis of $H$ which induces the given sequence of bases $\{H_n\}_{n\geq 0}$ .

PROPOSITION 2.4. Let $M$ be a strongly S-straight module. Then every submodule of $M[p]$ supports a pure S-straight submodule $H$ of $M$ such that $M/H$ is S-straight.

PROOF. Let $U$ be a submodule of $M[p]$ . Set $U_n = (p^nM)[p] \cap U$ . Let $V_n$ be a complementary summand of $U_n$ in $(p^nM)[p]$ . Let $C_n$ be a basis of $U_n$ and $D_n$ a basis of $V_n$ . Then $C_n \cup D_n$ is a basis of $(p^nM)[p]$. Let $B$ be an S-normal straight basis of $M$ such that:
$p^n(B(n)) = C_n \cup D_n$ . Let $B_n' = \{b \in B(n) \mid p^n b \in C_n\}$ , and

$B''_n = \{b \in B(n) \mid p^n b \in D_n\}$. Finally let $B' = \bigcup_{n \geq 0} B'_n$ and $H$ the submodule generated by $B'$. We show first that $B'$ is an $S$-normal straight basis of $H$. Let $b \in B'(n) = B'_n$, then: $pb = \sum_{i \in I} a_i b_i + \sum_{j \in J} a_j b_j$, where $b_i \in B'_{n-1}$, and $b_j \in B''_{n-1}$, and $a_i, a_j \in S$. We obtain: $p^n b = \Sigma a_i p^{n-1} b_i + \Sigma a_j p^{n-1} b_j$. But $p^n b \in C_n \subset U_n$, $p^{n-1} b_i \in C_{n-1} \subset U_{n-1}$ and $U_n \subset U_{n-1}$, therefore: $\Sigma a_j p^{n-1} b_j = p^n b - \Sigma a_i p^{n-1} b_i \in U_{n-1} \cap V_{n-1} = 0$. It follows that $p$ divides $a_j$ for every $j \in J$. This means that $a_j = 0$ for all $j \in J$ since $S$ is a system of representatives of $R$ modulo $pR$ containing $0$. Thus we have $pb = \sum_{i \in I} a_i b_i$. It remains to show that every element of $H$ can be written uniquely as a linear combination of elements of $B'$ with coefficients from $S$. We proceed by induction. Clearly those elements of $H$ which are linear combinations of elements of $B'_0$ can be written uniquely with coefficients from $S$. Suppose that all elements which belong to the submodule generated by $B'_0 \cup \ldots \cup B'_{n-1}$ have a unique representative as linear combinations of only the elements of $B'_0 \cup \ldots \cup B'_{n-1}$ with coefficients from $S$. Let $h$ be in the submodule generated by $B'_n$, say $h = \Sigma c_i b_i$, $b_i \in B'_n$, $c_i \in R$. Now for each $c_i$ there exist $a_i \in S$ and $k_i \in R$ such that $c_i = a_i + k_i p$, thus $h = \Sigma a_i b_i + \Sigma k_i p b_i$. But $\Sigma k_i p b_i$ is in the submodule generated by $B'_{n-1}$. By induction $h$ admits a representation involving only the elements of $B'_0 \cup \ldots \cup B'_n$, and coefficients from $S$. This representation is unique as $B'$ is a subset of the straight basis $B$. Therefore $B'$ is an $S$-normal straight basis of $H$ and $(p^n H)[p] = U_n$, $n \geq 0$. Knowing this, it is easy to verify that the elements of $H[p]$ have the same height in $H$ as in $M$ and accordingly $H$ is pure in $M$. We leave it to the reader to check that the set $\{b+H \mid b \in \bigcup_{n\geq 0} B''_n\}$ , is an S-normal straight basis of M/H , and M/H is also S-straight.

We are now ready to give more information on the structure of strongly S-straight modules.

THEOREM 2.5. Let M be a strongly S-straight module. Then M is the direct sum of a divisible module and a separable strongly S-straight module.

PROOF. From proposition 2.4 the socle of $M^1 = \bigcap_{n=0}^{\infty} p^n M$ supports a pure submodule D of M . It is then wellknown that D must be divisible and equal to $M^1$ . Therefore $M = D \oplus H$ , where H is separable (that is: $H^1 = 0$) . From proposition 2.3, H is strongly S-straight.

THEOREM 2.6. Direct sums of cyclic p-primary R-modules are strongly S-straight for every system S of representatives of R modulo pR containing 0 .

PROOF. The result follows from the fact that if two primary modules have isometric socles and one of them is direct sum of cyclic modules the other is also direct sum of cyclic modules and every isometry between the socles extends to an isomorphism of such modules. Therefore by theorem 2.2 these modules are strongly S-straight.

THEOREM 2.7. Torsion complete p-primary R-modules are strongly S-straight for every system S of representatives of R modulo pR containing 0 .

PROOF. Recall that an R-module is torsion complete if and only if its socle with the p-adic valuation is an injective valued vector space (see [3] theorem 45). Thus if M is torsion complete and f is an isometry between M[p] and N[p] where N is a p-primary R-module f extends to an isomorphism between a basic submodule of M and a basic submodule of N . This extension in turn extends to an isomorphism of M onto N since N is necessarily torsion complete and thus pure injective among torsion modules. Therefore by theorem 2.2 M is strongly S-straight for every S containing 0 .

The converse of theorem 2.5 is not true in general as we show in the nex section. We conclude this section with the following partial converse:

THEOREM 2.8. Let M be a torsion complete R-module and D a divisible R-module then $M \oplus D$ is strongly S-straight for every system S of representatives of R module pR containing 0 .

PROOF. Let f be an isometry between $(M\oplus D)[p]$ and N[p] for a module N . Then $T = f(M[p])$ is a subsocle of N , such that: $T \oplus N^1[p] = N[p]$. Let K be an $N^1$-high submodule of N containing T , then $K[p] = T$ and K is a pure submodule of N . Since K is pure f restricted to M[p] is an isometry between M[p] and K[p] therefore K is torsion complete. It follows from the pure injectivity of torsion complete modules that K is in fact a summand of N . That is $N = K \oplus D'$ and D' is the divisible submodule of N . Clearly f extends to an isomorphism of $M \oplus D$ and $N$ and by theorem 2.2, $M \oplus D$ is strongly S-straight.

## 3. FURTHER RESULTS AND EXAMPLES

We begin this section with the following unexpected result:

PROPOSITION 3.1. Let $M$ be an S-straight p-primary R-module where $S$ is a system of representatives of $R$ modulo $pR$ containing $0$ and closed under addition and multiplication. Then $M$ is strongly S-straight.

PROOF. Let $B = \bigcup_{n=0}^{\infty} B(n)$ be an S-normal straight basis of $M$ and let $C_n$ be a basis of $p^nM[p]$ , $n \geq 0$ . We set $B'(0) = C_0$ and we suppose inductively that for $0 \leq k \leq n-1$ , $B'(k)$ have been constructed such that:

(i) $p^k(B'(k)) = C_k$ , $0 \leq k \leq n-1$ .

(ii) For every $b \in B'(k)$ , $pb = \Sigma a_i b_i$ where $a_i \in S$ and $b_i \in B(k-1)$ .

(iii) For every $x \in B'(k)$ , $x = \Sigma d_i b_i$ , $b_i \in B(k)$ and $d_i \in S$ .

We construct $B'(n)$ . Let $c_n \in C_n$ , then $c_n = \Sigma a_i c_{n-1i}$ , $a_i \in S$ , $c_{n-1i} \in C_{n-1}$ . Let $b_i' \in B'(n-1)$ such that $p^{n-1}b_i' = c_{n-1i}$ and consider the element $z_n = \Sigma a_i b_i'$ . From (iii) and the fact that $S$ inclosed under addition and multiplication it follows that $z_n$ can be written as

(I) $$z_n = \Sigma r_j b_j \text{ , where } r_j \in S \text{ and } b_j \in B(n-1) .$$

In the one hand we note that in fact $z_n$ is the unique element that can be written in the above manner and satisfy: $p^{n-1}z_n = c_n$ . In the other hand, $p^n(B(n))$ is a basis of $p^nM[p]$ , therefore $c_n = \Sigma\, d_i p^n b_i$ , $b_i \in B(n)$ , $d_i \in S$ . Let:

(II) $$y_n = \Sigma\, d_i b_i \ .$$

Now: $py_n = \Sigma\, d_i pb_i$ . Since $B$ is S-normal and $S$ is closed under sums and products, $py_n$ can be written as: $py_n = \Sigma\, s_j b_j$ , $s_j \in S$ , $b_j \in B(n-1)$ . However $p^{n-1}(py_n) = c_n$ . Therefore $py_n = z_n$ . Now, to each $c_n \in C_n$ we associate $y_n$ constructed as above and form $B'(n)$ with these $y_n$ . Clearly $p^n(B'(n)) = C_n$ and from (I) and (II) we see that $B'(n)$ satisfy (ii) and (iii). It follows by induction that $B' = \cup B'(n)$ is an S-normal straight basis of $M$ which induces the given sequence of basis $\{C_n\}_{n\geq 0}$ and that $M$ is strongly S-straight.

<u>COROLLARY 3.2</u>. Over the ring of polynomials $K[t]$ where $K$ is a field, a t-primary module is K-straight if and only if it is strongly K-straight.

<u>COROLLARY 3.3</u>. Over the ring $K[t]$ where $K$ is a field there exists t-primary modules which are not K-straight.

<u>PROOF</u>. The Prüfer t-primary $K[t]$-module defined by generators $\{x_i\}_{i\geq 0}$ and relations $\{tx_o = 0$ and $t^n x_n = x_o$ , $n \geq 1\}$ is not K-straight in view of the preceeding corollary and theorem 2.5.

In contrast with the preceeding result we have the following astonishing fact:

PROPOSITION 3.4. The reduced p-primary Prüfer $\mathbb{Z}$-module P is S-straight for $S = \{0,1,\dots,p-1\}$.

PROOF. Let $P = \langle x_o, x_1, \dots, x_n, \dots\rangle$ with $px_o = 0$, $p^n x_n = x_o$, $n \geq 1$. Let $B(0) = \{x_o, x_1 - px_2, px_2 - p^2 x_3 \dots, p^n x_{n+1} - p^{n+1} x_{n+2}, \dots\}$ and let $0 < r,s < p$ such that $r+s = p$. Define $B(n)$, $n \geq 1$ as:

$B(n) = \{px_{n+1} + r(px_{n+2} - x_{n+1}), s(px_{n+2} - x_{n+1}), p^2 x_{n+3} - px_{n+2}, p^3 x_{n+4} - p^2 x_{n+3}, \dots\}$.

Note that $px_{n+1} + r(px_{n+2} - x_{n+1}) + s(px_{n+2} - x_{n+1}) = p^2 x_{n+2}$ therefore $\langle p^n(B(n))\rangle$ contains $x_o$ and $p^n x_{n+1} - p^{n+1} x_{n+2}$ etc... therefore $p^n(B(n))$ generates $p^n P[p]$. In fact it is easy to see that, $p^n(B(n))$ is a basis of $p^n P[p]$. To verify the normality we write the first elements of $B(n-1)$. We have: $px_n + r(px_{n+1} - x_n), s(px_{n+1} - x_n), p^2 x_{n+2} - px_{n+1}$, $p^3 x_{n+3} - p^2 x_{n+2}, \dots$. Clearly p times the third element in $B(n)$ is the fourth element in $B(n-1)$ and the normality is verified from the third element on. Now p times the second element of $B(n)$ is s-times the third element of $B(n-1)$, finally for the first element we have:

$$p[px_{n+1} + r(px_{n+2} - x_{n+1})] = p^2 x_{n+1} + r(p^2 x_{n+2} - px_{n+1})$$

which can be shown to be the sum of the first two elements of $B(n-1)$ with r times the third.

COROLLARY 3.5. If $p \neq 2$ then the Prüfer $\mathbb{Z}$-module P contains $2^{\aleph_o}$ different S-normal straight basis where $S = \{0,1,\dots,p-1\}$.

PROOF. Upon close examination of the construction in the preceeding proof we realize that in each $B(n)$ we can use a different pair of elements $0 < r,s < p$ such taht $r+s = p$. Thus for every sequence of choices we obtain a new normal straight basis.

REMARK 3.6. The reader can easily verify that the given construction in proposition 3.4 gives a normal straight basis of P . But both authors spent countless hours in unfruitfull attempts to show that the Prüfer group is not straight and changed their minds a dozen times about what should be true. They still cannot believe how simple the construction really was.

COROLLARY 3.7. There exists an S-straight p-primary $\mathbb{Z}$-module which is not strongly S-straight where $S = \{0,1,\ldots,p-1\}$.

COROLLARY 3.8. The class of strongly S-straight p-primary $\mathbb{Z}$-modules is not closed under direct sums where $S = \{0,1,\ldots,p-1\}$ .

PROOF. The Prüfer $\mathbb{Z}$-module P has an isometric socle with the socle of a direct sum of cyclic modules and $\mathbb{Z}_{p^\infty}$ . If the class of strongly S-straight p-primary Z-modules were closed under direct sums the latter module would be strongly S-straight and isomorphic to P by theorem 2.2 but P is reduced.

REMARK 3.9. It was pointed out to us by S.A. Khabbaz that every p-primary module M over any principal ideal domain R possesses a straight basis B such that p(B(n)) is included in the submodule generated by B(n-1) for every $n \geq 1$ . However, there exists primary modules over polynomial rings which are not straight (Corollary 3.3). This shows that the restriction on the coefficients in definition 1.5 plays a crucial role.

REMARK 3.10. A module $M$ is said to be pure complete if every submodule of the socle of $M$ is the socle of some pure submodule of $M$. Theorem 2.4 shows that strongly $S$-straight modules are pure complete. It can be shown that for every principal ideal domain $R$ which is not a field there exists primary $R$-modules which are not pure complete and have no elements of infinite height (see [1], p. 327). Therefore there exists separable modules which are not strongly straight. We do not know however, if there exists separable straight modules which are not strongly straight.

We conclude with a list of questions and problems.

1. Are there primary $\mathbb{Z}$-modules which are not $S$-straight for $S = \{0,1,\dots,p-1\}$.

2. For which principal ideal domains the class of $S$-straight modules is the same as the class of strongly $S$-straight modules for some $S$ or for all $S$'s ?

3. If $S$ and $S'$ are two systems of representatives of $R$ modulo $pR$ are $S$-straight modules also $S'$-straight?

4. Characterize strongly $S$-straight $p$-groups where $S = \{0,1,\dots,p-1\}$.

## REFERENCES

[1] BENABDALLAH, K. and IRWIN, J. "Pure N-high subgroups, p-adic topology and direct sums of cyclic groups". *Can. J. Math.*, vol. XXVI, no 2, 322, 1974.

[2] BENABDALLAH, K. and HONDA, K. "Straight bases of abelian p-groups". *Lecture notes in mathematics*, no 1005, 556, Springer-Verlag, 1983.

[3] FUCHS, L. "*Abelian p-groups and mixed groups*". Séminaire de Mathématiques Supérieures, vol. 70, Les Presses de l'Université de Montréal, Montréal 1980.

[4] FUCHS, L. "*Infinite Abelian groups*". Vol. 1, Academic Press, New York, 1970.

# A BASIS THEOREM FOR SUBGROUPS OF BOUNDED ABELIAN GROUPS

K. BENABDALLAH AND S. A. KHABBAZ

UNIVERSITÉ DE MONTRÉAL AND LEHIGH UNIVERSITY

§0. In this paper we present a generalization of the main aspects of the basis theorem for pairs $(G,H)$ in which $H$ is a subgroup of a finite abelian group $G$, see [4], to the case in which $G$ is an abelian group of bounded order. The proof we outline here is actually simpler than the one sketched in [4]. We remark that the assumption of the well ordering principle which occurs in the statement of the theorem is not an essential feature of it.

§1. The Main Theorem.

We use $o(g)$ to denote the order of a group element $g$, $\exp(g)$ to denote its exponent $\log_p o(g)$, and $V(n)$ to denote the $p$-order of the integer $n$, i.e. the exponent of the highest power of $p$ dividing $n$. For other notations and terms see [3].

Theorem: Given a bounded p-group G, and a subgroup H of G, there exists a basis $\{g_j\}_{j\in J}$ of G, and a basis $\{h_i\}_{i\in I}$ of H where I and J are some well ordered indexing sets, such that

(a) $\exp(g_t) \leq \exp(g_s)$ if $t \leq s$.

(b) Normalization: For each i, let $h_i = \sum_{j\in J} c_{i,j} g_j$, with finitely many $c_{i,j} \neq 0$. Let $c_{i,j_i}$ be the first nonzero coefficient, and let $c_{i,k(i)}$ be the last nonzero coefficient. Then $c_{i,j_i}$ is a power of p.

(c) Echelon form: the "row-length function" k(i) appearing in (b) is properly increasing, with the last coefficient, in which $j = k(i)$, not zero and indeed a power of p. (We will refer to $c_{i,k(i)}$ as the diagonal term);

(d) Blocks on the diagonal: If $g_j$ and $g_{k(i)}$ are of the same order, with j not equal to k(i), then $c_{i,j}$ is zero;

(e) Ordering within blocks: If $g_{k(i)}$ and $g_{k(j)}$ are of the same order, with $i < j$, then $V(c_{i,k(i)}) \leq V(c_{j,k(j)})$;

(f) The diagonal term determines the order of the row:

$$\exp(h_i) = \exp(c_{i,k(i)}g_{k(i)})$$

$$> \exp(h_i - c_{i,k(i)}g_{k(i)});$$

(This is equivalent to the condition that

$$V(c_{i,k(i)}) - V(c_{i,j})$$

$$< \exp(g_{k(i)}) - \exp(g_j)$$

for all $j < k(i)$ for which $c_{i,j}$ is not zero.)

Furthermore, the diagonal is unique: In each block, and for each $n$, the number of diagonal entries equal to $p^n$ is independent of our construction, as long as (a)-(f) are satisfied.

§2. The Construction.

The proof is by induction on the number $m$ of distinct orders among the orders of the generators of $G$. Note that the coefficients $c_{i,j}$, form an $I \times J$ matrix in which each row has only finitely many nonzero elements. If $m = 1$, the familiar basis theorem for a torsion-free module over a principal ideal ring applies. It gives us a diagonal matrix whose rank is the rank of $H$, augmented by additional zero columns for a total number of columns equal to the rank of $G$. Condition (a) of the theorem is trivial. Condition (b) can be achieved by multiplying the basis elements of $H$ by units. In (c) the row-length function is $k(i) = i$; and since there is only one nonzero entry in each row, when we made the first a power of $p$ we did the same with the last. (d) is trivial since the matrix is diagonal. (e) can be achieved by permuting the basis elements of $G$, and simultaneously doing the same with the basis elements of $H$, to arrange the diagonal in non-decreasing order. Condition (f) is trivial for diagonal matrices.

Now suppose that $m > 1$ and that $p^n G = 0$ but $p^{n-1}G \neq 0$. Let $E = (p^{n-1}G) \cap H$. Then $E$ equals its own socle $S(E)$. Since $G$ is bounded, $E$ has a purification $H_2$ in $H$, that is, $H_2$ is pure in $H$ and $S(E) = S(H_2)$. Since $H$ is bounded, we may write $H$ as a direct sum $H = H_1 \oplus H_2$. Now let $G_1 \subset G$ be a subgroup of $G$ which is maximal with respect to containing $H_1$ and being disjoint from $p^{n-1}G$. Clearly $H_1 = G_1 \cap H$. Also let $G_2$ denote a purification (in $G$) of $p^{n-1}G$. Then it is easy to veryify that $G = G_1 \oplus G_2$. But whereas $G_1 \supset H_1$ and $S(G_2) \supset E$, it is not necessarily true that $G_2 \supset H_2$. In any case note that $S(H) = S(H \cap G_1) \oplus S(H \cap G_2)$.

By the induction hypothesis $H_1$ and $G_1$, have bases $\{h_i\}_{i \in I'}$ and $\{g_j\}_{j \in J'}$ which satisfy the conditions of the theorem for $G = G_1$ and $H = H_1$. The most essential part of the theorem is condition (f). After this is proved, the rest of the theorem is easy to take care of.

In order to prove (f) we show the existence of a basis $\{\bar{g}_\alpha\}_{\alpha\in A}$ of $G/G_1$ and of a basis $\{\bar{h}_\alpha\}_{\alpha\in B\subseteq A}$ of $(H+G_1)/G_1$ which are related by the following condition: For each $\alpha \in A$ there is a representative $g_\alpha$ in $G$ of $\bar{g}_\alpha$, and for each $\alpha \in B$ there is a representative $h_\alpha$ (from $H$) of $\bar{h}_\alpha$ with the property that $h_\alpha = f_\alpha + p^{n_\alpha} g_\alpha$ where $f_\alpha \in G_1$ and $o(f_\alpha) < o(p^{n_\alpha} g_\alpha)$. The rest of the argument proceeds as follows.

Because of the maximum order property of the elements $g_\alpha$, it is not hard to see that the set $\{g_\alpha\}_{\alpha\in A}$ forms a basis for an absolute direct summand $G_2$ of $G$ with $G = G_1 \oplus G_2$. This $G_2$ is clearly a purification of $p^{n-1}G$. For the bases whose existence is asserted in the theorem we may take $\{h_i\}_{i\in B\cup I'}$ for $H$ and $\{g_i\}_{i\in A\cup J'}$ for $G$, so that $I = B\cup I'$ and $J = A\cup J'$. The inequality $o(f_\alpha) < o(p^{n_\alpha} g_\alpha)$ will guarantee that condition (f) of the theorem holds. Elements of $J'$ precede those of $A$.

To show that bases satisfying the condition exist note that the case $m = 1$ of the induction hypothesis yields bases $\{\bar{g}_\alpha\}_{\alpha\in A}$ of $G/G_1$ and $\{\bar{\bar{h}}_\alpha\}_{\alpha\in B}$ of $(H+G)/G_1$ which are related as in the condition. Let $g_\alpha$ be any element of $G$ with $g_\alpha + G_1 = \bar{g}_\alpha$, and let $\bar{h}_\alpha$ be any element of $H$ with $\bar{h}_\alpha + G_1 = \bar{\bar{h}}_\alpha$. Clearly the elements $\{g_\alpha\}_{\alpha\in A}$ form a basis for a subgroup $G_2$ of $G$ which is a purification of

$p^{n-1}G$ and is such that $G = G_1 \oplus G_2$. Since $\overline{h}_\alpha + G_1 = p^{n_\alpha}(g_\alpha + G_1)$ we may write $\overline{h}_\alpha = f'_\alpha + p^{n_\alpha} g_\alpha$ where $p^{n_\alpha} g_\alpha \in G_2$ and $f'_\alpha \in G_1$. We will show that $\overline{h}_\alpha$ can be chosen so that $o(f'_\alpha) < o(p^{n_\alpha} g_\alpha)$. We will do this by adding to $f'_\alpha$ an appropriate element of $H_1$. The proof that this choice can be made is divided into two cases.

Case 1. $p^c = o(f'_\alpha) > o(p^{n_\alpha} g_\alpha)$. In this case the element $p^{c-1}\overline{h}_\alpha$ equals $p^{c-1}f'_\alpha$ and hence is in $G_1 \cap H = H_1$. Since $H_1$ is a direct summand of $H$ and hence is pure in H, there is an element $f^\alpha$ of $H_1$ with $p^{c-1}f'_\alpha = p^{c-1}\overline{h}_\alpha = p^{c-1}f^\alpha$. Now $(f'_\alpha - f^\alpha) + p^{n_\alpha} g_\alpha$ certainly represents $\overline{\overline{h}}_a$, $f'_a - f^\alpha \in G_1$ and $o(f'_\alpha - f^\alpha) < o(f'_\alpha)$. This process may be repeated until a replacement $(f'_\alpha - f^\alpha)$ for $f'_\alpha$ has been found with $o(f'_\alpha - f^\alpha) \leq o(p^{n_\alpha} g_\alpha)$. This puts us in the next case.

Case 2. $o(f'_\alpha) = o(p^{n_\alpha} g_\alpha) = p^{n-n_\alpha}$. Now $\overline{h}_\alpha = f'_\alpha + p^{n_\alpha} g_\alpha \in H$ with $f'_\alpha \in G_1$, $g_\alpha \in G_2$. Since $p^{n-n_\alpha-1}\overline{h}_\alpha \in S(H)$, it follows from the uniqueness of the decomposition of $\overline{h}_\alpha$ as a sum of an element of $G_1$ and an element of $G_2$ that $(p^{n-n_\alpha-1} f'_\alpha, p^{n-n_\alpha-1}(p^{n_\alpha} g_\alpha))$ is the unique decomposition of $p^{n-n_\alpha-1}\overline{h}_\alpha$ as a sum corresponding to the direct sum decomposition of $S(H)$ given by $S(H) = S(H \cap G_1) \oplus S(H \cap G_2)$. Since $H = H_1 \oplus H_2$ with $S(H_1) = S(H \cap G_1)$ and $S(H_2) = S(H \cap G_2)$, it follows that the height of $p^{n-n_\alpha-1} f'_\alpha$ in $H_1$, is not less than that of $p^{n-n_\alpha-1}(f_\alpha+p^{n_\alpha}g_\alpha)$ in H. This means that there is an element $c_\alpha \in H_1$, with $p^{n-n_\alpha-1} c_\alpha = p^{n-n_\alpha-1} f'_\alpha$. Now we can make the needed and final alteration in $f'_\alpha$ referred to earlier. We add to $f'_\alpha$ the element $(-c_\alpha)$. Since $c_\alpha \in G_1 \cap H$ clearly $(f'_\alpha-c_\alpha) + p^{n_\alpha}g_\alpha$ represents $\bar{\bar{h}}_\alpha$, $f'_\alpha - c_\alpha \in H$, and $o(f'_\alpha-c_\alpha) \leq p^{n-n_\alpha-1} < o(p^{n_\alpha}g_\alpha) = p^{n-n_\alpha}$. For the required elements $f_\alpha$ and $h_\alpha$ we may thus take $f_\alpha = f'_\alpha - c_\alpha$ and $h_\alpha = f_\alpha + p^{n_\alpha}g_\alpha$ of the diagonal. Consider next the question of invariance. The proof of the invariance of the diagonal elements depends in an essential way on part (f) of the theorem. The invariance can be deduced by observing that in the block corresponding to the basis elements of G of order $p^u$ and for a given integer u-v, the number of diagonal entries equal to $p^{u-v}$ equals the number of basis elements $h_i$ of H of exponent v for which $p^{v-1}h_i$ has height u-1 in G.

Alternatively, this number equals the rank of a direct summand Q of H which is of maximum cardinality subject to being a direct sum of cyclic groups of order $p^v$ each, and having the property that S(Q) is the socle of a direct summand of G consisting of cyclic groups of order $p^u$ each.

When H is finite this number is given by $R(u-1,v-1) - R(u,v-1) - R(u-1,v) + R(u,v)$, where $R(u,v) = \operatorname{rank}(p^uG \cap p^vH)$. This formula which is proved in [4] for a finite G clearly holds if only H is assumed to be finite. The proofs of the remaining parts of the theorem are similar to the proofs of their counterparts in the finite case. For details see [4].

§3. In this section we indicate how to extend the theorem of §1 to the case of a pair (G,H) in which G is a direct sum of cyclic groups whose torsion subgroup $G_T$ is bounded.

Let $H_T$ denote the torsion subgroup of H and let $G_p$ and $H_p$ denote the p-primary components of $G_T$ and $H_T$ respectively. Let $\{g_\alpha\}_{\alpha \in J_p}$ and $\{h_\alpha\}_{\alpha \in I_p}$ denote bases for $G_p$ and $H_p$ respectively, which are related as in the theorem. Also let $\{\bar{g}_\alpha\}_{\alpha \in J_0}$ and $\{\bar{h}_\alpha\}_{\alpha \in I_0}$ denote bases for the free groups $G/G_T$ and $(H+G_T)/G_T$ which are related as follows: $J_0$ is a well ordered set containing $I_0$; and for each $\alpha \in I_0$, $\bar{h}_\alpha = n_\alpha \bar{g}_\alpha$ where $n_\alpha$ is a positive integer, and $n_\alpha$ divides $n_\beta$ if $\alpha < \beta$. Now for each $\alpha \in J_0$ let $g_\alpha$ be an element of $G$ representing the coset $\bar{g}_\alpha$ and for each $\alpha \in I_0$ let $h_\alpha$ be an element of $H$ representing $\bar{h}_\alpha$. Set $J = J_0 \cup (\bigcup_{p \text{ prime}} J_p)$ and $I = I_0 \cup (\bigcup_{p \text{ prime}} I_p)$, and order $J$ and $I$ by letting the elements of $J_x$ precede those of $J_y$ if $x < y$ or if $x \neq y$ and $y = 0$. In this way one obtains bases $\{g_\alpha\}_{\alpha \in J}$ for $G$ and $\{h_\alpha\}_{\alpha \in I}$ for $H$ which are related by a "triangular" matrix, see [4], which has in it all the main features of the theorem of §1. In particular, the diagonal is unique.

## BIBLIOGRAPHY

1. Donna Beers, R. Hunter, and E. Walker, Finite Valued p-Groups, Abelian Group Theory, Lecture Notes in Mathematics 1006, Springer-Verlag, New York 1983.

2. S. D. Berman and Z. P. Zilinskaza, On simultaneous direct decompositions of a finitely generated abelian group and a subgroup, translated in Sov. Math. Dokl. 14(1973), pp. 833-837.

3. Laszlo Fuchs, Infinite Abelian Groups, vol. 1, Academic Press New York and London, 1970.

4. S. A. Khabbaz and G. Rayna, A Basis Theorem for Subgroups of Finite Abelian Groups, Abelian Group Theory, Lecture Notes in Mathematics 1006, Springer-Verlag, New York 1983.

# ON A SPECIAL CLASS OF ALMOST COMPLETELY DECOMPOSABLE TORSION FREE ABELIAN GROUPS I

Rolf Burkhardt
Universität Würzburg

Lady[1] introduced the concept of a regulating subgroup of an almost completely decomposable group which turned out to be a completely decomposable subgroup of minimal index i(G). Every two regulating subgroups are isomorphic but he showed that there can exist two regulating subgroups with nonisomorphic factorgroups. For an almost completely decomposable group G let $\mathfrak{Z}(G)$ denote the set of factorgroups $\{G/W \mid W$ a regulating subgroup of $G\}$. One can ask the question: Is it possible that $\mathfrak{Z}(G)$ contains all groups of a given order? The answer is yes and that is the class of groups we want to consider in this note. We restrict ourselves to the case $|G/W| = p^n$ for some prime p. The group G of Lady's example has two regulating subgroups W and W' such that G/W is cyclic of order $p^2$ and G/W' is elementary-abelian. It is rather simple to generalize this example to a group G where each group of order $p^n$ occur as a factorgroup of G over one of its regulating subgroups ( see example 1 ). The general structure of these groups is described in the Theorem.

At the beginning we want to make some definitions and notations. If $\tau$ is a type then $G(\tau)$ is the subgroup of G consisting of those x with $t(x) \geq \tau$ . $G^*(\tau)$ denotes the pure subgroup of G generated by all $G(\sigma)$ for $\sigma > \tau$ . A regulating subgroup of an almost completely decomposable group G is a completely decomposable subgroup C with finite index in G such that, for each type $\tau$, $G(\tau) = C_\tau \oplus G^*(\tau)$ where $C_\tau$ is a maximal $\tau$-homogeneous summand of C. Denote the set of regulating subgroups of G by reg(G) and the intersection of all regulating subgroups by R(G) and call this intersection the regulator of G. For an almost completely decomposable group G let $T(G) = \{\tau \mid C_\tau \neq 0$ for a regulating subgroup $C\}$. Define a typegraph $\mathfrak{T}(G)$ with vertex set $\{\sigma \mid \sigma \in T(G)\}$ where $\sigma$ is connected with $\tau$ if and only if $\tau > \sigma$. $l(\mathfrak{T}(G))$ denotes the number of vertices in a longest chain in $\mathfrak{T}(G)$. Let $T_1 = \{\sigma \mid \sigma \in T(G)$, $\sigma$ maximal$\}$, $T_{i+1} = \{\sigma \mid \sigma \in T \setminus T_1 \cup \ldots \cup T_i$, $\sigma$ maximal$\}$ for $i \geq 1$ and $W_i = \oplus W_\tau$ with $\tau \in T_1 \cup \ldots \cup T_i$.

Example 1: For $n \in \mathbb{N}$ let

$$G(n) = \langle p_1^{-\infty} a_1,\ q_1^{-\infty} b_1,\ p_1^{-\infty} p_2^{-\infty} a_2,\ p_1^{-\infty} q_2^{-\infty} b_2,\ \ldots$$

$$\ldots,\ p_1^{-\infty} p_2^{-\infty} \cdots p_n^{-\infty} a_n,\ p_1^{-\infty} \cdots p_{n-1}^{-\infty} q_n^{-\infty} b_n,\ p_o^{-1}(a_1+b_1),$$

$$p_o^{-1}(a_2+b_2),\ \ldots,\ p_o^{-1}(a_n+b_n) \rangle .$$

Let I be any subset of $\{1,\ldots,n-1\}$. Define a regulating subgroup $W_I$ by

$$W_I = \bigoplus_{i=1}^{n} \langle b_i \rangle_* \oplus \bigoplus_{i \in I} \langle a_i + p_o^{-1}(a_{i+1}+b_{i+1}) \rangle_* \oplus \bigoplus_{i \notin I} \langle a_i \rangle_*$$

Let $\mathfrak{m} = (m_1,\ldots,m_r)$ be a tuple with $n \geq m_1 \geq \ldots \geq m_r \geq 1$, $m_1+\ldots+m_r = n$ and $\mathfrak{C}_\mathfrak{m}$ the complement of $\{m_1, m_1+m_2, \ldots, m_1+\ldots+m_{r-1}\}$ in $\{1,\ldots,n-1\}$. Then

$$G(n)/W_{\mathfrak{C}_\mathfrak{m}} \simeq Z_{p_o^{m_1}} \oplus \ldots \oplus Z_{p_o^{m_r}}$$

Hence $\mathfrak{J}(G(n))$ contains all groups of order $p^n$. The typegraph $\mathfrak{T}(G(n))$ has the following form

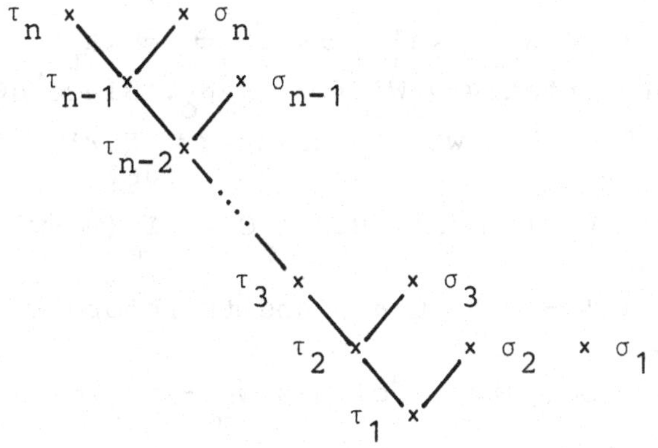


with $\tau_i \sim p_1^{-\infty} \cdots p_i^{-\infty}$ and $\sigma_i \sim p_1^{-\infty} \cdots p_{i-1}^{-\infty} q_i^{-\infty}$

It will turn out that a group G with the property that $\mathfrak{G}(G)$ contains all groups of order $p^n$ is similar to the group in the example.

Lemma 1: Let $W = \oplus W_\tau$ be a regulating subgroup of G. Then

$$R(G) = \oplus e_\tau W_\tau$$

with $e_\tau = \text{l.c.m.}\ \{\exp(G^*(\tau)/M) \mid M \in \text{reg}(G^*(\tau))\}$ .

Proof: a) Let W' be any regulating subgroup of G. Then $W_\tau \subset W'_\tau \oplus G^*(\tau)$ and therefore $e_\tau W_\tau \subset W'_\tau \oplus \bigcap_{\sigma > \tau} W'_\sigma \subset W'$, i.e. $\bigoplus_{\tau \in T} e_\tau W_\tau \subset R(G)$.

b) Let g be an element of R(G) which is not contained in the direct sum of the $e_\tau W_\tau$ , $\tau \in T := T(G)$. Without loss of generality we can assume that $g = \sum_{\sigma \in I \subset T} w_\sigma$ , $w_\sigma \in W_\sigma \setminus e_\sigma W_\sigma$ .
Let $\sigma_o$ be maximal in the type set I. There exists an element a of $G^*(\sigma_o)$ whose order mod $R(G^*(\sigma_o))$ is $e_{\sigma_o}$ . Let $W_{\sigma_o} = \langle a_1 \rangle_* \oplus \cdots \oplus \langle a_r \rangle_*$ . We can further assume that $\chi(a) \geqq \chi(a_i)$ for all i. In the expression $w_{\sigma_o} = q_1 a_1 + \ldots + q_r a_r$,

$q_i \in Q$ , not all $q_i$ can be divisible by $e_{\sigma_0}$ , say $e_{\sigma_0} \nmid q_1$ .
Let $a_1' = a_1 + a$ , $W'_{\sigma_0} = \langle a_1' \rangle_* \oplus \langle a_2 \rangle_* \oplus \quad \oplus \langle a_r \rangle_*$ and choose arbitrary representatives $W'_\sigma$ for $\sigma > \sigma_0$. Since $g \in R(G) \subset$ $\subset W' := \bigoplus_{\sigma \geqq \sigma_0} W'_\sigma \oplus \bigoplus_{\sigma \ngeqq \sigma_0} W_\sigma$ we can write $g = \sum_{\sigma \in T} w'_\sigma$ with $w'_\sigma \in W'_\sigma$ $(\sigma \geqq \sigma_0)$ and $w'_\sigma \in W_\sigma$ $(\sigma \ngeqq \sigma_0)$. Hence $0 = \sum_{\sigma \in T} (w_\sigma - w'_\sigma)$ , i.e.
$w_{\sigma_0} - w'_{\sigma_0} = \sum_{\sigma \neq \sigma_0} (w'_\sigma - w_\sigma) \in G(\sigma_0)$ and therefore $w_\sigma = w'_\sigma$ for $\sigma \ngeqq \sigma_0$. From $w_\sigma = 0$ $(\sigma > \sigma_0)$ follows $w_{\sigma_0} - w'_{\sigma_0} = \sum_{\sigma > \sigma_0} w'_\sigma$ .

If we write $w'_{\sigma_0}$ as a linear combination of the basic elements of $W'_{\sigma_0}$ , i.e. $w'_{\sigma_0} = q_1'a_1' + q_2'a_2 + \ldots + q_r'a_r$ we get the expression $w'_{\sigma_0} - w_{\sigma_0} = \sum_i (q_i' - q_i)a_i + q_1'a \in \bigoplus_{\sigma > \sigma_0} W'_\sigma$
in particular $\sum (q_i' - q_i)a_i \in W_{\sigma_0} \cap G^*(\sigma_0) = \{0\}$. We deduce $q_i' = q_i$ for $1 \leqq i \leqq r$ and $q_1 a \in \bigoplus_{\sigma > \sigma_0} W'_\sigma$ . Since $q_1$ was independent of the choice of $W'_\sigma$ $(\sigma > \sigma_0)$ , we arrive at $q_1 a \in R(G^*(\sigma_0))$ which is a contradiction. □

We should mention a necessary and sufficient condition for a completely decomposable subgroup of finite index to be the regulator $R(G)$.

<u>Lemma 2</u>: Let $U$ be a completely decomposable subgroup of $G$ with finite index. Let $\varepsilon_\tau = \exp(G^*(\tau)/U^*(\tau))$ and assume
(i) there exists a decomposition $U = \oplus U_\tau$ , $U_\tau$ homogeneous such that $\varepsilon_\tau^{-1} U_\tau \subset G$
(ii) $\varepsilon_\tau G(\tau) \subset U$ .
Then $\bigoplus_{\tau \in T} \varepsilon_\tau^{-1} U_\tau \in \operatorname{reg}(G)$ .

<u>Proof</u>: We have only to show that $G(\tau) = G^*(\tau) \oplus \varepsilon_\tau^{-1} U_\tau$.
If $g \in G(\tau)$ then (ii) forces $\varepsilon_\tau g \in U^*(\tau) \oplus U_\tau$ , i.e. $g = u_1 + u_2$ with $u_1 \in U^*(\tau)$ and $u_2 \in U_\tau$. From (i) we know that $u_2$ has an

$\varepsilon_\tau$-th root. Therefore the same is true for $u_1$, hence
$g = \varepsilon_\tau^{-1}u_1 + \varepsilon_\tau^{-1}u_2 \in G^*(\tau) \oplus \varepsilon_\tau^{-1}U_\tau$. □

Lemma 3: Let conditions (i) and (ii) of lemma 2 be satisfied. Then $U = R(G)$.

Proof: We first show that $\varepsilon_\tau \mid e_\tau$ for all $\tau$. If $\tau \in T_1$ than obviously $\varepsilon_\tau = e_\tau = 1$. Assume that $\varepsilon_\tau \mid e_\tau$ for all $\tau \in T_1 \cup \ldots \cup T_i$ and let $\tau \in T_{i+1}$. By lemma 1 $R(G)^*(\tau) = \bigoplus_{\sigma>\tau} e_\sigma \varepsilon_\sigma^{-1} U_\sigma \subset$

$$\subset \bigoplus_{\sigma>\tau} U_\sigma = U^*(\tau) \quad \text{and therefore } \varepsilon_\tau = \exp(G^*(\tau)/U^*(\tau))$$

divides $\exp(G^*(\tau)/R(G)^*(\tau)) = e_\tau$. In particular $R(G) \subset U$. Let $\tau$ be maximal with $\varepsilon_\tau < e_\tau$. Then $R(G)^*(\tau) = U^*(\tau)$ which contradicts the definition of $\varepsilon_\tau$. It follows $\varepsilon_\tau = e_\tau$ for all $\tau$ and $R(G) = U$. □

The following example shows that G has not necessarily to be decomposable if $\mathfrak{Z}(G)$ contains more than one isomorphism type.

Example 2: Let $G = \langle 2^{-\infty}3^{-\infty}a,\ 2^{-\infty}5^{-\infty}b,\ 7^{-\infty}c,\ 11^{-\infty}d,\ 2^{-\infty}e,\ 13^{-1}(a+b),\ 13^{-1}(b+c+d),\ 13^{-1}(c+e)\rangle$. Then $\{Z_p \oplus Z_p \oplus Z_p,\ Z_{p^2} \oplus Z_p\} \subset \mathfrak{Z}(G)$ and G is indecomposable.

We will see that G has to be decomposable as far as possible if $\mathfrak{Z}(G)$ contains an elementary abelian and a cyclic group.

Lemma 4: Let G be an almost completely decomposable group with $G = H \oplus V$, V completely decomposable and $W \in \operatorname{reg}(G)$. Then there exists a subgroup V' with $V \simeq V' \leq W$ and $G = H \oplus V'$.

Proof: We can restrict ourselves to the case that V is homogeneous. Let $t(V) = \tau$, $W = \bigoplus_{\sigma \in T} W_\sigma$.

The subgroup $W_\tau \cap H$ is pure in $W_\tau$ and therefore a direct summand. $W_\tau = (W_\tau \cap H) \oplus U_\tau$ . $H(\tau) = G(\tau) \cap H = (W_\tau \oplus G^*(\tau)) \cap H = (W_\tau \oplus H^*(\tau)) \cap H = H^*(\tau) \oplus (W_\tau \cap H)$ . Hence rank $U_\tau$ = rank $V$ and $U_\tau \cap H = 0$. We know $V \subset G(\tau) = G^*(\tau) \oplus W_\tau = H^*(\tau) \oplus W_\tau \subseteq H \oplus U_\tau$ so that $G = H \oplus U_\tau$. □

<u>Lemma 5</u>: Let G be almost completely decomposable with $G = H \oplus V$ , V completely decomposable. Then $\mathfrak{J}(G) = \mathfrak{J}(H)$ .

<u>Proof</u>: It is an easy consequence of lemma 4. □

In view of lemma 5 it is only necessary to look at groups which do not have a completely decomposable summand. We can therefore assume in the following that $W_\tau$ has rank 1 for all $\tau \in T(G)$.

<u>Lemma 6</u>: Let G be almost completely decomposable, $W = \oplus W_\tau$ a regulating subgroup and G/W cyclic of order $p^n$. If $G = \langle W, p^{-n}(\Sigma w_\tau)\rangle$ has no completely decomposable summand then $h_p(w_\tau) > h_p(w_\sigma)$ for all $\tau < \sigma$ .

<u>Proof</u>: Let $\tau < \sigma$ and $h_p(w_\sigma) \geq h_p(w_\tau)$ . Let $m_o = \min\{m \mid \chi(mw_\sigma) \geq \chi(w_\tau)\}$ . Choose $x, y \in Z$ with $1 = xm_o + yp^n$. Then

$$xm_o p^{-n}(\sum_{\tau \in T} w_\tau) = p^{-n}(\sum_{\rho \neq \tau} xm_o w_\rho + (1-yp^n)w_\tau)$$

$$= p^{-n}(\sum_{\rho \neq \tau} xm_o w_\rho + w_\tau) - yw_\tau = h - yw_\tau .$$

$$G = \langle W, p^{-n}(\sum_{\tau \in T} w_\tau)\rangle = \langle W, xm_o p^{-n}(\sum_{\tau \in T} w_\tau)\rangle = \langle W, h\rangle .$$

Hence $G = \langle w_\sigma\rangle_* \oplus (\bigoplus_{\tau \neq \rho \neq \sigma} W_\rho \oplus \langle w_\tau + xm_o w_\sigma\rangle)_*$ which is a contradiction. □

From lemma 6 follows that $l(\mathfrak{T}(G)) \leq n$ .

<u>Lemma 7</u>: Let G be almost completely decomposable without completely decomposable summands, W and W' regulating sub-

groups of G and G/W' elementary abelian. Then $p^i(W_i)_* \subset W_i$ .

Proof: We prove the lemma by induction on i. For i=1 there is nothing to prove. $p^i(W_i)_* = p^i(W_i')_* = p^{i-1}p(W_i')_* \subset \subset p^{i-1}W_i' \subset W_i$ . ( Use $W_i' \subset W_i + (W_{i-1})_*$ ) . □

Corollary 1: Let G be almost completely decomposable without completely decomposable summands. If $\mathfrak{J}(G)$ contains the cyclic and the elementary abelian group of order $p^n$ then $l(\mathfrak{T}(G)) = n$ .

Remarks: a) Under the hypothesis of lemma 7 we can show with a similar argument

$$\exp((W_{i+1})_*/W_{i+1}) \mid p\cdot\exp((W_i)_*/W_i) .$$

b) As a consequence we get under the hypothesis of the corollary: $(W_i)_*/W_i$ is cyclic of order $p^i$ for $1 \leq i \leq n$.

c) In contrast to the length of the type graph $\mathfrak{T}(G)$ its point independence number is not influenced by the richness of the set $\mathfrak{J}(G)$. If in example 1 the types $\sigma_i$ form a chain and also $\sigma_1$ is smaller than $\tau_n$ then $\mathfrak{J}(G)$ does not change and the point independence number is 2.

Lemma 8: Let G satisfy the hypothesis of corollary 1. Then for all i, $1 \leq i \leq n$ , there exists $\tau_i \in T_i$ with $e_{\tau_i} = p^{i-1}$.

Proof: Assume that $e_\tau \mid p^{i-2}$ for all $\tau \in T_i$. Then $p^{i-2}W_i' \subset W_i$ and therefore $\exp((W_i)_*/W_i) \mid p^{i-1}$. But that is a contradiction. □

Lemma 9: Let G satisfy the hypothesis of corollary 1. Let

$$G = \langle W, p^{-n}( \sum_{\tau\in T^0} w_\tau + \ldots + p^{n-1} \sum_{\tau\in T^{n-1}} w_\tau ) \rangle \text{ with}$$

$w_\tau \in W_\tau$ and $h_p(w_\tau) = 0$. Then $T^i \cap T_{i+1} \neq \emptyset$ for all i.

Proof: From lemma 6 we know that $T^i \subset T_1 \cup \ldots \cup T_{i+1}$ . Assume that $T^i \subset T_1 \cup \ldots \cup T_i$ . Then

$i(\bigoplus_{j=o}^{i} \bigoplus_{\tau\in T^j} W_\tau)_* \geq p^{i+1}$ forces $|(W_i)_*/W_i| \geq p^{i+1}$ which is a contradiction.

<u>Lemma 10</u>: Let G satisfy the hypothesis of corollary 1. Then there exists $\tau \in T^i$ with $e_\tau = p^i$.

<u>Proof</u>: Assume that $e_\tau \mid p^{i-1}$ for all $\tau \in T^i$. Then

$$p^i(\sum_{\tau\in T^o \cup..\cup T^i} G(\tau))_* = p^i(\sum_{\tau\in T^o\cup..\cup T^i} W'_\tau)_* \subset \sum_{\tau\in T^o\cup..\cup T^i} p^{i-1}W'_\tau$$

$$\subset \sum_{\tau\in T^o\cup..\cup T^i} W_\tau$$ which is a contradiction. □

<u>Lemma 11</u>: Let G satisfy the hypothesis of corollary 1. Then for all i, $0 \leq i \leq n-2$, there exists $\tau \in T^{i+1}$ such that $\sigma > \tau$ for all $\sigma \in T^i$.

<u>Proof</u>: We first show that $H_i := (\bigoplus_{\tau\in T^o\cup..\cup T^i} W_\tau)_*$ does not have a completely decomposable summand. Assume that $U_\tau$ is a summand of $H_i$ with $\tau\in T^j$. Let $g_j = p^{-(j+1)}(\sum_{\tau\in T^o} w_\tau + ..$ $.. + p^j \sum_{\tau\in T^j} w_\tau)$, $S = \{\sigma \mid \sigma \ngeq \tau\}$. Then $g_j \notin U_\tau \oplus G(S)_*$ but $g_j \in (U_\tau \oplus G(S))_*$. Therefore $U_\tau \oplus G(S)_*$ is not pure. But that contradicts lemma 8 of Lady's[1] paper. Hence a maximal completely decomposable summand of $(W_i)_*$ is isomorphic to $\bigoplus_{\substack{\tau\in T_1\cup..\cup T_i \\ \tau\notin T^o\cup..\cup T^{i-1}}} W_\tau$. Let $\tau\in T^{i+1}$ with $e_\tau = p^{i+1}$. From corollary 7 of Lady's[1] paper follows $\{\sigma \mid \sigma \ngtr \tau\} \subset (T_1\cup..\cup T_{i+1}) \setminus (T^o\cup..\cup T^i)$ and therefore $T^o\cup..\cup T^i \subset \{\sigma \mid \sigma > \tau\}$. □

Theorem: Let G satisfy the hypothesis of corollary 1. Then $G = G_1 \oplus \ldots \oplus G_n$ with

(i) $T(G_j)$ is rigid for all j

(ii) $i(G_j) = p$ for all j

(iii) For all j, $1 \leq j \leq n-1$, there exists $\tau \in T(G_{j+1})$ with $\tau < \sigma$ for all $\sigma \in T(G_j)$.

Proof: Let $G = \langle W, p^{-n}(\sum_{\tau \in T^o} w_\tau + \ldots + p^{n-1} \sum_{\tau \in T^{n-1}} w_\tau) \rangle$.

We can assume that $\chi(w_\sigma) \geq \chi(w_\tau)$ for all $\sigma > \tau$. For all i, $1 \leq i \leq n-1$, there exists $\tau_i \in T^i$ with $\tau_i < \sigma$ for all $\sigma \in T^{i-1}$. Let $w'_{\tau_i} = w_{\tau_i} + p^{-i}(\sum_{\tau \in T^o} w_\tau + \ldots + p^{i-1} \sum_{\tau \in T^{i-1}} w_\tau)$ for these $\tau_i$ and $w'_\sigma = w_\sigma$ for $\sigma \notin \{\tau_1, \ldots, \tau_{n-1}\}$. Then $W' := \bigoplus_{\tau \in T(G)} \langle w'_\tau \rangle_*$ is a regulating subgroup and

$$G = \langle W', p^{-1}(\sum_{\tau \in T^i} w'_\tau) \mid 0 \leq i \leq n-1 \rangle = \bigoplus_{i=o}^{n-1} (\bigoplus_{\tau \in T^i} \langle w'_\tau \rangle)_* .$$

The theorem follows with $G_{i+1} = (\bigoplus_{\tau \in T^i} \langle w'_\tau \rangle)_*$. □

Corollary 2: Let G be an almost completely decomposable group such that $\mathfrak{Z}(G)$ contains a cyclic and an elementary abelian group of order $p^n$. Then $\mathfrak{Z}(G)$ contains all groups of order $p^n$.

Proof: Let I be any subset of $\{1, \ldots, n-1\}$. Define a regulating subgroup $W'_I$ by

$$W'_I = \bigoplus_{i \in I} (\langle w'_{\tau_{n-i}} - p^{-1}(\sum_{\tau \in T^{n-i-1}} w'_\tau) \rangle_* \oplus \bigoplus_{\substack{\tau \in T^{n-i} \\ \tau \neq \tau_{n-i}}} \langle w'_\tau \rangle_*)$$

$$\oplus \bigoplus_{i \notin I} \bigoplus_{\tau \in T^{n-i}} \langle w_\tau \rangle_*$$

with $\tau_{n-i} \in T^{n-i}$ such that $\tau_{n-i} < \sigma$ for all $\sigma \in T^{n-i-1}$. ( The notation $w'_\tau$ and $w_\tau$ according to the theorem )
Let $\mathfrak{m} = (m_1, \ldots, m_r)$ be a tuple with $n \geqq m_1 \geqq \ldots \geqq m_r \geqq 1$, $\Sigma m_i = n$ and $\mathfrak{C}_\mathfrak{m}$ the complement of $\{m_1, m_1+m_2, \ldots, m_1+\ldots+m_{r-1}\}$ in $\{1, \ldots, n-1\}$ Then $G/W'_{\mathfrak{C}_\mathfrak{m}}$ is isomorphic to the direct sum of the cyclic groups of order $p^{m_i}$, $1 \leqq i \leqq r$. □

## Reference:

1. Lady ,E.L., Almost completely decomposable torsion free abelian groups, *Proc. Amer. Math. Soc.*, 45, 41, 1974

# CLASSIFICATION OF ALMOST COMPLETELY DECOMPOSABLE GROUPS

K.-J. KRAPF and O. MUTZBAUER
UNIVERSITÄT WÜRZBURG

## INTRODUCTION

A torsion-free abelian group of finite rank is said to be almost completely decomposable, if there is a completely decomposable subgroup of finite index. A completely decomposable subgroup of an almost completely decomposable group of minimal index is called regulating subgroup by Lady [4]. The intersection of all regulating subgroups of A is the regulator $R = R(A)$. Burkhardt [2] proved, that the regulator is completely decomposable. The regulator and the regulator quotient are invariants of an almost completely decomposable group.

Normally almost completely decomposable groups are given in the form

$$A = \left(\bigoplus_{i=1}^{n} S_i x_i\right) + \sum_{j=1}^{r} \mathbb{Z} g_j \ , \qquad (*)$$

where $1 \in S_i \leq \mathbb{Q}$ , $g_i = \sum_{j=1}^{n} a_{ij} x_j$ and $a_{ij} \in \mathbb{Q}$ .

In the first section some properties of completely decomposable groups are deduced. Then an almost completely decomposable group will be described by the finite group $\overline{A} = A/R \leq h^{-1}R/R = \overline{R}$ together with the embedding, where $h = \exp A/R$ with the regulator R of A. If A is given in the form (✳), then R can be explicitely calculated, so that $\overline{A}$ is given. If $\overline{A}, \overline{B} \leq \overline{R}$, where $\overline{A}$ and $\overline{B}$ are descriptions of groups A and B with regulator R and regulator quotient $\overline{A} \cong \overline{B}$, then the isomorphism problem is reduced to a numerical problem. Moreover, given $\overline{A} \leq \overline{R}$, the structure of A can be numerically determined. The background of this complete solution is a number-theoretical fact. We want to express our gratitude to R. Pierce who proved lemma 1.2 during the conference in Udine in April 84.

We follow the notations of Fuchs [3].

## 1. COMPLETELY DECOMPOSABLE GROUPS

LEMMA 1.1. Let R be completely decomposable of finite rank. Let h be a natural number and t a type. Let $R \to \overline{R} = R/hR$ be the canonical homomorphism. If $\overline{X}$ is a subgroup of $\overline{R}$ with $\overline{X} \oplus \overline{R^*(t)} = \overline{R(t)}$, then there is a subgroup X of R such that $(X+hR)/hR = \overline{X}$ and $X \oplus R^*(t) = R(t)$.

Proof. Let Z be the complete inverse image of $\overline{X}$ relative to the canonical homomorphism $R(t) \to R(t)/hR(t)$. Then $Z \cap R^*(t) \leq hR(t) \leq Z$. Intersection with $R^*(t)$ shows that $hR^*(t) = Z \cap R^*(t)$ is pure in Z. Moreover $Z/hR^*(t) \cong R(t)/R^*(t)$ is completely decomposable and homogeneous of type t and all elements of $Z \setminus hR^*(t) \subseteq R(t) \setminus R^*(t)$ have type t, so that $hR^*(t)$ is a direct summand of Z by the lemma of Baer [3;86.5]. A complement X of $hR^*(t)$ in Z has all claimed properties.

LEMMA 1.2. [R. PIERCE] Let $h \neq 1$ be a natural number and A be a $n \times n$ - matrix with integer entries and determinant congruent to 1 modulo h. Then there is a $n \times n$ - matrix B with integer entries such that $A + hB$ has determinant 1 .
Proof. A is non-singular and there exist $n \times n$ - matrices P and Q with integer entries and $|\det P| = |\det Q| = 1$ and $PAQ = D = \mathrm{diag}(d_1, \ldots, d_n)$. D is the so called Smith normal form. Multiplication of the first rows of P and Q with -1 , if necessary, leads to $\det P = \det Q = 1$ .Hence $d_1 \ldots d_n = \det D = \det A \equiv 1 (\mathrm{mod}\ h)$. It suffices to prove the statement for the diagonal matrix D.Let $d_1 \ldots d_n = \det D = 1+ha$. Then $d_1, \ldots, d_n$ are relatively prime to h, and it is possible to solve the congruence $xd_2 \ldots d_n \equiv a (\mathrm{mod}\ h^{n-1})$; say $b,c \in \mathbb{Z}$ satisfy $bd_2 \ldots d_n = a + ch^{n-1}$. Define

$$C = \begin{bmatrix} -b & o & \ldots & o & (-1)^{n-1}c \\ 1 & o & \ldots & \ldots & o \\ o & \ddots & & & \vdots \\ \vdots & & \ddots & & \vdots \\ o & \ldots & o & 1 & o \end{bmatrix}$$

Then $\det(A+hP^{-1}CQ^{-1}) = \det(D+hC) = (d_1 - hb)d_2 \ldots d_n + hch^{n-1} =$ $= d_1 \ldots d_n - hbd_2 \ldots d_n + c\,h^n = 1$ ,proving the lemma.

DEFINITION. Let $h \neq 1$ be a natural number, t a type and $\mathbb{Q}(t:t)$ the uniquely determined ring of rationals of type t:t with $1 \in \mathbb{Q}(t:t)$. Let $h_t := |\mathbb{Q}(t:t) : h\mathbb{Q}(t:t)|$ and $U_t$ the group of units of $\mathbb{Q}(t:t)$. Then

$$\omega : \mathbb{Q}(t:t) \to \mathbb{Z}/h_t\mathbb{Z}$$

defined by $\omega(\frac{x}{y}) := (x+h_t\mathbb{Z})(y+h_t\mathbb{Z})^{-1}$ is a well defined ring homomorphism and

$$\mathbb{Z}^*(h,t) := \omega(U_t)$$

a subgroup of the units of $\mathbb{Z}/h_t\mathbb{Z}$. The group $\mathbb{Z}^*(h,t)$ is generated by $-1+h_t\mathbb{Z}$ and by all $p+h_t\mathbb{Z}$ with $p \in U_t$, p a prime.

THEOREM 1.3. Let R be a completely decomposable group of finite rank. Let $h \neq 1$ be a natural number. Let $\overline{R} = R/hR = \oplus \overline{R_t}$, where $\overline{R(t)} = \overline{R_t} \oplus \overline{R^*(t)}$. Let $\overline{\alpha} \in \mathrm{Aut}\,\overline{R}$ be given as a matrix $(\overline{\alpha_{st}})$ with $\overline{\alpha_{st}} \in \mathrm{Hom}(\overline{R_t}, \overline{R_s})$. $\overline{\alpha}$ is induced by an automorphism $\alpha$ of R if and only if $\overline{\alpha_{st}} = 0$ for $t \nleq s$ and $\det \overline{\alpha_{tt}} \in \mathbb{Z}^*(h,t)$ for all t with $\overline{R(t)} \neq \overline{R^*(t)}$.

Proof. By lemma 1.1 all decompositions of $\overline{R} = \oplus \overline{R_t}$ are induced by decompositions of $R = \oplus R_t$. Let an automorphism $\alpha$ of R be given by a matrix $(\alpha_{st})$, where $\alpha_{st} \in \mathrm{Hom}(R_t, R_s)$, then the conditions on $(\overline{\alpha}_{st})$ are obviously necessary.

To show the converse note that the restriction map $\mathrm{Hom}(R_t, R_s) \to \mathrm{Hom}(\overline{R_t}, \overline{R_s})$ is epic if $t < s$. Therefore it suffices to prove, that the restriction map $\mathrm{Aut}\, R_t \to \{\overline{\alpha} \in \mathrm{Aut}\,\overline{R_t} \mid \det \overline{\alpha} \in \mathbb{Z}^*(h,t)\}$ is epic. We may assume that $\overline{R_t} = (R_t + hR)/hR$. There is a base $x_1, \ldots, x_n \in R_t$ such that $R_t = \oplus \langle x_i \rangle_*$ and $\overline{R_t} = \oplus \langle \overline{x}_i \rangle$, where $\overline{x}_i = x_i + hR$. Let $\overline{\alpha}$ be given by the matrix $A'$ with integer entries relative to the base $\overline{x}_1, \ldots, \overline{x}_n$ and $\det A' = \overline{u} + ha$, where $\overline{u} \in \mathbb{Z}^*(h,t)$ and a is an integer. Let $u \in U_t$ be an invere image of $\overline{u}$, then the new base $u^{-1}x_1, x_2, \ldots, x_n$ of $R_t$ has the same properties as the base $x_1, \ldots, x_n$, but the new matrix description A of $\overline{\alpha}$ with integer entries satisfies $\det A \equiv 1 (\text{modulo } h_t)$. By lemma 1.2 there is a matrix B with integer entries such that $\det(A + h_t B) = 1$; and $\overline{\alpha}$ is in fact induced by the automorphism $\alpha$ of $R_t$, which is described relative to the base $u^{-1}x_1, \ldots, x_n$ of $R_t$ by the matrix $A + h_t B$. This proves the theorem.

LEMMA 1.4. Let R be a homogeneous completely decomposable group of finite rank and let h be a natural number such that $\overline{R} = R/hR$ is of exponent h. If $\overline{R} = \overline{X} \oplus \overline{Y}$, then there are $X, Y \leq R$ with $R = X \oplus Y$ and $(X + hR)/hR = \overline{X}$ and $(Y + hR)/hR = \overline{Y}$.

Proof. Let $x_1,\dots,x_n$ be a base of $R=\oplus\langle x_i\rangle_*$ such that $\overline{R}=\oplus\langle\overline{x}_i\rangle$ ,where $\overline{x}_i=x_i+hR$. Let $\overline{R}=\oplus\langle\overline{y}_i\rangle$ and $\overline{y}_i=\Sigma_j a_{ij}\overline{x}_j$ for all i with integers $a_{ij}$. Hence $\det(a_{ij})=u+h\mathbb{Z}$ with unit u. If $\overline{y}_1$ is replaced by $a\overline{y}_1$ with $au\equiv 1 \pmod h$ ,then the corresponding determinant is congruent 1 modulo h. By lemma 1.2 there is a matrix $B=(b_{ij})$ with $\det(A+hB)=1$. Now if $z_i:=\Sigma_j(a_{ij}+hb_{ij})x_j$ for all i, then $z_1,\dots,z_n$ is a base of $R=\oplus\langle z_i\rangle_*$ and $\langle z_i+hR\rangle=\langle\overline{z}_i\rangle=\langle\overline{y}_i\rangle$. This proves the lemma.

LEMMA 1.5 Let B be a subgroup of A, both completely decomposable with finite index $|A:B|$. There is a chain of completely decomposable subgroups

$$B=C_1<C_2<\dots<C_n=A$$

such that the indices $|C_i:C_{i-1}|$ are primes.

Proof. If $p^{-1}B\cap A$ ,p prime, is shown to be completely decomposable, then it is enough to assume A/B elementary abelian. Now $(p^{-1}B\cap A)^*(t)_*=p^{-1}B^*(t)_*\cap A^*(t)_*=p^{-1}B^*(t)\cap A^*(t)$ by [1;2.5] and $(p^{-1}B\cap A)^*(t)=A\cap p^{-1}B^*(t)$. But $p^{-1}B^*(t)\cap A\le A^*(t)_*=A^*(t)$, so that for all types t

$$(p^{-1}B\cap A)^*(t)_*=(p^{-1}B\cap A)^*(t)$$

and again by [1;2.5] $p^{-1}B\cap A$ is completely decomposable.

We show now that $|A:B|=p$ is a prime, if A and B are completely decomposable, A/B is elementary abelian p-group and there is no completely decomposable group C with $B<C<A$. Let t be a type such that $B(t)\neq A(t)$ but $B^*(t)=A^*(t)$. Let $x\in A(t)\setminus B(t)$, i.e. $px=b+c$ ,where $B(t)=B_t\oplus B^*(t)$ and $b\in B_t\setminus\{0\}$, $c\in B^*(t)\setminus\{0\}$. $B_t$ is pure in A by the condition on B , so that $h_p^A(b)=h_p^A(c)=0$. If $h_q^A(b)>h_q^A(c)$ ,then choose an integer f such that $fq=1+up$. So we get $p(fx-uq^{-1}b)=q^{-1}b+fc$ ,and we can assume $\chi(b)\le\chi(c)$. If $B_t=\langle b\rangle_*\oplus H$ then $B_t\le\langle x\rangle_*\oplus H=C_t$ and $B<C_t\oplus B^*(t)\oplus(\oplus\{B_s\mid s\ngeq t\})\le A$. Therefore $|A:B|=p$ ,and the proof is finished.

REMARK 1.6. If a completely decomposable group A is given in the form (✱), then lemma 1.5 and the choice of the element x in the proof show that a decomposition base of A can be numerically calculated.

## 2. INVARIANTS

REMARK 2.1. The regulator R, i.e. the intersection of all completely decomposable subgroups of minimal index is a unique subgroup of an almost completely decomposable group A, therefore its isomorphism type and the isomorphism type of the regulator quotient $F \cong A/R$ are invariants. We have only to classify in the finite set of groups A, where

$$R \leq A \leq h^{-1}R \quad , \quad R(A) = R \quad , \quad A/R \cong F$$

with finite group F of exponent h. If two groups A and B in this set are isomorphic with isomorphism $\varphi : A \to B$, then $\varphi$ maps the regulator of A onto the regulator of B, i.e. $\varphi$ is an automorphism of R and Aut R restricted to $h^{-1}R/R$ operates on this finite set of groups. This procedure is the background of the classification.

CONSTRUCTION. Let R be completely decomposable of finite rank and F finite abelian of exponent h such that $F \lesssim R/hR$. $\mathfrak{H}(R,F)$ is the set of subgroups $\overline{A} \leq \overline{R} := h^{-1}R/R$ isomorphic to F such that

$$e_t^{-1}h_t\overline{X} \leq \overline{A} \cap \overline{R(t)} \leq e_t^{-1}h_t\overline{R(t)}$$

for all t, where $\overline{X}$ is a suitable complement of $\overline{R^*(t)}$ in $\overline{R(t)}$, $h_t = \exp \overline{R(t)}$ and $e_t := \exp(\overline{A} \cap \overline{R^*(t)})$. The set $\mathfrak{H}(R,F)$ can be empty. Let be defined

$$\text{Aut } R|_{\overline{R}} := \{\alpha|_{\overline{R}} \mid \alpha \in \text{Aut } R\} \leq \text{Aut } \overline{R} .$$

LEMMA 2.2. If A is almost completely decomposable of finite rank with regulator R, then

$$\bar{A} = A/R \in \mathfrak{H}(R,A/R).$$

If conversely R is completely decomposable of finite rank, F a finite abelian group of exponent h, such that $F \cong\leq R/hR$ ,then the inverse image of $\bar{A} \in \mathfrak{H}(R,F)$ relative to the canonical homomorphism $R \to h^{-1}R/R$ is an almost completely decomposable group with regulator R and regulator quotient F.

Proof. The kernel of the proof is a characterisation of the regulator by Burkhardt [2]. If A is almost completely decomposable and if C is a completely decomposable subgroup of finite index, then C is the regulator of A if and only if there is a homogeneous direct summand $C_t$ of C with type t and of maximal rank, so that for all types t

$$C_t \leq e_t A(t) \leq C(t)$$

where $e_t = \exp(A^*(t)_*/C^*(t))$ .

Straightforward calculations show that the image $\bar{A}$ of such a group A with regulator R is in $\mathfrak{H}(R,A/R)$. Conversely lemma 1.1 has to be used to show that the inverse image A of $\bar{A} \in \mathfrak{H}(R,F)$ has regulator R and regulator quotient F.

REMARK. Lemma 2.2 makes sure that the set $\mathfrak{H}(R,F)$ contains descriptions for all almost completely decomposable groups with regulator R and regulator quotient F and conversely that all elements of the set $\mathfrak{H}(R,F)$ are indeed descriptions of such groups.

There remains the classification problem to give conditions for two such descriptions to describe isomorphic groups.

THEOREM 2.3. Let R be completely decomposable of finite rank and F finite abelian of exponent h. Let be $\bar{R} = h^{-1}R/R$. Then Aut $R|_{\bar{R}}$ operates on $\mathfrak{H}(R,F)$ by $\bar{\alpha}\bar{A} = \overline{\alpha A}$ ,where $\alpha \in$ Aut R has the restriction $\alpha|_{\overline{R}} = \bar{\alpha}$.

Let conversely be A and B almost completely decomposable groups of finite rank with regulator R and regulator quotient F. The groups A and B are isomorphic if and only if $\overline{A} = A/R$ and $\overline{B} = B/R$ are in the same orbit of $\mathfrak{H}(R,F)$ with operating group Aut $R|_{\overline{R}}$.

Proof. The construction together with remark 2.1 show all the statements.

REMARK 2.4. The orbits of $\mathfrak{H}(R,F)$ are precisely the invariants of almost completely decomposable groups with regulator R and regulator quotient F. The number of orbits equals the number of isomorphism types of such groups.

LEMMA 2.5. Let A be an almost completely decomposable group with regulator R. Then Aut A is a subgroup of Aut R and the orbit of a description $\overline{A} \in \mathfrak{H}(R,A/R)$ of A has length

$$|\text{Aut } R : \text{Aut } A| = |\text{Aut } R|_{\overline{R}} : \text{Aut } A|_{\overline{R}}|.$$

Proof. An automorphism of A can be restricted to an automorphism of the characteristic regulator R. Therefore Aut A $\leq$ Aut R and Aut $A|_{\overline{R}} \leq$ Aut $R|_{\overline{R}}$ . The kernel of the homomorphism Aut R $\to$ Aut $R|_{\overline{R}}$ is contained in Aut A , so that the claimed equality of the indices is shown. The length of the orbit of $\overline{A} = A/R$ ,following theorem 2.3, equals $|\text{Aut } R|_{\overline{R}} : \text{Stab } \overline{A}|$ but the stabilizer of $\overline{A}$ is Stab $\overline{A}$=Aut $A|_{\overline{R}}$.

## 3. STRUCTURE

It will be shown in this section that the knowledge of a description $\overline{A} \in \mathfrak{H}(R,F)$ of a group A is sufficient to deal with a lot of structure problems by numerical calculation. Moreover it is possible if A is explicitely given in the form (✱) to calculate $\overline{A} \in \mathfrak{H}(R,F)$, so that the treatment of almost completely decomposable groups is not restricted to a special form of description. It is only necessary to be able to see that A is in fact almost completely decomposable.

THEOREM 3.1. Let A be an almost completely decomposable group with regulator R and description $\overline{A} \in \mathfrak{H}(R,A/R)$.

A is direct decomposable if and only if there are complements $\overline{S_t}$ of $\overline{R^*(t)}$ in $\overline{R(t)}$ and decompositions $\overline{S_t} = \overline{X_t} \oplus \overline{Y_t}$ for all types t, such that for $\overline{X} = \oplus\overline{X_t}$ and $\overline{Y} = \oplus\overline{Y_t}$ :

$$\overline{A} = (\overline{A} \cap \overline{X}) \oplus (\overline{A} \cap \overline{Y}).$$

Proof. If $A = B \oplus C$ is a direct decomposition of A, then $R = (R \cap B) \oplus (R \cap C)$ because the regulator R is fully invariant [2]. Consequently there is a homogeneous decomposition $R = \oplus\{S_t \mid t \in T\}$ such that for all $t \in T$: $S_t = X_t \oplus Y_t$ where $X_t \leq B$, $Y_t \leq C$. Using lemma 2.2 to get the corresponding descriptions, all occuring groups are mapped like $A \to \overline{A} = A/R$ and the claimed direct decomposition of $\overline{A}$ follows.

If conversely the direct decomposition of $\overline{A}$ is given like above, then the lemmas 1.1 and 1.4 show that there is a direct decomposition of $R = X \oplus Y$ such that the inverse image A of $\overline{A}$ satisfies: $A = (A \cap h^{-1}X) \oplus (A \cap h^{-1}Y)$ ,i.e. A is decomposable and the proof is complete.

If A is an almost completely decomposable group of finite rank given explicitely in the form (✱) then all completely decomposable subgroups with $C \geq |A:S| \cdot S$ and $S = \oplus S_i x_i$ can be calculated by remark 1.6, together with a decomposition base. Moreover the finite quotients A/C can be calculated. Now by a theorem of Lady [4;theorem 1] all regulating subgroups and therefore the regulator of A contain $|A:S| \cdot S$ ,and consequently all regulating subgroups and the regulator can be calculated with decomposition bases and all the quotients can be determined numerically. We therefore assume all almost completely decomposable groups A of finite rank to be given by $\overline{A} \leq \overline{R} = h^{-1}R/R$ relative to their regulator R with regulator quotient A/R of exponent h.

The restricted automorphism group $\mathrm{Aut}A|_{\overline{R}}$ is explicitely given by theorem 1.3. If $\overline{A}, \overline{B} \in \mathfrak{H}(R,F)$, then it suffices to compare $\overline{\alpha}\overline{A}$ with $\overline{B}$ to decide if A and B are isomorphic or not; but $\overline{\alpha}$ is explicitely described as matrix with well defined properties which can be numerically checked.

Using theorem 3.1 it is possible to determine all direct decompositions of a group A given by $\overline{A} \in \mathfrak{H}(R,F)$. By the way, here the problem is included to calculate explicitely a complement $R_t$ of $R^*(t)$ in $R(t)$ with given $\overline{R_t}$. This can be done by connection of the methods used in the proofs of lemma 1.4 and 1.5 to change decomposition bases.

The restriction $\mathrm{Aut}\, A|_{\overline{R}}$ of the automorphism group of A to $\overline{R}$ obviously can be calculated as a subgroup of $\mathrm{Aut}\, R|_{\overline{R}}$ so that the length of the orbit of $\overline{A}$ in $\mathfrak{H}(R,F)$ is determined.

If R and F are fixed the set $\mathfrak{H}(R,F)$ can be numerically presented, even the number of orbits can be specified, i.e. the number of isomorphism types of almost completely decomposable groups with regulator R and regulator quotient F.

The fact that R and F don't fit together is equivalent to $\mathfrak{H}(R,F)$ to be empty. Instead of $\mathfrak{H}(R,F)$ even more interesting is the set

$$\{\overline{A} \in \mathfrak{H}(R,F) \mid A \text{ is indecomposable}\}$$

of indecomposable groups with regulator R and regulator quotient F.

We present solutions of the classification problem and of the main structure problems, which could be verified numerically. This shows that our method is useful. But this is no replacement for theoretical insight, and in this area there are a lot of unsolved problems, for instance:

(1) The dependence between the isomorphism types of the regulator, the regulator quotient and indecomposability.

(2) The structure of groups with certain sets of quotients of regulating subgroups (see [2]).

(3) Formulas for numbers of isomorphism types, if the regulator, the regulator quotient and may be additionally indecomposability are prescribed.

(4) Numbers of non-isomorphic direct decompositions of decomposable almost completely decomposable groups.

## REFERENCES

[1] Arnold, D.,Pure subgroups of finite rank completely decomposable groups, Proceedings of Abelian Group Theory (Oberwolfach), Lecture Notes 874 (1981),1-31.

[2] Burkhardt, R.,On a special class of almost completely decomposable torsion free abelian groups, to appear.

[3] Fuchs, L.,Infinite Abelian Groups I+II ,Academic Press, New York (1970,1973).

[4] Lady, E.L.,Almost completely decomposable torsion free abelian groups, Proc. A.M.S. 45 (1974),41-47.

# THE DIVISIBLE AND E-INJECTIVE HULLS OF A TORSION FREE GROUP

C. Vinsonhaler
University of Connecticut
Storrs, Connecticut 06268

Recently there has been considerable interest in the structure of torsion free abelian groups G as modules over their endomorphism rings E = End(G). In particular, homological properties of the E-module G have been the focus of a number of papers (see (2), (7), (8)). In (2) the structure of finite rank groups G such that G was E-projective was determined, while in (7) the injective hull of G as an E-module was investigated. In the latter paper the question was asked: For which groups G is the E-injective hull of G equal to QG, the divisible hull? This is equivalent to the question: When is QG injective as a QE-module? In this note we obtain a partial answer by imposing an additional condition on G. We examine groups G for which QG is injective as a QE-module and $Hom_{QE}(QG,QG)$ = Q(center E). This property, called aEqi, is equivalent to a condition weaker than quasi-injectivity on ${}_EG$.

The strongly indecomposable finite rank aEqi groups G turn out to be exactly those with a commutative, local, self-injective quasi-endomorphism ring for which rank E(G) = rank G (Proposition 1.3). It is also shown (Corollary 1.4) that any finite dimensional, local, commutative self-injective rational algebra arises in this way. The arbitrary finite rank aEqi groups are classified in Theorem 2.8, giving a large class of groups G such that QG is injective as a QE-module.

## 0. Definitions and preliminaries.

Throughout, the letter G always denotes a group, where group means torsion free abelian group. The endomorphism ring of the group G is written E(G) or E, and C is the center of E. If X is a group or ring, the symbol QX represents the group or ring $Q \otimes_Z X$, and X is always regarded as a subgroup or subring of QX. This is possible since X is torsion free. The ring QE(G) is called the quasi-endomorphism ring of G, and G, respectively QG, is regarded as a left module over E, respectively QE.

We employ the standard definitions of quasi-isomorphism, quasi-equality and strongly indecomposable.

Our initial proposition, stated for reference, is due to J.D. Reid.

Proposition 0.1. Let $G$ be a finite rank group. The following are equivalent:

1) $G$ is strongly indecomposable,
2) $QE(G)$ is local,
3) Any endomorphism of $G$ is monic or nilpotent.

Define a group $G$ to be almost E-quasi-injective (aEqi) provided that given any E-submodule $H$ of $G$, and E-homomorphism $f:H \to G$, there is a positive integer $n$ and E-homomorphism $g:G \to G$ such that $g$ lifts $nf$.

Note that the map $g$ may be regarded as an element of $C$, the center of $E$. The next lemma restates the definition of aEqi in more familiar terms.

Lemma 0.2. A group $G$ is aEqi if and only if

(a) $QG$ is quasi-injective as a QE-module, and

(b) $\mathrm{Hom}_{QE}(QG,QG) = QC$.

Proof. ($\Rightarrow$) (a) Let $K$ be a QE-submodule of $QG$ and $f:K \to QG$ a QE-homomorphism. Choose a Q-basis $\{x_i \mid i \in I\}$ for $K$ such that for each $i$, $x_i \in G$ and $f(x_i) \in G$. Then $K' = \Sigma_{i \in I} Ex_i$ is an E-submodule of $G$ and $f(K') \subset G$. Furthermore, the restriction of $f$ to $K'$ is an E-map and

$K/K'$ is torsion. Since $G$ is aEqi, there is a $g$ in $C$ = center $E$, and $n > 0$ such that $g$ lifts the restriction of $nf$ to $K'$. Then $(1/n)g \in QC$ is a lifting of $f$.

(b) Clearly $QC \subset Hom_{QE}(QG,QG)$. To show the reverse inclusion, let $f$ be an element of $Hom_{QE}(QG,QG)$. As in (a), choose a Q-basis $\{x_i\}$ for $QG$ such that $x_i \in G$ and $f(x_i) \in G$ for each $i$. Then $K = \Sigma Ex_i$ is an E-submodule of $G$ and the restriction of $f$ to $K$ is an E-map into $G$. Thus, there exist $g \in C$, and $n > 0$, such that $g$ lifts $nf$ restricted to $K$. This implies that $(1/n)g \in QC$ must equal $f$ since $G/K$ is torsion. Therefore $f \in QC$.

($\Leftarrow$) Given (a) and (b), let $K$ be an E-submodule of $G$, and $f:K \to G$ an E-map. Then $f$ extends uniquely to a QE-map $f:QK \to QG$, which lifts to an element $g$ of $QC$ by (a) and (b). Thus there exists $n > 0$ such that $ng \in C$, and $ng$ lifts $nf$.

<u>Corollary 0.3</u>. If $G$ and $H$ are quasi-isomorphic groups, then $H$ is aEqi if and only if $G$ is aEqi.

Proof. Assume without loss of generality that $G$ is quasi-equal to $H$. Then $QG = QH$ and $QE(G) = QE(H)$, and the result follows directly from Lemma 0.2.

<u>Corollary 0.4</u>. If $G$ is a finite rank aEqi group, then $QG$ is injective as a QE-module, and hence is the injective

hull of G as an E-module.

Proof. Under the hypotheses, QE is left and right Artinian. By Lemma 0.2, QG is quasi-injective as a QE-module. Since QG is faithful as a QE-module, QG is injective (see (3), p.68). It follows that QG is the injective hull of G as an E-module, since G is essential in QG.

The converse to Corollary 0.4 is false. If G is a finite rank group with rank greater than one, and E(G) is a subring of Q, then QG is injective as a QE-module, but G is not aEqi.

## 1. Strongly indecomposable finite rank aEqi groups.

Lemma 1.1. If G is a strongly indecomposable aEqi group, then QG is indecomposable as a QE-module.

Proof. If QG = A ⊕ B as QE-modules, then projection onto A gives a central idempotent in QE, by Lemma 0.2. This in turn induces a quasi-summand of G, a contradiction unless A is 0 or QG.

Lemma 1.2. If G is a strongly indecomposable finite rank aEqi group, then E = E(G) is commutative.

Proof. Note that under the hypotheses, QJ, the Jacobson radical of QE, is nilpotent. Also, if $r \notin QE - QJ$, then $r$ is a unit in QE, since QE is local. Define

$H_i = \{x \in QG \mid QJ^i x = 0\}$ $(H_0 = 0)$, and

$\text{ann}H_i = \{r \in QE \mid rH_i = 0\}$ for $i = 0,1,\ldots$

Note that $H_i$ is a QE-submodule of QG and $\text{ann}H_i$ is an ideal in QE. Since QJ is nilpotent, $H_i = QG$ and $\text{ann}H_i = 0$ for some $i$. Hence it suffices to show, by induction, that for each $i$, $QE = QC + \text{ann}H_i$.

For $i = 0$ the result is trivial since $\text{ann}H_0 = QE$. Assume $QE = QC + \text{ann}H_i$ for an arbitrary $i \neq 0$, and let $r \in \text{ann}H_i$. We will show $r \in QC + \text{ann}H_{i+1}$, and hence that $QC + \text{ann}H_i = QC + \text{ann}H_{i+1}$.

Let $M = H_{i+1}/H_i$, a semisimple QE-module. Write $M = QE\bar{x}_1 + \ldots + QE\bar{x}_m$, where for each $j$, $x_j \in H_{i+1}$, $\bar{x}_j = x_j + H_i$ and $QE\bar{x}_j$ is a simple (non-zero) QE-module. Note that $H_{i+1} = QEx_1 + \ldots + QEx_m + H_i$, and define $f_r : H_{i+1} \to H_{i+1}$ by $f_r(s_1x_1 + \ldots + s_mx_m + h) = s_1rx_1 + \ldots + s_mrx_m$ for all choices of $s_1,\ldots,s_m \in QE$, $h \in H_i$. To show that $f_r$ is well-defined, suppose that $s_1x_1 + \ldots + s_mx_m + h = 0$. By the definition of the $x_j$, $s_1\bar{x}_1 + \ldots + s_m\bar{x}_m = 0$, so that $s_jx_j \in H_i$ for each $j$. Hence $s_j \in QJ$ for each $j$, since QE is local. If $i = 0$, then $s_jrx_j = 0$ for each $j$ since $rx_j \in H_{i+1} = H_1$. If $i > 0$, for each $j$ use the induction hypothesis to write $s_j = c_j + a_j$, where $c_j \in QC$, $a_j \in \text{ann}H_i$. Then $s_j r x_j = (c_j + a_j) r x_j = r c_j x_j + a_j r x_j$. However, $c_j = s_j - a_j \in QJ$, so that $c_j x_j \in H_i$ and $r c_j x_j = 0$ since $r \in \text{ann}H_i$. Furthermore, $r \in QJ$ since $r \in \text{ann}H_i$, so $rx_j \in H_i$ and $a_j(rx_j) = 0$ since $a_j \in \text{ann}H_i$. Thus $s_j r x_j = 0$ for all $j$ and $f_r$ is well-defined.

Obviously $f_r$ is a QE-map, so by the QE-injectivity of QG, there exists $c \in QC$ such that $(f_r - c)H_{i+1} = 0$. To complete the proof we show $r-c \in \text{ann}H_{i+1}$. Let $x = s_1x_1 + \dots + s_mx_m + h$ be any element of $H_{i+1}$. Then $(r-c)x = (r-f_r)x = rs_1x_1 + \dots + rs_mx_m - s_1rx_1 - \dots - s_mrx_m$, since $rH_i = 0$. In particular, $(r-c)x_j = 0$ for each $j$, so that if $i = 0$, $r-c \in QJ \quad \text{ann}H_1 = \text{ann}H_{i+1}$. If $i > 0$, write $s_j = c_j + a_j$ with $c_j \in C$, $a_j \in \text{ann}H_i$, as above. The expression for $(r-c)x$ becomes $ra_1x_1 + \dots + ra_mx_m - a_1rx_1 - \dots - a_mrx_m$, which is $0$ since both $ra_j$ and $a_jr$ annihilate $H_{i+1}$. Thus $r-c \in \text{ann}H_{i+1}$ and the proof is complete.

We can now characterize the strongly indecomposable finite rank aEqi groups $G$ in terms of the quasi-endomorphism ring QE.

Proposition 1.3. The following are equivalent for a finite rank group $G$:

(a) $G$ is a strongly indecomposable aEqi group.

(b) QE is commutative, local and self-injective, and rank E = rank G.

Proof. (a) => (b). By Lemma 1.2, QE is commutative, and by the hypotheses is a finite dimensional algebra over Q. Since G is strongly indecomposable, QE is local. By Corollary 0.4 and Lemma 1.1, QG is injective and indecomposable as a QE-module. Hence $QG^* = Hom_Q(QG,Q)$ is projective and indecomposable as a right QE-module, by Morita duality. Since QE is local, $QG^* = QE$ as right QE-modules. Therefore $QG = QG^{**} = QE$ as left QE-modules. It follows that QE is self-injective and rank E = rank G.

(b) => (a). If QE is local, then G is strongly indecomposable. Since QG is a finitely generated faithful QE-module, and QE is Artinian, there is a QE-embedding of QE into $(QG)^n$, for some $n > 0$. However, QE is injective and hence is a summand of $(QG)^n$. Since QE is local, it is indecomposable as a QE-module, and therefore is a summand of QG by the Krull-Schmidt Theorem. The fact that rank E = rank G implies that QE = QG as QE-modules. Thus, QG is injective as a QE-module and $Hom_{QE}(QG,QG) = Hom_{QE}(QE,QE) = QE = QC$ since QE is commutative. By Lemma 0.2, G is aEqi.

<u>Corollary 1.4</u>. Let A be a commutative, local, self-injective and finite dimensional Q-algebra. Then A = QE(G)

for some finite rank aEqi group $G$ with rank $G$ = rank $A$.

Proof. Let $R$ be a Z-order in $A$. That is, $R$ is a full subring of $A$, finitely generated as a Z-module. By Zassenhaus (9), there exists a finite rank group $G$ such that $R = E(G)$ and rank $G$ = rank $R$. Then $QE(G) = A$ and the conditions of Proposition 1.3(b) are satisfied, so $G$ is aEqi.

2. Finite rank aEqi groups.

Throughout this section $G$ is a finite rank aEqi group. It is well known that $G$ admits a "unique" quasi-decomposition $G \doteq G_1 \oplus \dots \oplus G_k$, where each $G_i$ is strongly indecomposable. Since the aEqi property is invariant under quasi-isomorphism by Corollary 0.3, we can assume that if $G_i$ and $G_j$ are quasi-isomorphic then they are isomorphic. The next lemma, actually an application of Morita equivalence, allows a further simplification of $G$.

Lemma 2.1. Let $H_1,\dots,H_r$ be mutually non-quasi-isomorphic groups, and $m(1),\dots,m(r)$ positive integers. Then $G = H_1 \oplus \dots \oplus H_r$ is aEqi if and only if $G' = H^{m(1)} \oplus \dots \oplus H^{m(r)}$ is aEqi.

Proof. It is easy to see that there is a 1-1 correspondence between E(G) submodules $K = K_1 \oplus \dots \oplus K_r$ of $G$ and $E(G')$-submodules $K' = K^{m(1)} \oplus \dots \oplus K^{m(r)}$ of $G'$, noting that submodules are invariant under projection and shifting maps. Furthermore, there is a 1-1 correspondence between E(G)-maps $f = f_1 \oplus \dots \oplus f_r$ from K into G and E(G′)-maps $f' = f^{m(1)} \oplus \dots \oplus f^{m(r)}$ from K′ into G′. In particular, if $K_i = H_i$ for each i, there is a "natural" isomorphism center E(G) = center E(G′). These 1-1 correspondences together imply the lemma.

As a consequence of Lemma 2.1, it suffices to assume $G = G_1 \oplus \dots \oplus G_r$, where the $G_i$'s are strongly indecomposable and mutually non-quasi-isomorphic.

A final simplifying assumption is that G is not quasi-decomposable into a sum of two non-trivial E-submodules. If $G \doteq H_1 \oplus H_2$ as E-modules, then by Cor. 0.3, we can assume $G = H_1 \oplus H_2$ as E-modules. In this case both $H_1$ and $H_2$ are aEqi.

<u>Summary</u>. In the sequel G is always a finite rank aEqi group, $G = G_1 \oplus \dots \oplus G_k$, where the $G_i$ are strongly indecomposable, mutually non-quasi-isomorphic, and G does not quasi-decompose into two non-trivial E-submodules.

<u>Lemma 2.2</u>. Let G be as in the Summary. Then

(a) If $f \in C$, then f is either monic or nilpotent.

(b) G does not contain the direct sum of two non-trivial E-modules.

Proof. (a) Since $f$ commutes with projection onto $G_i$, $f(G_i) \subset G_i$ for each $i$. If $f_i$ is the restriction of $f$ to $G_i$, then by Proposition 0.1, $f_i$ is monic or nilpotent for each $i$. Choose a positive integer $n$ so that $f_i^n$ is 0 or monic for each $i$, and let $H_1 = \oplus\{G_i \mid f_i^n(G_i) = 0\}$, $H_2 = \oplus\{G_i \mid f_i$ is monic$\}$. Then $G = H_1 \oplus H_2$, and since $G$ is not a direct sum of E-modules, either $H_1 = 0$, $H_2 = 0$, or there exists a non-zero homomorphism between $H_1$ and $H_2$, say $g: H_1 \to H_2$. In the latter case, $f^n g(H_1) \neq 0$, while $gf^n(H_1) = 0$, contradicting $f \in C$. Thus, either $H_1$ or $H_2$ is 0, and $f$ is monic or nilpotent.

(b) If $K = K_1 \oplus K_2 \subset G$, with $K_1$ and $K_2$ E-submodules of $G$, then projection of $K$ onto $K_1$ is an E-map. By the aEqi property, some positive multiple of this projection map lifts to $f \in C$. However $f$ is then neither monic nor nilpotent, contradicting part (a) unless $K_1$ or $K_2$ is 0.

Corollary 2.3. The socle, $S$, of $QG$ as a QE-module is simple.

Proof. Apply Lemma 2.2 to $S \cap G$.

Corollary 2.4. If $S$ is the socle of $QG$, then $S \cap G \subset G_i$ for some $i$.

Proof. Let $p_j$ be the projection of $G$ onto $G_j$, $1 \leq j \leq k$. Then for some $i$, $p_i S \neq 0$. Suppose that also $p_j S \neq 0$

for some $j \neq i$. Let $0 \neq x_1 \in p_i(S \cap G) \subset G_i \cap S$ and $0 \neq x_2 \in p_j(S \cap G) \subset G_j \cap S$. Since $S$ is simple as a QE-module, there exist $r, s \in QE$ such that $rx_1 = x_2$ and $sx_2 = x_1$. Choose non-zero integers $a, b$ such that $ar$, $bs \in E$. Then $(p_2ar)(p_1bs): G_j \to G_j$ and $(p_1bs)\cdot(p_2ar): G_i \to G_i$ are not nilpotent, hence are monic by Proposition 0.1. This implies $G_i$ and $G_j$ are quasi-isomorphic, a contradiction.

By Corollary 2.4, we may assume without loss of generality that $S \cap G \subset G_1$. Denote $E_1 = E(G_1)$, $J_1$ = nil radical of $E_1$, $C_1$ = center $E_1$. Note that $QJ_1$ = Jacobson radical of $QE_1$ is nilpotent, so that $J_1$ is nilpotent. Define inductively (as in the proof of Lemma 1.5)

$$H_1 = S \cap G \subset G_1, \quad H_{j+1} = \{x \in G_1 \mid J_1 x \subset H_j\}, \text{ and}$$
$$\mathrm{ann} H_j = \{r \in E_1 \mid rH_j = 0\} \quad \text{for } j = 1, 2, \ldots.$$

<u>Lemma 2.5</u>. For all $j$, $H_j$ is an $E_1$-submodule of $G_1$ and $H_1 = \{x \in G_1 \mid J_1 x = 0\}$.

Proof. In fact, $H_1$ is an E-submodule of $G$. Since $J_1$ is a right ideal in $E_1$, $H_j$ is an $E_1$-submodule of $G_1$.

For the second part of the lemma, note that $EJ_1H_1 = p_1Ep_1J_1H_1 \subset E_1J_1H_1 \subset J_1H_1$, where $p_1$ is the projection of $G$ onto $G_1$. Thus, $J_1H_1$ is an E-submodule of $G$. There are therefore two possibilities: (1) $QJ_1H_1 = 0$, or (2) $QJ_1H_1 = S = QH_1$. However, (2) is impossible since $J_1$ is nilpotent, so that $H_1 \subset \{x \in G_1 \mid J_1x = 0\}$. To show the reverse containment, suppose there is an $x \in G_1$ such that $J_1x = 0$ but $x \notin H_1$. Since $QH_1$ = socle $QG$, there is $r \in E$ such that $0 \neq rx \in H_1$. But then $0 \neq p_1rp_1x \in H_1$, so assume $r \in E_1$. If $r$ is monic, then $nx \in H_1$ for some positive integer $n$. This implies $x \in H_1$, a contradiction. On the other hand, if $r$ is nilpotent then $r \in J_1$ so that $rx = 0$, another contradiction. Since $r$ must be either monic or nilpotent (Proposition 0.1) the proof is complete.

Lemma 2.6. If $K$ is an $E_1$-submodule of $G_1$, and $f:K \to G_1$ is an $E_1$-map, then there exists $n > 0$ and $g \in C$ such that $g|_K = nf$.

Proof. The proof is by induction on $m$ such that $K \subset H_m$. First note that for each $j$, $f(H_j \cap K) \subset H_j$. This can be seen inductively: $f(H_1 \cap K) \subset H_1$ because $J_1f(H_1 \cap K) = f(J_1(H_1 \cap K)) = 0$, and $J_1f(H_{j+1} \cap K) = f(J_1(H_{j+1} \cap K)) \subset f(H_j \cap K) \subset H_j$ by induction, so that $f(H_{j+1} \cap K) \subset H_{j+1}$ by definition of $H_{j+1}$.

If $K \subset H_1$, then $K$ is in fact an E-submodule of $G$. Hence the desired $n$ and $g$ exist by the aEqi property. Inductively, assume the result for $K \subset H_{m-1}$, and let $K \subset H_m$. Then $0 \neq K \cap H_{m-1}$ is an $E_1$-submodule of $K$ and $f(K \cap H_{m-1}) \subset H_{m-1}$. Hence by the induction hypothesis there exists $n > 0$ and $g_1 \in C$ such that $(g_1 - nf)(K \cap H_{m-1}) = 0$.

That is, $J_1(g_1-nf)(K) = (g_1-nf)(J_1K) \subset (g_1-nf)(K \cap H_{m-1}) = 0$, so that $(g_1-nf)(K) \subset H_1$.

Let $K' = EK \cap (1-p_1)G = (1-p_1)EK$, and define a map $f_1 : EK \to G$ by $f_1|_K = g_1-nf$ and $f_1|_{K'} = 0$. Clearly, $f_1|_K$ is an $E_1$-map and $f_1|_{K'}$ is a $(1-p_1)E(1-p_1)$-map. Let $r = (1-p_1)rp_1 \in E$. Then $r(K) \subset K'$, and $f_1r(K) \subset f_1(K') = 0$, while $rf_1(K) \subset r(H_1) = 0$. Similarly, let $s = p_1s(1-p_1) \in E$. Then $s(K') = p_1s(1-p_1)Ep_1K$. However, $p_1s(1-p_1)Ep_1 \subset J_1$, since $E_1$ is local and $G_1$ is not quasi-isomorphic to $G_i$ for $i \neq 1$. Hence, $f_1s(K') \subset f_1(J_1K) = 0$, while $sf_1(K') = 0$ by the definition of $f_1$. It follows that $f_1$ is an E-map. Therefore, by the aEqi property, there exist $n' > 0$ and $g_2 \in C$, such that $n'f_1 = g_2|_{EK}$. In particular, $(n'g_1-n'nf-g_2)(K) = 0$, and $g = n'g_1-g_2 \in C$ is a lifting of $n'nf$.

<u>Corollary 2.7</u>. The group $G_1$ is aEqi.

Proof. This corollary follows immediately from Lemma 2.6, since the $g \in C$ obtained in that lemma provides $p_1gp_1 \in C_1$ = center $E_1$ to satisfy the definition of aEqi for $G_1$.

We can now prove the structure theorem for finite rank aEqi groups.

<u>Theorem 2.8</u>. Let G be a finite rank aEqi group. Then G is aEqi if and only if G is quasi-isomorphic to a direct

sum of fully invariant subgroups of the form,

(1) $A = G_1 \oplus M$, where

(2) $G_1$ is a strongly indecomposable finite rank aEqi group,

(3) $M$ is isomorphic to a subgroup of $G_1^k$ for some $k > 0$,

(4) $Q(\text{center } E(A))$ is isomorphic to $QE(G_1)$.

Proof. As in the Summary at the beginning of this section, we may assume, without loss of generality, that $G = A$ and $G$ is a direct sum of strongly indecomposable, mutually non-quasi-isomorphic groups. Denote by $p_1$ the projection of $G$ onto $G_1$.

Suppose $G = A$ has the form (1) satisfying (2), (3), and (4). We first remark that conditions (1) - (3) imply that the isomorphism in (4) is given by $f \to p_1 f p_1$, for $f \in Q(\text{center } E(A))$. Indeed, if $0 \neq f$, choose $x \in G$ such that $f(x) \neq 0$. Then there exists $r \in E$ such that $0 \neq rf(x) \in G_1$. But $rf(x) = p_1 rf(x) = p_1 f(rx) \neq 0$. This implies $p_1 f \neq 0$ and hence that $p_1 f p_1 \neq 0$.

Let $K$ be any E-submodule of $G$. Then $K = K_1 \oplus M_1$, where $K_1$ is an $E_1$-submodule of $G_1$ and $M_1 \subset M$, since $E$ contains the projections onto $G_1$ and $M$. If $f:K \to G$ is an E-map, then $f(K_1) \subset G_1$ and $f(M_1) \subset M$, since $f$ commutes with projections. Since $G_1$ is aEqi, there exist $n > 0$, and $g_1 \in \text{center } E_1 = E_1$ (Lemma 1.2), such that $(nf-g_1)(K_1) = 0$. By (4), $mg_1 = p_1 g p_1$ for some $m > 0$ and

$g \in C$. We show $g$ is a lifting of $mnf$ by showing $(mnf-g)(M_1) = 0$. Suppose there exists an $x \in M_1$ such that $(mnf-g)(x) \neq 0$. By (3), there exists $h:M \to G_1$ such that $h((mnf-g)(x)) \neq 0$. Both $g$ and $f$ commute with $h$ (regarded as an element of $E$), so that $(mnf-g)(hx) \neq 0$. However, $hx \in K_1$, so we have contradicted $(mnf-g)(K_1) = 0$. Thus $G$ is aEqi.

Conversely, suppose $G$ is aEqi. Assuming $G$ is as in the Summary, apply Corollaries 2.4 and 2.7 to write $G = G_1 \oplus M$, where $G_1$ is strongly indecomposable aEqi (conditions (1) and (2)) and the QE-socle of $QG$ is contained in $QG_1$. In particular, for each $x \neq 0$ in $M$, there is an element $f$ of $E$ such that $0 \neq f(x) \in G_1$. This fact, together with the finite rank of $G$ gives (3). That the center of $QE(G)$ is isomorphic to $QE_1$ (condition (4)) follows from Lemma 2.6 and the observation that if $f \in C =$ center $E$ satisfies $f(G_1) = 0$, then $f = 0$. The proof of the theorem is completed by a straightforward transition from the $G$ of the Summary to the general case.

<u>Corollary 2.9</u>. Let $G$ be a torsion free finite rank aEqi group. Then Q(center $E(G)$) is a finite dimensional, local, self-injective Q-algebra.

This corollary follows directly from Theorem 2.8 and Proposition 1.3.

## References


(1) D. Arnold, Finite Rank Torsion Free Abelian Groups and Rings, Lecture Notes in Math. No. 931, Springer-Verlag, 1982, Berlin-Heidelberg.

(2) Arnold, Pierce, Reid, Vinsonhaler and Wickless, Torsion free abelian groups of finite rank projective over their endomorphism rings, J. Algebra, Vol. 71, No. 1, (1981), 1-10.

(3) C. Faith, Algebra II, Ring Theory, Springer-Verlag, 1976, Berlin-Heidelberg.

(4) G.D. Poole and J.D. Reid, Abelian groups quasi-injective over their endomorphism rings, Can. J. Math., Vol. 24 No. 4, (1972), 617-621.

(5) J.D. Reid, On the ring of quasi-endomorphisms of a torsion-free abelian group, Topics in Abelian Groups, Scott Foresman, Chicago, 1963.

(6) P. Schultz, The endomorphism ring of the additive group of a ring, J. Australian Math. Soc. 15 (1973), 60-69.

(7) C. Vinsonhaler and W. Wickless, Injectivive hulls of torsion free abelian groups as modules over their endomorphism rings, J. Algebra, Vol.5, No.1,(1979), 64-69.

(8) C. Vinsonhaler and W. Wickless, Torsion free abelian groups quasi-projective over their endomorphism rings, Pac. J. Math. Vol. 68, No. 2, (1977), 527-535.

(9) H. Zassenhaus, Orders as endomorphism rings of modules of the same rank, J. London Math. Soc. 42(1967),180-182.

# E-UNISERIAL TORSION-FREE ABELIAN GROUPS OF FINITE RANK

Jutta Hausen

University of Houston
University Park

An abelian group A is said to be E-uniserial if the lattice of fully invariant subgroups of A is a chain.

In an earlier paper[4] we characterized the E-uniserial abelian groups up to torsion-free reduced E-uniserial direct summands. The purpose of this note is to give a complete description of all torsion-free reduced E-uniserial groups of finite rank.

Except where noted, the word "group" will be used to mean "reduced torsion-free abelian group of finite rank", and G and H are such groups.

We list some earlier results[4]:

LEMMA 1. If G is E-uniserial then

(i) G is p-local for some prime p;

(ii) G is E-cylic;

(iii) G is strongly irreducible (cf. Reid[7]);

(iv) $G \doteq H^m$ where H is a strongly irreducible strongly indecomposable group and m is an integer (cf. Reid[8]).

LEMMA 2. For H a group and $m \geq 1$ a cardinal, $H^m$ is E-uniserial if and only if H is E-uniserial.

THEOREM 3. Let H be a strongly indecomposable group. Then H is E-uniserial if and only if H supports an E-ring with totally ordered ideal lattice.

E-rings were introduced by P. Schultz as the rings R such that every endomorphism of $R^+$ is the left multiplication by some ring element. Note that E-rings are commutative[9].

Numerous properties of abelian groups which are of interest in the E-module setting are preserved under quasi-isomorphism (e.g. strong irreducibility, strong indecomposability, being finitely E-generated, etc.). If this were the case for E-uniseriality, the results above would yield a complete description of the E-uniserial groups. However, this is not the case as can be seen from the following example which is due to Charles I. Vinsonhaler.

EXAMPLE 4. Let p be a prime and let $f(x) = x^5 - x^3 - p$. Then f is irreducible in $Z[x]$. If $\lambda$ is a root of f, then $\lambda$ generates a maximal ideal in $Z[\lambda]$ so that the localization $R = (Z[\lambda])_{(\lambda)}$ at $(\lambda)$ is a discrete valuation ring in the sense of Kaplansky[5]. Hence $R^+$ is a torsion-free E-uniserial group of rank 5. One verifies that $R/pR$ has rank 3 which, by a result of J. D. Reid[6], implies that $R^+$ is strongly indecomposable. By Bowshell and Schultz[3], R is an E-ring. Let $G = R + Z_{(p)} \cdot \frac{1}{p} \subseteq Q \otimes R$ where $Z_{(p)}$ denotes the integers localized at p. Then G is quasi-equal to the E-uniserial group $R^+$. Every endomorphism of G can be shown to be the multiplication with an element of the form $pr + q$ with $r \in R$ and $q \in Z_{(p)}$ . It follows that $G$ is E-cyclic but not E-uniserial.

This example points to the heart of the matter: if we let $S = \text{End } G$ , the endomorphism ring of $G$ , then

$$S \doteq \text{End } R^+ \simeq R$$

so that $S^+$ is strongly indecomposable but $S$ is not a valuation ring since $G$ is not E-uniserial. In fact, one can show

PROPOSITION 5. Let $R$ and $S$ be two unital rings with totally ordered ideal lattices whose additive groups are finite rank torsion-free and strongly indecomposable. Then $R^+ \doteq S^+$ implies $R \simeq S$ as rings.

It follows that two strongly indecomposable quasi-isomorphic E-uniserial groups are, in fact, isomorphic. This is reminiscent of the situation for D. M. Arnold's strongly homogeneous groups. Every p-local strongly homogeneous group is E-uniserial, and two strongly homogeneous quasi-isomorphic groups are isomorphic. A strongly homogeneous group is a direct sum of copies of the same strongly homogeneous strongly indecomposable group, and the strongly indecomposable strongly homogeneous groups are precisely the additive groups of certain subrings of algebraic number fields[1]. We have the exact same situation for E-uniserial groups except that in our case the subrings of algebraic number fields that occur are the strongly indecomposable E-rings with totally ordered ideal lattice, i.e. the strongly indecomposable discrete valuation rings. To see this, let $G$ be an E-uniserial group. By Reid[8], $G \doteq H^m$ where $H$ is strongly indecomposable and strongly irreducible. Hence

$$\text{End } G \doteq \text{End } H^m \simeq \text{Mat}_m(\text{End } H) ,$$

and Reid[8] implies that End H is an E-ring with

$$H \doteq (\text{End } H)^+ .$$

Consequently, End H is commutative,

$$\text{End } H \simeq \text{Cent}[\text{Mat}_m(\text{End } H)] ,$$

and $\text{Mat}_m(\text{End } H)$ is finitely generated as module over is center. (We use $\text{Mat}_m(R)$ and $\text{Cent}(R)$ to denote the ring of all $m \times m$ matrices over $R$ and the center of $R$, respectively.) Let

$$C = \text{Cent}(\text{End } G) .$$

Quasi-isomorphic torsion-free rings of finite rank must have quasi-isomorphic centers and, if one of them is finitely generated over its center, then so is the other one.
Thus,

$$C \doteq \text{End } H$$

is an E-ring (cf. Bowshell and Schultz[3]), and End G is a finitely generated C-module. Since $(\text{End } H)^+ \doteq H$, C is strongly indecomposable which implies C is a domain[3]. For each $\xi \in C$, both the kernel of $\xi$ and the image, $\text{im}\xi$, of $\xi$ are fully invariant subgroups. Since G is strongly irreducible and E-cyclic, it follows that G is a finitely generated torsion-free C-module; the map

$$\xi C \rightarrow \text{im}\xi$$

is an order isomorphism from the poset of principal ideals of C into the lattice of fully invariant subgroups of G . The additive group of C is reduced and torsion-free of finite rank. Hence C is a discrete valuation ring in the sense of Kaplansky[5] and, in particular, a principal ideal domain. It follows that G is a free C-module, $C \doteq H$ strongly indecomposable.

We have established

THEOREM 6. Let G be a reduced torsion-free abelian group of finite rank. Then G is E-uniserial if and only if the center C of the endomorphism ring of G is a strongly indecomposable discrete valuation ring[5] and G is a free C-module.

Note that strongly indecomposable discrete valuation rings are E-rings[3].

Analogous to Arnold's results on strongly homogeneous groups[1] we have

COROLLARY 7. Let G be an E-uniserial reduced torsion-free abelian group of finite rank.

(a) $G \simeq H^m$ where H is E-uniserial and strongly indecomposable.

(b) G is indecomposable iff G is strongly indecomposable.

(c) If B is a direct summand of G then $B \simeq H^n$ for some $n \leq m$.

(d) If K is another E-uniserial reduced torsion-free abelian group then $G \simeq K$ iff rank G = rank K and Cent(End G) $\simeq$ Cent(End K).

Suppose G and K are two E-uniserial groups and $G \doteq K$. Then the centers of their endomorphism rings are quasi-isomorphic strongly indecomposable valuation rings. Proposition 5 implies that Cent(End G) and Cent(End K) are isomorphic. Combining these results with (d) of Corollary 7 proves

PROPOSITION 8. Two E-uniserial reduced torsion-free abelian groups of finite rank are quasi-isomorphic if and only if they are isomorphic.

By earlier results[4], a torsion-free abelian group $G$ of finite rank is E-uniserial if and only if $G = Q^n \oplus K$ with $K$ reduced and E-uniserial and $n$ any non-negative integer.

Let $\mathcal{F}$ be a complete and irredundant set of representatives of the isomorphism classes of strongly indecomposable finite rank torsion-free E-uniserial groups (note that $Q$ is in $\mathcal{F}$). Then, by Proposition 7, no two groups in $\mathcal{F}$ are quasi-isomorphic. Also, since the endomorphism ring of each group $C$ in $\mathcal{F}$ is a principal ideal domain, C-projective groups are C-free (cf. Arnold and Lady[2] or Corollary 7, (c)). Thus, Corollary 4.3 of Arnold and Lady[2] implies yet another analogon with the class of strongly homogeneous groups[1]:

COROLLARY 9. Let $\mathcal{S}$ be the class of finite direct sums of E-uniserial torsion-free abelian groups of finite rank.

(a) If $G$ is in $\mathcal{S}$ and $B$ is a summand of $G$ then $B$ is in $\mathcal{S}$.

(b) Every $G$ in $\mathcal{S}$ has the Krull-Schmidt property (i.e. any two decompositions of $G$ into direct sums of indecomposable groups are equivalent).

Example 4 shows that the class of local strongly homogeneous groups is properly contained in the class of E-uniserial groups.

REFERENCES

1. D. M. Arnold, Strongly homogeneous torsion-free abelian groups of finite rank, Proc. Amer. Math. Soc. 56 (1976), 67-72.

2. D. M. Arnold and E. L. Lady, Endomorphism rings and direct sums of torsion free abelian groups, Trans. Amer. Math. Soc. 211 (1975), 225-237.

3. R. A. Bowshell and P. Schultz, Unital rings whose additive endomorphisms commute, Math. Ann. 228 (1977), 197-214.

4. J. Hausen, Abelian groups which are uniserial as modules over their endomorphism rings, Abelian Group Theory, Lecture Notes in Mathematics, Vol. 1006, Springer-Verlag, Berlin 1983, pp. 204-208.

5. I. Kaplansky, "Infinite Abelian Groups", Revised Edition, The University of Michigan Press, Ann Arbor 1969.

6. J. D. Reid, On the ring of quasi-endomorphisms of a torsion-free group, Topics in Abelian Groups, Scott, Foresman, Chicago 1963, pp. 51-68.

7. J. D. Reid, On rings on groups, Pacific J. Math. 53 (1974), 229-237.

8. J. D. Reid, Abelian groups finitely generated over their endomorphism rings, Abelian Group Theory, Lecture Notes in Mathematics Vol. 874, Springer-Verlag, Berlin 1981, pp. 41-52.

9. P. Schultz, The endomorphism ring of the additive group of a ring, J. Austral. Math. Soc. 15 (1973), 60-69.

# THE EXISTENCE OF RIGID SYSTEMS OF MAXIMAL SIZE

Rüdiger Göbel
Universität Essen

## §1 Introduction

In [CG] we have not been able to answer a question on the maximal size of (generalized) rigid systems of R-modules over commutative rings; see Remark (5.3) in [CG]. A (generalized) rigid system at the cardinal $\rho$ will be a set $G_i$ $(i \in \rho)$ of R-modules of power $\lambda$ such that the endomorphism algebra End $G_i$ coincides with a prescribed R-algebra A modulo some "inessential" endomorphisms which form an ideal Ines $G_i$. Furthermore Hom$(G_i, G_j)$ consists of inessential homomorphisms only for any $i \neq j \in \rho$. Naturally, we want $\rho$ to be as large as possible which is $\rho = 2^\lambda$. In all "classical cases" we derived $\rho = 2^\lambda$, but it would be much nicer to obtain $\rho = 2^\lambda$ without any restrictions as assumed in [CG], Theorem 5.2(b). The following theorem will settle this problem which will be our main result.

<u>THEOREM</u>: Let A be a $\lambda^{\kappa}$-representable topological R-algebra then we have R-modules $G^J$ for each $J \subseteq \lambda$ and

$$\mathrm{Hom}(G^{J_1}, G^J) = \begin{cases} A \oplus \mathrm{Ines}(G^{J_1}, G^J) & (J_1 \subseteq J) \\ \mathrm{Ines}(G^{J_1}, G^J) & (J_1 \not\subseteq J) \end{cases}$$

where A is embedded as a topological subalgebra of the finitely topologized End $G^J$ for any $J_1 = J$.

This is a result on torsion-free, torsion or mixed R-modules discussed in detail in [CG]. The distinction of the three cases depends on the unexplained notions of $\lambda^{\kappa}$-representable algebras and Ines G. For the convenience of the reader we will repeat briefly these natural definitions from [CG] in the next paragraph. In §3 we will give the proof of our theorem. The most interesting case will be a result on torsion-free modules and A equipped only with the discrete topology. Therefore we will restrict our attention to this case. The extension of the proof to mixed and torsion modules including the topology can then easily be carried out using the extra load of terminology and Lemmata from [CG].

The fixed non-zero commutative ring R with 1 will have a fixed and countable multiplicatively closed set S with $1 \in S$, containing no zero-divisors such that $\bigcap_{s \in S} Rs = 0$. The S-topology on an R-module M is generated by $\{Ms : s \in S\}$ and M is Hausdorff (= reduced) if $\bigcap_{s \in S} Ms = 0$. The letter $\hat{M}$ denotes the completion of M in the S-topology and $M[s] = \{m \in M : ms = 0\}$ is the s-socle of M. The module M is torsion-free if $M[s] = 0$ for all $s \in S$ and M is torsion if $M = \sum_{s \in S} M[s]$. We say that A is representable if A is a complete hausdorff topological R-algebra with a set $\mathfrak{N}$ of right ideals N such that A/N is reduced and torsion-free. Furthermore we require

(a) For the construction of torsion-free or mixed modules: $\mathfrak{N}$ is a basis of neihgbourhoods of 0 in A.

(b) Construction torsion-modules: There exists $p \in S$ such that $S = \{p^k : k \in \omega\}$ and $\{p^k A + N : k \in \omega, N \in \mathfrak{N}\}$ is a basis of neighbourhoods of 0 in A.

Now A is $\lambda^\kappa$-representable if $\lambda, \kappa$ are any cardinals such that $\kappa \geq \max\{\aleph_0, |\mathfrak{N}|\}$ and $\lambda \geq |A|^\kappa$ with the further cardinal-requirements.

(c) Constructing torsion or mixed modules: There is a pure and dense R-submodule D of A such that

 (i) if $\mathfrak{N} \neq \{0\}$, then D has at most $\kappa$ generators

 (ii) if $\mathfrak{N} = \{0\}$, then $D = \bigoplus_{i \in I} D_i$ and each $D_i$ has at most $\kappa$ generators.

Requirement (c) serves only for the construction of modules of small cardinal and (a), (b) are most natural. For a discussion see [CG] §1.

Remark: We also dropped the condition $\lambda^\kappa = \lambda$ in [CG] because an additional ultra filter argument of [S] is applicable to to replace $\lambda^{\kappa} = \lambda \geq |A|$ by $\lambda \geq |A|^{\kappa}$; see [S]. If we restrict to the discrete toplology and torsion-free modules the theorem simply becomes the

<u>COROLLARY</u>: Let S be a countable multiplicatively closed subset of an (S-)reduced and (S-)torsion-free commutative ring R with $1 \neq 0$. If A is any reduced and torsion-free algebra and $\lambda$ any cardinal with $\lambda \geq |A|^{\aleph_0}$, then we have for each $J \subseteq \lambda$ a reduced and torsion-free R-module $G^J$ such that

$$\mathrm{Hom}(G^{J_1}, G^J) = \begin{cases} A \oplus \mathrm{Ines}(G^{J_1}, G^J) & (J_1 \subseteq J) \\ \mathrm{Ines}(G^{J_1}, G^J) & (J_1 \not\subseteq J) \end{cases}$$

where $\mathrm{Ines}(G^{J_1}, G^J) = \{\varphi \in \mathrm{Hom}(G^{J_1}, G^J) : G^{J_1}\varphi \subseteq G^J\}$.

Now it is obvious that $\rho = 2^{\lambda}$, using suitable subsets of $\lambda$. The corollary will be derived by a modification of the construction of the modules in [CG]. Many of the results in [CG] remain valid under this change. Therefore we will refer heavily to proofs in [CG] and will carry out all basic changes in details.

Finally we want to present a positive solution of Problem 7.22 (p.84) from the Kourovka Notebook [K]:

> "Suppose G is a finite group acting as a group of automorphisms on some torsion-free abelian group. Do there exist for each cardinal number m, $2^m$ non isomorphic torsion-free abelian groups A of cardinal m with G as its group of automorphisms?"

The problem is attributed to S.F. Kozhukov, but at least parts of it may also be found in [HH].

It is easy to see that G must be the group of units of some countable ring, A with free additive group ; see e.g. [DG] p.334. If $m = \aleph_0$ , apply [C] to get $2^{\aleph_0}$ desired abelian groups of cardinality $\aleph_0$ . If $m \geq 2^{\aleph_0}$ , the existence of a rigid system of $2^m$ torsion-free abelian groups $G_i$ $(i \in 2^m)$ with $Aut(G_i) = G$ follows from the Corollary above. Since $Hom(G_i, G_j) = Ines(G_i, G_j)$ for $i \neq j$, and $Ines(G_i, G_j) = 0$ for any free $A^+$ from [CG], Theorem 6.3, we have $Hom(G_i, G_j) = A \cdot \delta_{ij}$ $(i,j \in 2^m)$. In particular $G_i$ and $G_j$ are not isomorphic for $i \neq j$. On the other hand, End $G_i = A$, and the automorphism group is the group of units, hence Aut $G_i = G$.

In view of the above results, the following problem 4.43 in [K] attributed to V.T. Nagrebetskii does not make sense any more:

> "Describe the torsion-free abelian groups with only a finite number of automorphisms".

We also want to mention that Kulikov's Problem 1.66 in [K] has an obvious positive solution:

> "Let T be a torsion abelian group and $\kappa$ a cardinal $> \aleph_0$. Find an abelian group U such that for abelian group A of cardinal $\leq \kappa$ we have $A \subseteq U$ iff $Ext(A,T) = 0$."

Choose $U = \oplus \{B : |B| \leq \kappa, Ext(B,T) = 0\}$ and apply functorial properties of $Ext(.,T)$.

## §2 The suitable form of Shelah's Black Box

Let $\lambda$ be a cardinal of the corollary (§1). We will assume $\lambda^{\aleph_0} = \lambda$, the case $\lambda^{\aleph_0} > \lambda \geq |A|^{\aleph_0}$ is similar and needs only a combinatorial extra step; we refer to [S]. Consider the tree $T = {}^{\omega>}\lambda$ of all finite sequences $\tau : n \to \lambda$ $(n \in \omega)$. Then $l(\tau) = n$ will be the length of $\tau$ and the order $\subseteq$ on T will be containment. If $X \subseteq T$, then Br(X) denotes all branches in X. If $v = \{v_n = v \restriction n : n \in \omega\}$ is a branch, let $\alpha_l^v$ $(l \in \omega)$ be a set of ordinals which satisfy $\alpha_l^v < v(l)$ at each level $l \in \omega$. The object $\{(v \restriction n) \cup (v \restriction n-1)^\wedge \langle\alpha_{n-1}^v\rangle : n \in \omega\}$ is called the branch v with leaves $\alpha_l^v$. Fix an (S-)reduced R-module $B = \bigoplus_{\tau \in T} B_\tau$ and assume $\tau \in B_\tau \setminus \{0\}$. Every element g of the (S-)completion $\hat{B}$ of B is expressible as a convergent sum $g = \sum_{\tau \in T} g_\tau$ with $g_\tau \in B_\tau$; see §1. Then $[g] = \{\tau \in T, g_\tau \neq 0\}$ denotes the <u>support</u> of g and similarly $[X] = \bigcup_{g \in X} [g]$ for all $X \subseteq B$. In order to define a <u>norm</u> of g, we fix a strictly increasing continuous map $\rho : cf(\lambda) + 1 \to \lambda + 1$ with $\rho(0) = 0$ and $(cf(\lambda)) = \lambda$. The norm of g is now $\|g\| = \min\{\nu \leq cf(\lambda) : [g] \subseteq {}^{\omega>}\rho(\nu)\}$ and $\|X\| = \sup_{g \in X}\|g\|$. In §3 we will define a set $\mathcal{C}$ of canonical submodules of B which have the following properties

(a) If $P \in \mathcal{C}$, $X \subseteq B$ a countable set, then there exists $P' \in \mathcal{C}$ such that $P \cup X \subseteq P'$.

(b) $P \in \mathcal{C}$ has at most countably many generators.

(c) $\mathcal{T} \neq \emptyset$ is closed under countable ascending unions.

Obviously $\|P\| < cf(\lambda)$ for all $P \in \mathcal{T}$. We also fix a <u>coding set</u> $\mathcal{R}$ of cardinal $\leq \lambda$.

<u>Definition 2.1</u>: A tuple $(f,a,P,\varphi,r)$ is called a trap if $f : {}^{\omega>}\omega \to {}^{\omega>}\lambda = T$ is a tree-embedding, $P \in \mathcal{T}$, $a \in \hat{P}$, $\varphi \in \operatorname{End} P$ and $r \in \mathcal{R}$ such that the following holds:

(a) $\operatorname{Im} f \subseteq P$

(b) $[P] \subseteq P$ and $[P]$ is a subtree of $T$.

(c) $cf\|P\| = \omega$

(d) $\|v\| = \|P\|$, for all $v \in B_r(\operatorname{Im} f)$

(e) $\|a\| < \|P\|$.

Then we have the following

<u>Black Box 2.2</u>: For some ordinal $\lambda^*$ there exists a sequence $(f_\alpha,a_\alpha,P_\alpha,\varphi_\alpha,r_\alpha)$ $(\alpha < \lambda^*)$ of traps such that all $\alpha,\beta \in \lambda^*$ satisfy the following

(i) $\beta \leq \alpha \Rightarrow \|P_\beta\| \leq \|P_\alpha\|$

(ii) There are disjoint subsets $V,V',V''$ of $\omega$-limit ordinals in $cf(\lambda)$ such that $\|P_\alpha\| \in V \cup V'$ for all $\alpha$.

(iii) $\beta \neq \alpha \Rightarrow B_r(\operatorname{Im} f_\alpha \cap \operatorname{Im} f_\beta) = \emptyset$

(iv) $\beta + 2^{\aleph_0} \leq \alpha \Rightarrow B_r(\operatorname{Im} f_\alpha \cap [P_\beta]) = \emptyset$

(v) <u>The prediction</u>: If $X \subseteq B$ is a countable set, $P \in \mathcal{T}$, $\varphi \in \operatorname{End} \hat{B}$, $Y \in \{V,V'\}$, $a \in \hat{B}$ and $r \in \mathcal{R}$,

(a) there exists $\alpha < \lambda^*$ such that (*) holds:

$$(*) \begin{cases} P \cup X \subseteq P_\alpha , \ \|P \cup X\| < \|P_\alpha\| \in Y \\ a_\alpha = a, \ \varphi \restriction P_\alpha = \varphi_\alpha \text{ and } r_\alpha = r. \end{cases}$$

(b) if in addition $B_\varphi$ is bounded, i.e. there exists $\beta \in cf(\lambda)$ with $\|B_\varphi\| < \beta$, then there exists $\alpha < \lambda^*$ such that (*) holds and

$$(**) \begin{cases} \text{For every } v \in B_r(\mathrm{Im}\, f_\alpha) \text{ there are leaves } \alpha_1^v \\ (1 \in \omega) \text{ such that } (v \restriction 1)^\wedge \langle\alpha_1^v\rangle \in P_\alpha \text{ and} \\ \langle v \restriction 1\rangle\varphi^\alpha = (v \restriction 1 - 1)^\wedge \langle\alpha_{1-1}^v\rangle \text{ for all } 1 \in \omega. \end{cases}$$

<u>Proof</u>. (mixture of [CG] and [GS]).

## <u>§3 Proof of the Corollary</u>

First we want to construct the required modules $G^J$. Let $B_\tau = \tau A$ and $\mathrm{Ann}_A \tau = 0$ because we restrict to the construction of torsion-free modules in the discrete case. Then $B = \bigoplus_{\tau \in T} B_\tau$ and the set $\mathcal{C}$ of all <u>canonical</u> submodules of B will simply be all pure submodules of B which are generated (as pure submodules) by at most countably many elements. It will be convenient to use a special representation for limits in $\hat{B}$. Therefore let $S = \{s_n : n \in \omega\}$ be an enumeration with $s_o = 1$ and define $q_n = \prod_{i \leq n} s_n$. Hence elements g in $\hat{B}$ may be expressed by $g = \sum_{n \in \omega} g_n q_n$ with $g_n \in B$ respectively

$g = \sum_{\tau \in [g]} g_\tau \tau$ with $g_\tau \in \hat{A}$ and $|[g]| \leq \aleph_0$. A divisor chain $g^k (k \in \omega)$ of $g = g^o$ will be a sequence of elements $g^k \in \hat{B}$ such that $g^k - g^{k+1} q_{k+1} \in B$ and $[g^k] \subseteq [g]$. Naturally, each element h has a divisor chain and any branch v can easily be made into a divisor chain, if we let

$$v^k = \sum_{\sigma \in v, l(\sigma) \geq k} \sigma \frac{q_{l(\sigma)}}{q_k} \quad (k \in \omega).$$

The coding set will be $\mathfrak{R} = \lambda + 1$ and from the Black box we have

$$(f_\alpha, a_\alpha, P_\alpha, \varphi_\alpha, j_\alpha) \quad (\alpha < \lambda^*).$$

Let $\lambda^*_c = \{\alpha \in \lambda^* : \|P_\alpha\| \in V'\}$. First we define only one R-module $G = G^\lambda$. The module G will be an A-module and can be written as $\bigcup_{\mu < \lambda^*} G_\mu$ a union of a continuous ascending chain of A-modules, starting at the bottom with $G_o = B$. The construction of the layers $G_\mu$ is by induction. Simultaneously we define for each successor ordinal $\mu = \alpha + 1$ new generators $g^k_\alpha \in \hat{B}$ $(k \in \omega)$ and test elements $b_\alpha \in \hat{B} \cup \{\infty\}$ subject to the following conditions. Suppose $G^k_\alpha$ and $g_\beta$, $b_\beta (\beta < \alpha)$ have been constructed. If $\alpha \in \lambda^* \setminus \lambda^*_c$, we want a branch $v_\alpha \in B_r(\mathrm{Im}\, f_\alpha)$ and if $\alpha \in \lambda^*_c$ we want a branch $v_\alpha \in B_r(\mathrm{Im}\, f_\alpha)$ and leaves $\alpha^v_l (l \in \omega)$ with $(v_\alpha \upharpoonright l-1)^\wedge \langle \alpha^v_{l-1} \rangle \in P_\alpha$. Then we define

(*) $a_{\alpha n} = v_\alpha \upharpoonright n$ for $\alpha \in \lambda^* \setminus \lambda^*_c$ and

$a_{\alpha n} = (v_\alpha \upharpoonright n) - (v_\alpha \upharpoonright n-1)^\wedge \langle \alpha^v_{n-1} \rangle$ for $\alpha \in \lambda^*_c$

and choose $\delta_\alpha \in \{0,1\}$ such that

$$(**)\quad \begin{cases} g_\alpha^k = a_\alpha^k \delta_\alpha + \sum_{n\geq k} a_{\alpha n} \frac{q_n}{q_k} \\ G_{\alpha+1} = G_\alpha + \sum_{k\in\omega} g_\alpha^k A. \end{cases}$$

We will force in the construction

$$(***)\qquad b_\beta \notin G_\alpha \quad \text{for all } \beta < \alpha.$$

Suppose $\alpha \in \lambda_c^*$ and it is possible to find a branch $v_\alpha$ with leaves $\alpha_1^v$ and $b_\alpha = \infty$, $\delta_\alpha = 1$ such that (*), (**), (***) hold and also $a_{\alpha n} \varphi_\alpha = 0$ for all $n \in \omega$; then we say that $\alpha$ is <u>complete</u> and we make this choice. If $\alpha \in \lambda^* \setminus \lambda_c^*$ and it is possible to find a branch $v_\alpha$ and $b_\alpha = g_\alpha \varphi^\alpha$ which satisfy (*), (**), (***), then we choose these elements and say that $\alpha$ is <u>strong</u>. If $\alpha \in \lambda^* \setminus \lambda_c^*$ is neither complete nor strong, but we find $v_\alpha$, $\delta_\alpha$ and $b_\alpha = \infty$ to satisfy (*), (**), (***), then we fix such elements and call $\alpha$ <u>weak</u>. In the case that such choices are impossible, we simply let $g_\alpha^k = 0$, $b_\alpha = \infty$ and call $\alpha$ <u>useless</u>.

Induction completes the construction of G. In order to derive a rigid system, choose any $J \subseteq \lambda$ and let $J^\# = \{\alpha \in \lambda^* : j_\alpha \in J \cup \{\lambda\}\}$. The required module $G^J$ will be

$$G^J = B + \sum_{\alpha\in J^\#} \sum_{k\in\omega} g_\alpha^k A.$$

In particular we have $G^\lambda = G$.

For a better understanding of the construction we recall from [CG] that there are no useless ordinals; the proof of this result holds in this construction without any changes.

Since B is a dense submodule of any $G^J$, the given definition (§1) of inessential homomorphism $\varphi : G^{J_1} \to G^J$ is equivalent to say that $\hat{B}\varphi \subseteq G^J$. This will be used to show the crucial

Lemma 3.1: If $\varphi \in$ Ines G, we find some ordinal $\alpha < cf(\lambda)$ such that $\|B\varphi\| < \alpha$.

Proof: Suppose (3.1) does not hold, i.e. $\|B\varphi\| = cf(\lambda)$. Inductively we can choose sequences $\alpha_n < \alpha_{n+1} \in V''$, $b_n\varphi \in B$, $s_n | s_{n+1} \in S$ and $\sigma_n \in [b_n\varphi]^*$ such that $\alpha_{n-1} < \|\sigma_n\|$, $\|b_n\varphi\| < \alpha_n$, and

$$(b_n\varphi \restriction \sigma_n) s_n \not\equiv 0 \bmod A\, s_{n+1}.$$

Recall from [CG] the notation $[X]^*$ of the top of an element $x \in G$:

$$(+) \quad [x]^* = \begin{cases} \{\tau \in T : \|\tau\| = \|x\|,\ x \restriction \tau \neq 0\} & \text{if } \|x\| \text{ is not a limit} \\ \bigcup\limits_{i \leq n} \{{}_\nu[v_{\alpha_i}] : x = x' + \sum\limits_{i \leq n} g^k_{\alpha_i} a_i \ (a_i \neq 0 \\ \qquad [x'] \cap \sum\limits_{i \leq n} {}_\nu[v_{\alpha_i}] = \emptyset\} & \text{if } \|x\| \text{ is a limit.} \end{cases}$$

We used ${}_\nu X = \{x \in X, \|x\| > \nu\}$ ($\nu \in cf(\lambda)$). The choice of in (+) is always possible, because x is a linear combination of the generators $g^k_\alpha$. And the existence of the above sequences follows immediately from our assumptions $cf(\lambda) > \aleph_0$, (from $\lambda^{\aleph_0} = \lambda$), $\|B\varphi\| = cf(\lambda)$ and $\bigcap\limits_{s \in S} As = 0$.

Consider the element $f = \sum\limits_{i=1}^{\infty} b_i s_i \in \hat{B}$. Since $\varphi$ is inessential we have $f\varphi \in G$. A support argument will give the desired contradiction. Since $\alpha_n \in V''$, we derive $\alpha^* = \sup_{n\in\omega} \alpha_n \in V''$ and obviously $\alpha^* = \sup_{n\in\omega} \|\sigma_n\|$. From the definition of $\alpha^*$ and $f\varphi = \sum_{i=1}^{\infty} b_i \varphi s_i$ we derive $\|f\varphi\| \leq \alpha^*$. Since $\|f\varphi\| \in V \cup V'$ and $\alpha^* \in V''$, $(V \cup V') \cap V'' = \emptyset$, we have

(*) $\|f\varphi\| < \alpha^*$.

In order to contradict (*), it is enough to show $\sigma_n \in [f\varphi]$. This can be seen as follows

$$f\varphi \restriction \sigma_n = \left(\sum_{i=1}^{\infty} b_i s_i \varphi\right) \restriction \sigma_n = \sum_{i=1}^{\infty} (b_i \varphi \restriction \sigma_n) s_n$$

$$= (b_n \varphi \restriction \sigma_n) s_n + s_{n+1} \left( \sum_{i=n+1}^{\infty} (b_i \frac{s_i}{s_{n+1}}) \varphi \restriction \sigma_n \right)$$

$$= (b_n \varphi \restriction \sigma_n) s_n \not\equiv 0 \mod A\, s_{n+1}.$$

Hence $f\varphi \restriction \sigma_n \neq 0$ and $\sigma_n \in [f\varphi]$.

<u>Lemma 3.2</u>: If $\varphi \in$ Ines $G$, then $\hat{B}\varphi = G^{\emptyset}\varphi$.

<u>Proof</u>: (The proof is similar to [GS], Theorem 8.1, but much shorter). If $f \in \hat{B}$, we want to find some element $g \in G^{\emptyset}$ such that $g\varphi = f\varphi$. Immediately we apply Lemma 3.1 and the fact that there are no useless ordinals to the Black Box. Hence we obtain a <u>complete</u> ordinal $\alpha \in \lambda^*_c$ such that $(f_\alpha, a_\alpha, P_\alpha, \varphi_\alpha, r_\alpha) = (f_\alpha, f, P_\alpha, \varphi \restriction P_\alpha, \lambda)$.

Since $\alpha \in \emptyset^{\#}$, we have $g_\alpha \in G^{\emptyset}$ by definition of $G^{\emptyset}$. In view of the Black Box (v)(b) we derive $g_\alpha = a_\alpha + \sum_{n\in\omega} a_{\alpha_n} q_n = f + \sum_{n\in\omega} a_{\alpha_n} q_n$ and $g_\alpha \varphi = g_\alpha \varphi^\alpha = a_\alpha \varphi^\alpha + \sum_{n\in\omega} (a_{\alpha_n} \varphi^\alpha) q_n = f\varphi$. □

The last lemma can be used to derive the

Observation 3.3: If $J, J_1 \subseteq \lambda$ , then

$$\mathrm{Hom}(G^{J_1}, G^J) \cap \mathrm{Ines}\ G = \mathrm{Hom}(G^{J_1}, G^J) \cap \mathrm{Ines}\ G^J .$$

Proof: The right-hand side is always contained in the left-hand side; therefore let $\varphi \in \mathrm{Hom}(G^{J_1}, G^J) \cap \mathrm{Ines}\ G$ . From (3.2) we have $G^{\emptyset}\varphi = \hat{B}\varphi$ and since $G^{\emptyset} \subseteq G^{J_1}$ also $B\varphi = G^{\emptyset}\varphi \subseteq G^{J_1}\varphi \subseteq G^J$. Therefore $\varphi \in \mathrm{Ines}\ G^J$ and the above equality holds. □

Remark: It was this equation (3.3) which was derived from the unpleasant extracondition on inessential homomorphisms. Now the corollary and the theorem follow from [CG] without further assumptions. Observe that all proofs remain unchanged.

## REFERENCES

[C] A.L.S. Corner, Every countable reduced torsion free ring is an endomorphism ring, Proc. Lond. Math. Soc. (3) 13 (1963), 687-710

[CG] A.L.S. Corner, R.Göbel, Prescribing endomorphism algebras - A unified treatment, to appear in Proc. Lond. Math. Soc.

[DG] M. Dugas, R. Göbel, Every cotorsion-free ring is an endomorphism ring, Proc. Lond. Math. Soc. (3) 45 (1982), 319-336

[GS] R. Göbel, S. Shelah, Modules Over Arbitrary Domains II, to appear

[HH] J.T. Hallett and K.A. Hirsch, Torsion-free groups having finite automorphism groups, Journ. Algebra 2 (1965), 287-298.

[K] The Kourovka Notebook, Unsolved Problems in Group Theory, AMS Translations 121 (1983)

[S] S. Shelah, this volume.

The author would like to thank the Deutsche Forschungsgemeinschaft and MINERVA for financial support.

# PURE SUBGROUPS OF BUTLER GROUPS

Ladislav BICAN

MFF UK, Sokolovská 83

Praha 8-Karlín

Czechoslovakia

In [5] the class of Butler groups of arbitrary rank was introduced. Recently, in [6], it was shown that the class of Butler groups with ordered type set is closed under countable pure subgroups, while the general case remained open. In this note we shall show that the answer to the last question is negative, while the class of Butler groups with inversely well-ordered type set is closed under arbitrary pure subgroups. Moreover, we extend [5;Proposition 3.5] to arbitrary homogeneous groups by showing that a homogeneous torsionfree group is Butler if and only if it is completely decomposable. The main tool in this direction is a slight modification of Griffith's proof [9] of the freeness of Baer's groups (as presented in [8]).

By the "word" group we shall always mean an additively written abelian group. If $p$ is a prime and $g$ is an element of a group $H$ then $h_p^H(g)$ denotes the p-height of $g$ in $H$. Recall [5], that a torsionfree group $H$ is called a Butler group if $\mathrm{Bext}(H, T) = 0$ for all torsion groups $T$, where Bext is the subfunctor of Ext consisting of the equivalence classes of the balanced exact sequences, as defined by Hunter in [10]. Note, that a subgroup $K$ of a torsionfree group $H$ is said to be full if the quotient $H/K$ is a torsion group. A subgroup $K$ of a torsionfree group $H$ is called regular if every element of $K$ has in $K$ the same type as in $H$. For all other unexplained notations and terminology we refer to [8].

Lemma 1: There is an uncountable system $\{M_\lambda \mid \lambda \in \Lambda\}$ of infinite subsets of the set $\underline{N}$ of all positive integers such that $|M_\lambda \cap M_\mu| < \infty$ whenever $\lambda \neq \mu$.

Proof: By Zorn's lemma, among the systems with the desired property there is a maximal one. Assuming it countable, $\mathcal{M} = \{M_i \mid i \in \underline{N}\}$, select inductively the elements $x_i \in M_i \setminus \bigcup_{j=1}^{i-1} M_j$ and set $M = \{x_1, x_2, \ldots\}$. Then $M \cap M_i \subset \{x_1, x_2, \ldots, x_i\}$ and $M \notin \mathcal{M}$, which contradicts the maximality of $\mathcal{M}$.

Lemma 2: Let $H = \bigoplus_{\alpha \in A} J_\alpha$ be a completely decomposable torsionfree group, where $J_\alpha$, $\alpha \in A$, is of rank one and of the type $\hat{\tau}_\alpha$. If $A$ is uncountable and for any $\alpha, \beta \in A$, $\alpha \neq \beta$, it is $\hat{\tau}_\alpha || \hat{\tau}_\beta$ and $\hat{\tau}_\alpha \cap \hat{\tau}_\beta = \hat{\tau}$, then $H$ contains a homogeneous pure subgroup of the type $\hat{\tau}$, which is not completely decomposable.

Proof: See [3;Lemma 12].

Theorem 1: The class of Butler groups is not closed under pure subgroups.

Proof: Take an uncountable system $\mathcal{M} = \{M_\alpha | \alpha \in A\}$ of infinite subsets of $\underline{N}$ having the property stated in Lemma 1. Let $P = \{p_1, p_2, \ldots\}$ be the set of all primes and for each $\alpha \in A$ take the characteristic $\tau_\alpha$ having 1 at all primes $p_i$ with $i \in M_\alpha$ and 0 elsewhere. Denoting $J_\alpha$ a rank one torsionfree group of the type $\hat{\tau}_\alpha$, Lemma 2 yields the existence of a pure subgroup $K$ of $\bigoplus_{\alpha \in A} J_\alpha$ which is not completely decomposable, but homogeneous of the type $Z$. By [5;Proposition 3.5] $K$ is not Butler.

Proposition 2: The class of Butler groups is closed under full regular subgroups.

Proof: Let $K$ be a full regular subgroup of a Butler group $H$. Considering the commutative diagram

$$\begin{array}{lccccccccc} E: & 0 & \longrightarrow & T & \longrightarrow & X & \longrightarrow & K & \longrightarrow & 0 \\ & & & \| & & \downarrow & & \downarrow & & \\ F: & 0 & \longrightarrow & T & \longrightarrow & Y & \longrightarrow & H & \longrightarrow & 0 \end{array}$$

with exact row, $T$ torsion and $E$ balanced, the full regularity of $K$ in $H$ easily gives the balancedness of $F$. Thus the splitting of $F$ induces that of $E$ and $K$ is therefore Butler.

Remark: The converse of the preceding Proposition do not hold in general. To see this it suffices to take the group $K$ from the proof of Theorem 1, every maximal free subgroup of which satisfies the conditions of Proposition 1 (also the group $Z^N$ can be considered).

Theorem 2: Let $H$ be a Butler group with ordered typeset and $K$ be its pure subgroup with inversely well-ordered type set. Then $K$ is Butler.

Proof: Let $\{\hat{\tau}_\alpha | \alpha<\delta$, $\delta$ an ordinal$\}$ be the type set of $K$ such that $\hat{\tau}_\alpha < \hat{\tau}_\beta$ whenever $\alpha > \beta$. Choose arbitrarily a basis $M'_1$ of $H^*(\hat{\tau}_1)$, add a basis $N_1$ of $K(\hat{\tau}_1)$ and the linearly independent set $M'_1 \cup N_1$ extend by $M''_1$ to a basis $M'_1 \cup M''_1 \cup N_1 = M_1 \cup N_1$ of $H(\hat{\tau}_1)$. Continuing by the transfinite induction, let $M_\alpha \cup N_\alpha$ be a basis of $H(\hat{\tau}_\alpha)$ such that $N_\alpha$ is that of $K(\hat{\tau}_\alpha)$. There is a linearly independent set $M'_{\alpha+1}$ such that $M_\alpha \cup M'_{\alpha+1} \cup N_\alpha$ is a basis of $H^*(\hat{\tau}_{\alpha+1})$. Extending $N_\alpha$

to a basis $N_{\alpha+1}$ of $K(\hat{\tau}_{\alpha+1})$ we can extend the linearly independent set $M_\alpha \cup M'_{\alpha+1} \cup N_{\alpha+1}$ by adding $M''_{\alpha+1}$ to a basis $M_{\alpha+1} \cup N_{\alpha+1}$ of $H'\hat{\tau}_{\alpha+1})$. If $\alpha$ is limit, then setting $\bar{M}_\alpha = \bigcup_{\beta<\alpha} M_\beta$ and $\bar{N}_\alpha = \bigcup_{\alpha<\beta} N_\beta$, the union $\bar{M}_\alpha \cup \bar{N}_\alpha$ can be extended by $M'_\alpha$ to a basis $\bar{M}_\alpha \cup M'_\alpha \cup \bar{N}_\alpha$ of $H^*(\hat{\tau}_\alpha)$. Extending $\bar{N}_\alpha$ to a basis $N_\alpha$ of $K(\hat{\tau}_\alpha)$, the linearly independent set $\bar{M}_\alpha \cup M'_\alpha \cup N_\alpha$ can be extended by $M''_\alpha$ to a basis $M_\alpha \cup N_\alpha$ of $H(\hat{\tau}_\alpha)$. Putting $M' = \bigcup_{\alpha<\delta} M_\alpha$, $N = \bigcup_{\alpha<\delta} N_\alpha$, we can extend $M' \cup N$ to a basis $M \cup N$ of $H$ such that $N$ is a basis of $K$.

Obviously, $L = \langle M \rangle_*$ is a K-high subgroup of $H$ and we proceed to show that $\hat{\tau}^H(k+l) = \hat{\tau}^H(k) \wedge \hat{\tau}^H(l)$ whenever $k \in K$ and $l \in L$ are non-zero elements. Since the type set of $H$ is ordered, this equality is easily verified in case that $k$ and $l$ are of different types. If $\hat{\tau}^H(k) = \hat{\tau}^H(l) = \hat{\tau}$, then necessarily $\hat{\tau} = \hat{\tau}_\alpha$ for some $\alpha<\delta$ and so $k$ linearly depends on $N_\alpha$ and it has at least one non-zero component at $N_\alpha \setminus \bigcup_{\beta<\alpha} N_\beta$. Similarly, the element $l$ linearly depends on $M_\alpha$ and it has at least one non-zero component at $M''_\alpha$. Assuming that $\hat{\tau}^H(k+l) > \hat{\tau}_\alpha$, we have $k+l \in H^*(\hat{\tau}_\alpha)$ and consequently, by the above construction, $k+l$ linearly depends on $M_\alpha \cup N_\alpha$ with zero component at $(N_\alpha \setminus \bigcup_{\beta<\alpha} N_\beta) \cup M''_\alpha$. The contradiction obtained proves the desired equality.

The subgroup $A = K \oplus L$ is obviously full in $H$. Moreover, for every $a \in A$, $a = k+l$, $k \in K$, $l \in L$, we have $\hat{\tau}^H(a) = \hat{\tau}^H(k) \wedge \hat{\tau}^H(l) =$

$= \hat{\tau}^K(k) \wedge \hat{\tau}^L(l) \leqslant \hat{\tau}^A(k) \wedge \hat{\tau}^A(l) = \hat{\tau}^A(a) \leqslant \hat{\tau}^H(a)$ and $A$ is regular in $H$.

By Proposition 1, $A$ is Butler, and consequently, $K$ is so.

Corollary 1: The class of Butler groups with inversely well-ordered type set is closed under pure subgroups.

Now we proceed to show that homogeneous Butler groups are completely decomposable. In the rest of the paper for a fixed type $\hat{\tau}$ we use the notations $P_1 = \{p \in P \mid \tau(p) = \infty\}$ and $P_2 = P \setminus P_1$.

Lemma 3: There is a countable mixed group $N$ with $P_2$-primary torsion part $T$ such that $N/T \simeq Q$ and every element of $N$ of infinite order is of the type $\hat{\tau}$.

Proof: For each prime $p \in P_2$ denote $\tau(p) = k_p$ and for each $n \in \underline{N}$ take a cyclic group $\langle a_{pn} \rangle \simeq Z(p^{k_p+2n-1})$. Setting $T_p = \bigoplus_{n=1}^{\infty} \langle a_{pn} \rangle$ and $U_p = \prod_{n=1}^{\infty} \langle a_{pn} \rangle$, the element $u_p = (p^{k_p} a_1, p^{k_p+1} a_2, \dots, p^{k_p+1} a_n, \dots) \in U_p$ is of infinite order and has the p-height equal to $k_p$.

Now, for $T = \bigoplus_{p \in P_2} T_p$ and $U = \prod_{p \in P_2} U_p$ the element $u + T$, where $u = (u_p)_{p \in P_2}$, is obviously of infinite p-height for each prime $p \in P$, and consequently, the factor--group $U/T$ contains a subgroup $N/T \simeq Q$ containing $u + T$.

Obviously, the element $u \in N$ is of characteristic $\tau$. If $a \in N$

is an arbitrary element of infinite order, then $ma = nu$ for some non-zero integers $m, n$ and $a$ is of the type $\hat{\tau}$, $T$ being $P_2$-primary.

Lemma 4: Let $M_1$ be a $P_1$-divisible mixed group of finite rank with the $P_2$-primary torsion part $T_1$ and $M_1/T_1$ divisible such that every element of $M_1$ of infinite order is of the type $\hat{\tau}$ and every torsion-free subgroup of $M_1$ homogeneous of the type $\hat{\tau}$ is completely decomposable. If $M_2$ is the group from the preceding Lemma, then every torsionfree subgroup of $M = M_1 \oplus M_2$ homogeneous of the type $\hat{\tau}$ is completely decomposable.

Proof: Denote $M/T = M_1/T_1 \oplus M_2/T_2$, $T = T_1 \oplus T_2$ and let $\theta_i : M_i \longrightarrow M_i/T_i$, $\pi_i : M \longrightarrow M_i$, $i = 1, 2$, $\theta: M \longrightarrow M/T$ be the canonical projections. For a torsionfree subgroup $F$ of $M$ homogeneous of the type $\hat{\tau}$ set $H = \{a \in F \mid \pi_2 a \in T_2\} = \{a \in F \mid \theta a \in \theta_1 M_1\}$. For $a \in H \cap \operatorname{Ker} \pi_1$ we have $\pi_2 a \in T_2$, $\pi_1 a = 0$, hence $a \in T \cap F = 0$ and consequently $H \cap \operatorname{Ker} \pi_1 = H \cap M_2 = 0$. Thus the restriction $\pi_1 | H$ is monic and so $\pi_1 H \subseteq M_1$ is torsionfree. The mapping $\phi: F \longrightarrow M_2/T_2$ given by $\phi(m_1 + m_2) = m_2 + T_2$ has the kernel $H$ and so, if non-zero, $F/H$ is a rank one torsionfree group. Thus $H$, being pure in $F$, is homogeneous of the type $\hat{\tau}$ and so it is completely decomposable as an isomorphic copy of $\pi_1 H \subset M_1$.

To finish the proof it remains now to show that for $F/H \neq 0$ it is $\hat{\tau}(F/H) = \hat{\tau}$, since in this case the well-known Baer's lemma [8;Proposition 86.5] gives that $F \simeq H \oplus F/H$ is completely decomposable.

So, let $H = \bigoplus_{i=1}^{k} \langle u_i \rangle_*^H$ and $a = b + c \in F \setminus H$, $b \in M_1$, $c \in M_2$, be an element. It is easy to see that it can be supposed that $u_1, u_2, \ldots, u_k \in M_1$ and $h_p^F(a) = h_p^F(u_1) = \ldots = h_p^F(u_k) = k_p$ for all primes $p$. If $u = v + t \in H$, $v \in M_1$, $t \in M_2$, is arbitrary, then $mu = \sum_{i=1}^{k} \lambda_i u_i$ for suitable integers with $(m, \lambda_1, \lambda_2, \ldots, \lambda_k) = 1$ and $mt = mu - mv \in M_1 \cap M_2 = 0$. So, if $(p, m) = 1$ then $h_p^M(t) = \infty$ and from $a + u = (b + v) + (c + t)$ it immediately follows $h_p^F(a+u) \leq h_p^M(a+u) \leq h_p^M(c)$. Finally, if $p|m$ then $(p, \lambda_j) = 1$ for some $j \in \{1, 2, \ldots, k\}$ and consequently $h_p^F(\sum_{i=1}^{k} \lambda_i u_i) = k_p = h_p^F(mu)$. Thus $h_p^F(u) < k_p$ and again $h_p^F(a+u) = h_p^F(u) < k_p \leq h_p^M(c)$.

<u>Lemma 5:</u> Let $M_i$, $i \in I$, be groups of the form of Lemma 3. Then every torsionfree subgroup of $M = \bigoplus_{i \in I} M_i$ homogeneous of the type $\hat{\tau}$ is completely decomposable.

<u>Proof:</u> In view of the preceding Lemma the proof runs on the same lines as that of Lemma 101.4 in [8] (use [8;Theorem 98.2] for the $\aleph_1$-complete decomposability of $F$ and the projectivity of free groups replace by Baer's lemma [8;Theorem 86.5]).

<u>Proposition 2:</u> For every type $\hat{\tau}$ and every cardinal number $\mathfrak{m}$ there is a mixed group M with the $P_2$-primary torsion part T such that:

(i) M/T is a divisible group of rank $\mathfrak{m}$;

(ii) every element of M of infinite order is of the type $\hat{\tau}$;

(iii) every torsionfree subgroup of M homogeneous of the type $\hat{\tau}$ is completely decomposable.

<u>Proof:</u> It follows immediately from Lemmas 5 and 3.

<u>Theorem 3:</u> The following conditions are equivalent for a homogeneous torsionfree group H of the type $\hat{\tau}$:

(i) H is Butler;

(ii) H is completely decomposable;

(iii) Bext(H, T) = O for every $P_2$-primary group T.

<u>Proof:</u> (i) ⇒ (ii). Choose a suitable group M from Proposition 2 and an embedding $\phi: H \longrightarrow M/T$. Considering the commutative diagram

$$\begin{array}{rccccccccc} E: & 0 & \longrightarrow & T & \longrightarrow & X & \xrightarrow{\lambda} & H & \longrightarrow & 0 \\ & & & \Vert & & \downarrow{\scriptstyle\rho} & & \downarrow{\scriptstyle\phi} & & \\ & 0 & \longrightarrow & T & \longrightarrow & M & \xrightarrow{\pi} & M/T & \longrightarrow & 0 \end{array}$$

with exact rows, we know that the right square is a pullback diagram. Now it is easy to see that the sequence E is balanced and hence there is the splitting map $\mu: H \longrightarrow X$ Then $\pi\rho\mu = \phi\lambda\mu = \phi$, $\rho\mu: H \longrightarrow M$ is monic and H is completely decomposable by Proposition 2.

(ii) ⇒ (iii). Obvious.

(iii) ⇒ (i). If $E: 0 \longrightarrow T \longrightarrow X \longrightarrow H \longrightarrow 0$ is any balanced exact sequence with T torsion then the exact sequences $E_i : 0 \longrightarrow T \otimes Z_{p_i} \longrightarrow$ $\longrightarrow X \otimes Z_{p_i} \longrightarrow H \otimes Z_{p_i} \longrightarrow 0$, $i = 1, 2$, are obviously balanced. Moreover, $H \otimes Z_{p_1}$ is divisible and so $E_1$ splits by [2]. Further, $T \otimes Z_{p_2}$ is $p_2$-primary, $H \otimes Z_{p_2} \simeq H$ and $E_2$ splits by the hypothesis. Now it easily follows from the main result of [4] that E splits and H is therefore Butler.

## References


[1] L. Bican: Mixed abelian groups of torsionfree rank one, Czech. Math. J. 20 (95), (1970), 232-242.

[2] L. Bican: A note on mixed abelian groups, Czech. Math. J. 21 (96), (1971), 413-417.

[3] L. Bican: Completely decomposable abelian groups any pure subgroup of which is completely decomposable, Czech. Math. J. 24 (99), (1974), 176-191.

[4] L. Bican: A splitting criterion for abelian groups, Comment. Math. Univ. Carolinae 19, (1978), 763-774.

[5] L. Bican, L. Salce: Butler groups of infinite rank, Proc. Honolulu, Lecture Notes n. 1006, 171-189, Springer Verlag, Berlin 1983.

[6] L. Bican, L. Salce, J. Štěpán: A characterization of countable Butler groups (to appear).

[7] M. C. R. Butler: A class of torsionfree abelian groups of finite rank, Proc. London Math. Soc. 15, (1965), 680-698.

[8] L. Fuchs: Infinite abelian groups, Vol. I and II, Academic Press London, New York, 1971 and 1973.

[9] P. Griffith: A solution to the splitting mixed group problem of Baer, Trans. Amer. Math. Soc. 139, (1969), 261-270.

[10] R. H. Hunter: Balanced sequences of abelian groups, Trans. Amer. Math. Soc. 215, (1976), 81-98.

# ON COSEPARABLE COMPLETELY DECOMPOSABLE TORSIONFREE ABELIAN GROUPS

CLAUDIA METELLI

UNIVERSITA' DI PADOVA (*)

All groups in this Note are abelian.

Let G be a torsionfree group, L a pure subgroup of G. Following [1], we say L is <u>severable</u> in G if $G = X \oplus H \geq L \geq H$, where X is completely decomposable of finite rank. G is <u>coseparable</u> if all pure subgroups L of G such that G/ L is reduced of finite rank are severable in G.

In [1], the class of torsionfree coseparable groups is proved to be closed with respect to direct products and direct summands, but non closed with respect to infinite direct sums. Since torsionfree groups of rank 1 are coseparable, it is natural to investigate which (reduced) completely decomposable torsionfree groups are coseparable; this is the purpose of this Note.

(*) Lavoro eseguito nell'àmbito dei Gruppi di Ricerca CNR-GNSAGA.

We recall that a group G is <u>1-coseparable</u> if all pure subgroups U such that G/U is reduced of rank 1 are severable in G, and that the class of 1-coseparable groups properly contains the class of coseparable groups ([1]).

Terminology and notation will follow [F II]. Moreover, for i belonging to an index set, $n \in \mathbb{N} \cup \{0\}$, $k_{in} \in \mathbb{N} \cup \{0, \infty\}$, we will denote by $R_i = (k_{i1}, k_{i2}, \ldots)$ the <u>proper</u> subgroup of $\mathbb{Q}$ containing $\mathbb{Z}$ for which $\chi_{R_i}(1) = (k_{i1}, k_{i2}, \ldots)$. We recall that a subgroup of $\mathbb{Q}$ is a proper maximal subring of $\mathbb{Q}$ if it is divisible by all primes but one, say p; we will denote it by $\mathbb{Q}_p$.

<u>Lemma</u>. Let $G = \bigoplus_{i \in I} R_i x_i$, and for $n \in \mathbb{N}$ $I_n = \{i \in I \mid k_{in} < \infty\}$. If for some $m, n \in \mathbb{N}$, $m \neq n$, the set $I_m \cap I_n$ is infinite, G is not 1-coseparable, hence not coseparable.

<u>Proof</u>. Since by [1] summands of 1-coseparable groups are 1-coseparable, we need only show that G has a summand which is not 1-coseparable; thus w.l.o.g. let $I = I_m = I_n = \mathbb{N}$; moreover, let $k_{im} = 0$, $k_{in} = i$ $\forall i \in I$. For $f: G \to \mathbb{Q}_{p_m}$, $x_i \mapsto 1$ $\forall i \in I$, we have $f(G) = (k_1, k_2, \ldots)$ with $k_n = \infty$, but no finite rank summand of G can be mapped onto $f(G)$. Thus $U = \operatorname{Ker} f$ is not severable in G, hence G is not 1-coseparable.

<u>Proposition</u>. Let $G = \bigoplus_{i \in I} R_i$ be reduced homogeneous. Then G is coseparable iff I is finite; G is 1-coseparable iff either I is finite or $R_i \cong \mathbb{Q}_p$, a maximal subring of $\mathbb{Q}$.

<u>Proof</u>. Sufficiency is obvious in the first case; as for the

second, if G is $t(Q_p)$-homogeneous and for $U \leq_* G$ G/U is rank 1 reduced, it must be isomorphic to $Q_p$, thus U itself splits G.

To prove necessity, let I be infinite. If $R_i$ is not isomorphic to a maximal subring of $\mathbb{Q}$, for some $m,n \in \mathbb{N}$, $m \neq n$, we have $I = I_m = I_n$, thus $I_m \cap I_n$ is infinite, and by the Lemma G is not 1-coseparable. If $R_i \cong Q_p$ for some prime number p, being a free $Q_p$-module of infinite rank, G has a quotient $A \cong G/L$ which is an irreducible $Q_p$-module of rank, say, 2 (see e.g. Ex.5 p.125 of [F II], with the obvious adaptations). Now L cannot be severable in G, for otherwise we would have $A \cong G/L \cong X/(L \cap X)$, with X a free $Q_p$-module of finite rank, which is impossible. Therefore G is in this case 1-coseparable but not coseparable.

<u>Corollary</u>. Let $G = \bigoplus_{i \in I} R_i$ be reduced coseparable. Then I is countable.

<u>Proof</u>. From the Lemma we know that for $n,m \in \mathbb{N}$, $n \neq m$, the set $I'_n = \cup_m (I_m \cap I_n)$ is countable. By the Proposition, the set $I_n \setminus I'_n = \{i \in I_n \mid R_i \cong Q_{p_n}\}$ is finite; thus $I_n$ is countable, and, G being reduced, $I = \cup_{n \in \mathbb{N}} I_n$ is countable.

<u>Theorem</u>. The reduced group $G = \bigoplus_{i \in I} R_i x_i$ is coseparable iff $\forall n \in \mathbb{N}$ the set $I_n = \{i \in I \mid k_{in} < \infty\}$ is finite.

<u>Proof</u>. To prove sufficiency, let f be a surjection of G onto a finite rank reduced t.f. group A, L = Ker f. Given the sequence of types $t_1 = t((\infty, 0, 0, 0, \ldots))$, $t_2 = t((\infty, \infty, 0, 0, \ldots))$, $t_3 = t((\infty, \infty, \infty, 0, \ldots))$... the chain $A(t_1) \geq A(t_2) \geq A(t_3) \geq \ldots$ must vanish in a finite number, say r,

of steps. Since for $i\in I' = I\setminus(I_1\cup I_2\cup\ldots\cup I_r)$ we have $R_ix_i \leq G(t_r)$, thus $f(R_ix_i) \leq f(G(t_r)) \leq A(t_r) = 0$, there follows that $L = \mathrm{Ker}\, f \geq \bigoplus_{i\in I'} R_ix_i$ is severable in G.

To show necessity, let $I_n$ be infinite for some $n\in\mathbb{N}$; w.l.o.g. $n = 1$, and $I = I_1$, since by [1] we only need to show that G has a non (1-)coseparable summand. By choosing $k_{i1} = 0$ $\forall i\in I$, we ensure that $f: G \to Q_{p_1}$, $x_i \mapsto 1$, defines a homomorphism, and $U = \mathrm{Ker}\, f$ is such that $G/U$ is rank 1 reduced.

In our situation G cannot contain a homogeneous summand of infinite rank; thus $T(G) = \{t(R_i) \mid i\in I\}$ is infinite. Let T(G) contain - w.l.o.g. consist of - an infinite set of 2 by 2 non comparable types (= a horizontal chain). Then G contains no rank 1 summand except for the $R_ix_i$'s, and $f(x_i) \neq 0$ $\forall i\in I$ ensures that $U = \mathrm{Ker}\, f$ contains no rank 1 summand of G, hence is not severable in G; therefore G is not (1-)coseparable. If T(G) contains no infinite horizontal chain, it must contain - w.l.o.g. consist of - an infinite chain of types, and we may suppose them to be all strictly smaller than $t(Q_{p_1})$. If $\mathrm{Im}\, f = R \lneq Q_{p_1}$, for some $n\in\mathbb{N}$, $n \neq 1$ we would have $I_n = I_1 = I$ infinite, and by the Lemma G would not be (1-)coseparable. If $\mathrm{Im}\, f = Q_{p_1}$, $U = \mathrm{Ker}\, f$ cannot be severable: for otherwise $Q_{p_1} \simeq G/U \simeq X/(U\cap X)$, where X is finite rank completely decomposable, with T(X) a finite chain of types strictly smaller than $t(Q_{p_1})$, which is impossible. Thus again G is not (1-)coseparable, and this exhausts all possibilities.

<u>Corollary</u>. Let $G' = \bigoplus_{i\in K} R_i$ be reduced, $I = \{i\in K \mid R_i$ is not isomor-

phic to a maximal subring of $\mathbb{Q}\}$, $G = \bigoplus_{i\in I} R_i \leq G'$. Then G' is 1-coseparable iff G is coseparable.

Proof. Let $J = K\setminus I$, $G_\circ = \bigoplus_{i\in J} R_i$, so $G' = G_\circ \oplus G$. To prove sufficiency, we only need to show that $G_\circ$ is 1-coseparable. Now if $f: G_\circ \to R$ is a nonzero homomorphism of $G_\circ$ on a proper subgroup R of $\mathbb{Q}$, R must be isomorphic to a maximal subring $\mathbb{Q}_p$ of $\mathbb{Q}$, hence Ker f will contain all $R_i$'s non isomorphic to $\mathbb{Q}_p$; thus we fall back into the homogeneous case, which was proved in the Proposition. To prove necessity, we note that, since $G_\circ$ is 1-coseparable, w.l.o.g. we may suppose $G' = G$; and now the proof of necessity in the above Theorem applies verbatim to our case.

Observations. 1. Coseparable completely decomposable (t.f. abelian) groups, like separable vector groups, are particular instances of "biseparable" groups, i.e. t.f. abelian groups which are at the same time separable and coseparable; a class that might deserve a closer look.

2. The classical dual to the problem treated in this Note - i. e. the characterization of separable vector groups - is, in fact, still open. The characterizations given in 1961-62 by M. Król [2] and A.P.Mišina[3] are incorrect, as the following example shows: let $V = \prod_{n\in\mathbb{N}} R_n x_n$, where the $R_n$'s are defined inductively as follows:

$$R_1 = (1,0,1,0,1,0,1,0,1,0,1,0,1,0,1,0,\ldots)$$
$$R_2 = (\infty,1,\infty,0,\infty,1,\infty,0,\infty,1,\infty,0,\infty,1,\infty,0,\ldots)$$
$$R_3 = (\infty,\infty,\infty,1,\infty,\infty,\infty,0,\infty,\infty,\infty,1,\infty,\infty,\infty,0,\ldots)$$
$$R_4 = (\infty,\infty,\infty,\infty,\infty,\infty,\infty,1,\infty,\infty,\infty,\infty,\infty,\infty,\infty,0,\ldots)$$
..............................................................

Now although $T(V)$ is an infinite chain of non reduced types, $V$ is separable. In fact, let $v = (r_n x_n)_{n\in\mathbb{N}}$, $r_n \in R_n$, and w.l.o.g. $r_1 \neq 0$. Then $\chi(r_n x_n) \geq \chi(x_1)$ $\forall n>1$, and if $p_m$ is the maximum prime occurring in $r_1$ with a positive exponent, $\chi(r_n x_n) \geq \chi(r_1 x_1)$ $\forall n > m$. Let $v' = (s_n x_n)_{n\in\mathbb{N}}$, where $s_n = 0$ for $n \leq m$, $s_n = r_n$ for $n > m$. Then $\chi(r_1^{-1} v') \geq \chi(x_1)$, thus $x_1' = x_1 + r_1^{-1} v'$ is substitutable to $x_1$ in the given decomposition of $V$ (see e.g. Lemma 0 in [1]); therefore $v = r_1 x_1' + r_2 x_2 + \ldots + r_m x_m$ is separable in $V$, q.e.d.

## REFERENCES

[F II] - Fuchs, L. "Infinite Abelian Groups" Vol.II, Academic Press (1973).

[1] - Metelli, C. "Coseparable torsionfree Abelian Groups" (to appear).

[2] - Król, M. "Separable Groups" I, Bull. Acad. Polon. Sci. 9 (1961).

[3] Mišhina, A.P. "Separability of Complete Direct Sums of Torsion-free Abelian Groups of Rank 1", Mat. Sb. 57 (1962).

# THE NON-SLENDER RANK OF AN ABELIAN GROUP

Burkhard Wald

**Freie Universität Berlin**

For a family $(A_i)_{i\in I}$ of Abelian groups and a cardinal $\kappa$ we define the $\kappa$-product $\prod_{i\in I}^{(\kappa)} A_i$ to be the subgroup of the cartesian product $\prod_{i\in I} A_i$ consisting of all elements which support is less than $\kappa$. Let us write $A^{I(\kappa)}$ instead of $\prod_I^{(\kappa)} A$, $A^{(I)}$ instead of $A^{I(\omega)} = \bigoplus_I A$ and $A^{[I]}$ instead of $A^{I(\omega_1)}$. We are going to use the groups $Z^{[\kappa]}$ to introduce a new cardinal invariant for an abelian group.

Let G be an Abelian group. We say that G has <u>non-slender rank</u>, if there is a cardinal $\kappa$ and an embedding of $Z^{[\kappa]}$ into G but no embedding of $Z^{[\kappa^+]}$ into G. Such a $\kappa$ is unique determinate by G and we can define $nsr(G) = \kappa$. For an arbitrary Abelian group G let $nsr^+(G)$ be the smallest cardinal $\kappa$ for which there is no embedding of $Z^{[\kappa]}$ into G. Obviously, G has non-slender rank, if and only if $nsr^+(G)$ is a successor cardinal. In this case $nsr^+(G) = nsr(G)^+$. Later on, we see there is a group without non-slender rank.

1. Our first observation is that the non-slender rank is limited by the torsionfree rank $r_o$ of the group.

<u>Proposition</u>: Let G be an Abelian group

a) $nsr^+(G)$ is finite if and only if $r_o(G)$ is finite. Is this the case, G has non-slender rank and $nsr(G) = r_o(G)$

b) $r_o(G) < 2^\omega$ implies $nsr^+(G) \leq \omega$

c) $r_o(G) \geq 2^\omega$ implies $\sup\{\kappa^\omega : \kappa < nsr^+(G)\} \leq r_o(G)$

d) $nsr^+(G) \leq r_o(G)^+$.

e) Either $nsr^+(G) \leq r_o(G)$ or G has non-slender rank and $nsr(G) = r_o(G) = r_o(G)^\omega$.

Proof. a) is obvious. For b) notice that the size of $Z^\omega$ is $2^\omega$. In general, the size of $\mathbb{Z}^{[\kappa]}$ is $\kappa^\omega$ and hence $r_o(\mathbb{Z}^{[\kappa]}) = \kappa^\omega$, which implies c). d) and e) follows from a), b) and c).

We list a few facts about the computation of $\kappa^\omega$. $\omega^\omega = 2^\omega$, $(2^\kappa)^\omega = 2^\kappa$ and hence $\kappa^\omega = 2^\omega$ for $\kappa \leq 2^\omega$. $\omega = cf\kappa$ implies $\kappa < \kappa^\omega$ and for every $\mu$ between $\kappa$ and $\kappa^\omega$ we have $\mu^\omega = \kappa^\omega$. Furthermore every cardinal $\mu$ with $\mu < \mu^\omega$ lies between some $\kappa$ and $\kappa^\omega$ where $\omega = cf\kappa$; cf. T. Jech [J], §6. Under special set-theoretical assumptions, GCH (the Generalized Continuum Hypothesis) or "$0^\#$ does not exist", the arithmetic is much easier: For $\kappa > 2^\omega$ we have $\kappa^\omega = \kappa^+$ in case $cf\kappa = \omega$ and $\kappa^\omega = \kappa$ otherwise; cf. T. Jech [J], p. 357.

2. In order to prove theorems like $nsr^+(G_1 \oplus G_2) = nsr^+(G_1) + nsr^+(G_2)$ we restrict our investigation to the class of cotorsionfree groups. This is the class of all torsionfree, reduced (no divisible part) groups which contains no copy of the group p-adic integers for some prime p as a subgroup. By a theorem of R.J. Nunke [N], a group G is slender if and only if G is cotorsionfree and $nsr^+(G) \leq \omega$. Recall that G is called slender, if for every homomorphism from $\mathbb{Z}^\omega$ into G almost all of the elements $e_i$ are mapped onto zero; cf [F], §94. For $i \in I$ we denote by $e_i$ the element of $\mathbb{Z}^I$ defined by $e_i(j) = \delta_{ij}$, where $\delta_{ij}$ is the Kronecker symbol. The following theorem will be the essential tool of this paper.

Theorem: For an Abelian group and an infinite cardinal $\kappa$ the following is equivalent:

a) G is cotorsionfree and $nsr^+(G) \leq \kappa$ (i.e. there is no embedding of $\mathbb{Z}^{[\kappa]}$ into G).

b) For every homomorphism $\varphi$ from $\mathbb{Z}^{[\kappa]}$ into G the set $E = \{i \in \kappa : \varphi(e_i) \neq 0\}$ has size less than $\kappa$.

In the case of b) in addition $\varphi(\mathbb{Z}^{[\kappa \setminus E]}) = 0$.

Proof: Assuming b), $nsr^+(G) \leq \kappa$ is obvious. To prove that G is cotorsion-free, we have to show that G contains no algebraically compact subgroup $K \neq 0$; cf. [GW], Folgerung 4.2. If $0 \neq x \in K$ we choose a homomorphism $\varphi$ from the free group $\mathbb{Z}^{(\kappa)}$ to K such that $\varphi(e_i) = x \neq 0$ for all $i \in \kappa$. Because K is algebraically compact, $\varphi$ extends to a homomorphism from $\mathbb{Z}^{[\kappa]}$ to K. This contradicts b). By the well-known theorem of Balcerzyk [B] $\mathbb{Z}^{[\kappa\setminus E]}/\mathbb{Z}^{(\kappa\setminus E)}$ is algebraically compact and since G is cotorsionfree, $\varphi(\mathbb{Z}^{[\kappa\setminus E]}) = 0$. Now we come to the proof that a) implies b). In [GWW] we have already shown b) under the hypothesis $|G| < \kappa$, cf. [GWW], Theorem 4.5. For that we used the following fact, which is also useful for our stronger result:

If $\Psi$ is a homomorphism from $\mathbb{Z}^{\omega}$ into a cotorsionfree group G, for every $x \in G\setminus\{0\}$ the set $\{n \in \omega : \varphi(e_n) = x\}$ is finite.

For the proof of b) in case $\kappa = \omega$ we refer to R.J. Nunke [N], so that we can assume $\kappa > \omega$. Let $\varphi$ be a homomorphism from $\mathbb{Z}^{[\kappa]}$ into a cotorsionfree group G, satisfying $|E| = \kappa$. We will construct an injective map $i : \kappa \longrightarrow E$ and a map $F : \kappa \longrightarrow P(\kappa)$ with the following properties.

1) $\forall \alpha \in \kappa$: $|F(\alpha)| \leq |\alpha| \cdot \omega$ and $\varphi(\mathbb{Z}^{[\kappa\setminus F(\alpha)]}) \cap \mathbb{Z}\varphi(e_{i(\alpha)}) = 0$

2) $\forall \alpha, \beta \in \kappa$, $\alpha < \beta$: $F(\alpha) \subseteq F(\beta)$ and $i(\beta) \notin F(\alpha)$.

Let us assume that $i(\alpha)$ and $F(\alpha)$ are defined for all $\alpha$ less than some $\beta \in \kappa$. If $D = \bigcup_{\alpha<\beta} F(\alpha)$, we obtain $|D| \leq \sum_{\alpha<\beta} |\alpha| \cdot \omega = |\beta| \cdot \omega$. Because $|E| = \kappa$ and $|D| < \kappa$, we can choose $i(\beta)$ in $E \setminus D$. Now let T be a subset of $\mathbb{Z}^{[\kappa]}$ which is maximal with the properties that the supports of the elements of T are pairwise disjoint and $\varphi(a) \in \mathbb{Z}\varphi(e_{i(\beta)}) \setminus \{0\}$ for all $a \in T$. In this situation the result mentioned above applies and therefore $|T| \leq |\mathbb{Z}| = \omega$. If we define $F(\beta) = D \cup \bigcup_{a\in T} \operatorname{supp}(a)$ where supp(a) is the support of a, $|F(\beta)| \leq |\beta| \cdot \omega + \omega = |\beta| \cdot \omega$ and by the maximality of T, consequently $\varphi(\mathbb{Z}^{[\kappa\setminus F(\beta)]}) \cap \mathbb{Z}\varphi(e_{i(\beta)}) = 0$. Now it is easy to see that $\varphi$, restricted on $\mathbb{Z}^{[K]}$, is injective if K is the image of the map i. If $a \in \mathbb{Z}^{[K]}$, $a \neq 0$ and $\alpha \in \kappa$ is minimal such that $a(i(\alpha)) \neq 0$, then $\varphi(a) = 0$ implies

$$0 \neq a(i(\alpha))\varphi(e_{i(\alpha)}) = -\varphi(a - a(i(\alpha))e_{i(\alpha)}) \in \mathbb{Z}\varphi(e_{i(\alpha)}) \cap \varphi(\mathbb{Z}^{[\kappa\setminus F(\alpha)]}$$

which is a contradiction.

3. Using the above theorem, we can prove the following proposition.

Proposition: For an infinite cardinal $\kappa$ and a finite n we define $\kappa - n = \kappa$.

a) Let $0 \longrightarrow A \xrightarrow{\alpha} B \xrightarrow{\beta} C \longrightarrow 0$ be an exact sequence of Abelian groups, where C is cotorsionfree, then

$$nsr^+(A) \leq nsr^+(B) \leq nsr^+(A) + nsr^+(C) - 1$$

b) Let A and C be Abelian groups, where one of them is cotorsionfree. Then

$$nsr^+(A \oplus C) = nsr^+(A) + nsr^+(C) - 1$$

c) Let n be a natural number and $A_1, \ldots, A_n$ be cotorsionfree groups. Then

$$nsr^+\left(\bigoplus_{i=1}^{n} A_i\right) = \sum_{i=1}^{n} nsr^+(A_i) - (n-1)$$

Proof: $nsr^+(A) \leq nsr^+(B)$ is trivial. Also the second inequation is clear if $nsr^+(A)$ and $nsr^+(C)$ are both finite. Is one of them infinite, let $\varphi$ be an embedding of $\mathbb{Z}^{[\kappa]}$ into B, where $\kappa = nsr^+(A) + nsr^+(C) - 1$. By the theorem in 2., we can find a subset E of $\kappa$ such that $|E| < \kappa$ and $(\beta \circ \varphi)(\mathbb{Z}^{[\kappa \setminus E]}) = 0$. This induces a homomorphism $\Psi$ of $\mathbb{Z}^{[\kappa \setminus E]}$ into A, satisfying $\alpha \circ \Psi = \varphi|_{\mathbb{Z}^{[\kappa \setminus E]}}$. Because $\Psi$ is an embedding, too, and $|\kappa \setminus E| = \kappa$, we get a contradiction to $nsr^+(A) \leq \kappa$. Hence $nsr^+(B) \leq \kappa$ is shown.

An embedding of $Z^{[\nu]}$ into A together with an embedding of $Z^{[\mu]}$ into C induces an embedding of $Z^{[\nu+\mu]}$ into $A \oplus C$. c) follows by induction.

Notice that if C is an epimorphic image of B, $nsr^+(C)$ does not stand in any general relation to $nsr^+(B)$. For example every group is an epimorphic image of a free group and for free groups $nsr^+$ is less or equal to $\omega$.

4. It is not difficult to see, that whenever $\varphi$ is a homomorphism from a group $\prod_{i \in I}^{(\omega_1)} A_i$ into a cotorsionfree group G there is a subset E of I such that $|E| < nsr^+(G)$ and $\varphi\left(\prod_{i \in I \setminus E}^{(\omega_1)} A_i\right) = 0$.

Now we are able to compute the non-slender rank of Cartesian products.

Theorem: Let $(G_i)_{i \in I}$ be an infinite family of cotorsionfree groups $\neq 0$ and G a group satisfying $\prod_{i\in I}^{(\omega_1)} G_i \subseteq G \subseteq \prod_{i\in I} G_i$ then $nsr^+(G)$ depends only on $|I|$ and the cardinals $nsr^+(G_i)$ $(i \in I)$.

a) If $nsr^+(G_i) \leq |I|^+$ for all $i \in I$ then G has non-slender rank and $nsr(G) = |I|$.

b) If there is $i_o \in I$ such that $|I| < nsr^+(G_{i_o})$ and $nsr^+(G_i) \leq nsr^+(G_{i_o})$, for any $i \in I$, then $nsr^+(G) = nsr^+(G_{i_o})$.

c) If $\{nsr^+(G_i) : i \in I\}$ has no maximum and $|I| < \sup\{nsr^+(G_i) : i \in I\}$, then G has non-slender rank and $nsr(G) = \sup\{nsr^+(G_i) : i \in I\}$.

<u>Proof</u>: Of course, there is an embedding of $\mathbb{Z}^{[I]}$ into G and hence $|I| < nsr^+(G)$. Furthermore, if $\varphi_i$ is an embedding of $\mathbb{Z}^{[M_i]}$ into $G_i$ where $M_i$ is a set of cardinality less than $nsr^+(G_i)$ and the family $(M_i)_{i\in I}$ is pairwise disjoint, then an embedding of $\mathbb{Z}^{[M]}$ into G is induced where $M = \dot{\bigcup}_{i\in I} M_i$. This implies that in case a) $|I|^+ \leq nsr^+(G)$, in case b) $nsr^+(G_{i_o}) \leq nsr^+(G)$ and in case c) $\sup\{nsr^+(G_i) : i \in I\} < nsr^+(G)$. Now let $\kappa$ be an arbitrary cardinal (large enough) and $\varphi$ a homomorphism from $\mathbb{Z}^{[\kappa]}$ into G. Then for every $i \in I$ there is a subset $E_i$ of $\kappa$ such that $|E_i| < nsr^+(G_i)$ and $(\pi_i \circ \varphi)(Z^{[\kappa \setminus E_i]}) = 0$, where $\pi_i$ is the projection of G onto $G_i$. We build the union $E = \bigcup_{i\in I} E_i$ and obtain $\varphi(Z^{[\kappa \setminus E]}) = 0$ and $|E| < |I| \cdot \sup\{|E_i| : i \in I\}$. This implies in case a) $|E| < |I|^+$ and hence $nsr^+(G) \leq |I|^+$, in case b) $|E| < nsr^+(G_{i_o})$ and hence $nsr^+(G) \leq nsr^+(G_{i_o})$ and in case c) $|E| \leq \sup\{nsr^+(G_i) : i \in I\}$ and hence $nsr^+(G) \leq \sup\{nsr^+(G_i) : i \in I\}^+$. This completes the computation of $nsr^+(G)$. Because in case a) and c) $nsr^+(G)$ is a successor cardinal, G must have non-slender rank and we get the noted formulas for $nsr(G)$.

5. To handle direct sums we use the following generalization of a result of S.U. Chase [Ch].

<u>Lemma</u>: Let $(A_i)_{i\in I}$ and $(G_j)_{j\in J}$ be two families of Abelian groups, where the groups $G_j$ are torsionfree and reduced. Then for every homomorphism $\varphi$ from $\prod_{i\in I}^{(\omega_1)} A_i$ into $\bigoplus_{i\in J} G_j$ there are finite sets $E \subseteq I$ and $F \subseteq J$, such that

$$\varphi\Big(\prod_{i\in I\setminus E}^{(\omega_1)} A_i\Big) \subseteq \bigoplus_{j\in F} G_j .$$

We want to apply the result of Chase, who proved the lemma just in the case $|I| = \omega$. There is another generalization due independently by M.Dugus and B.Zimmermann-Huisgen [DZ] and A.V.Ivanov [I]. They prove the same assertion for the full product $\prod_{i\in I} A_i$, where the cardinality of I is less than all measurable cardinals.

<u>Proof</u>: Let $E = \{i \in I: \forall F \subseteq J \text{ finite } \varphi(A_i) \not\subseteq \bigoplus_{j\in F} G_j\}$ and $F = \bigcup\{\operatorname{supp}\varphi(a): a \in \prod_{i\in I\setminus E}^{(\omega_1)} A_i\}$ where $\operatorname{supp}\varphi(a)$ is the support of $\varphi(a)$ in J. Applying the special case of the lemma proved by Chase, we obtain that E is finite. If F is infinite, there is a subset T of $\prod_{i\in I\setminus E}^{(\omega_1)} A_i$, satisfying $|T| = \omega$ and $|\bigcup_{a\in T} \operatorname{supp}\varphi(a)| = \omega$. Because there is $X \subseteq I$, $|X| = \omega$, such that $T \subseteq \prod_{i\in X} A_i$, the theorem of Chase applies once more, which leads to a contradiction. Thus, F has to be finite, too.

<u>Theorem</u>: Let $(G_j)_{j\in J}$ be an infinite family of cotorsionfree groups $\neq 0$. Then

$$\operatorname{nsr}^+\Big(\bigoplus_{j\in J} G_j\Big) = \sup\{\operatorname{nsr}^+(G_j):\ j \in J\}\cdot\omega$$

<u>Proof</u>: It is only to show that $\operatorname{nsr}^+(\bigoplus G_j)$ is less or equal to $\kappa = \sup\{\operatorname{nsr}^+(G_j):\ j \in J\}\cdot\omega$. If $\varphi$ is a homomorphism from $\mathbb{Z}^{[\kappa]}$ into $\bigoplus G_j$, by the above lemma we find a finite subset F of J such that $\varphi(\mathbb{Z}^{[\kappa]}) \subseteq \bigoplus_{j\in F} G_j$. But by 3.c) we have $\operatorname{nsr}^+(\bigoplus_{j\in F} G_j) = \sum_{j\in F} \operatorname{nsr}^+(G_j) \leq \kappa$ and hence $\varphi$ cannot be injective.

We can see now, that every cardinal can be $\operatorname{nsr}^+(G)$ for some Abelian group G. In particular, it is allowed for a group G that $\operatorname{nsr}^+(G)$ is a limit cardinal. But this implies that G hasn't non-slender rank.

6. In the last three sections we continue our investigation of quotients of products $\prod_{i\in I} A_i$ started in [W1] and [W2]. Therefore we need the notion of saturated ideals. Let I be a set, $\mathbb{I}$ an ideal on I and $\kappa$ a

cardinal. $\mathbb{I}$ is called $\kappa$-saturated, if every subset T of $\mathbb{P}(I)\setminus\mathbb{I}$ which is pairwise almost disjoint (i.e. $X \cap Y \in \mathbb{I}$ for all $X \neq Y$ in T) has cardinality less than $\kappa$. sat($\mathbb{I}$) is the smallest cardinal $\kappa$, such that $\mathbb{I}$ is $\kappa$-saturated. It is clear, that $\mathrm{sat}(\mathbb{I}) \leq (2^{|I|})^+$.

Theorem: Let $(A_i)_{i\in I}$ be a family of Abelian groups and U a subgroup of $\prod_{i\in I} A_i$ such that $G = \prod_{i\in I} A_i/U$ is a cotorsionfree group. We define $\mathbb{I}_U$ to be the ideal $\{X\subseteq I: \prod_{i\in X} A_i \subseteq U\}$. Then $\mathrm{sat}(\mathbb{I}_U) \leq \mathrm{nsr}^+(G)$.

Proof: We have to show that $\mathbb{I}_U$ is $\mathrm{nsr}^+(G)$-saturated. Let T be a pairwise almost disjoint subset of $\mathbb{P}(I)\setminus\mathbb{I}_U$. For $X \in T$ we can choose some $a_X$ in $\prod_{i\in I} A_i\setminus U$. Now we are able to define a homomorphism $\varphi$ from $Z^T$ into G in the following way: For $s \in \mathbb{Z}^{[T]}$ let $\varphi(s) = a + U$, where $a(i) = s(X)a_X(i)$ if $X \in \mathrm{supp}(s)$ and $i \in X$ and $a(i) = 0$ if $i \notin X$ for all $X \in \mathrm{supp}(s)$. The problem by the definition of a(i) are those $i \in I$, which lie in more than one $X \in \mathrm{supp}(s)$. This is the set $D = \bigcup \{X \cap Y: X,Y \in \mathrm{supp}(s),\ X \neq Y\}$. Since G is cotorsionfree, $\mathbb{I}_U$ is $\omega_1$-complete (cf. [W2]) and therefore $D \in \mathbb{I}_U$. Thus, $\varphi$ is a homomorphism satisfying $\varphi(e_X) = a_X + U \neq 0$ for all $X \in T$ and consequently, by 2., $|T| < \mathrm{nsr}^+(G)$.

Of special interest are the groups $Q_\kappa = \mathbb{Z}^\kappa/\mathbb{Z}^{\kappa(\kappa)}$. It is well-known, that for the ideal $\mathbb{P}_\kappa(\kappa) = \{X\subseteq\kappa: |X| < \kappa\}$ of $\mathbb{P}(\kappa)$ the cardinal $\mathrm{sat}(\mathbb{P}_\kappa(\kappa))$ is greater than $\kappa^+$; cf. J.E. Baumgardner [Bg], Theorem 2.8. So we have the following Corollar.

Corollar: For every cardinal $\kappa$ satisfying $\omega < \mathrm{cf}\kappa$ we have $\kappa^+ < \mathrm{nsr}^+(Q_\kappa) \leq (2^\kappa)^+$ and in particular, if $\kappa^+ = 2^\kappa$, $Q_\kappa$ has non-slender rank and $\mathrm{nsr}(Q_\kappa) = 2^\kappa$.

Notice, that if $\omega = \mathrm{cf}\kappa$, $Q_\kappa$ is not cotorsionfree. Under certain assumption weaker as GCH (Generalized Continuum Hypothesis) $\mathrm{sat}(\mathbb{P}_\kappa(\kappa)) = (2^\kappa)^+$, too; cf. J.E. Baumgardner [Bg] and T. Jech and K. Prikry [JP], 5.1.1 and 6.1.1. Also in these cases we get $Q_\kappa$ has non-slender rank $2^\kappa$. For example, if $2^\omega < 2^{\omega_1}$ and $2^\omega < \omega_{\omega_1}$, then $Q_{\omega_1}$ has non-slender rank $2^{\omega_1}$.

7. Problems concerning saturation of ideals are of high interest in set theory; see [KM] §11,12,17 and [J] §27,35. From 1930 there is a fundamental paper of S. Ulam [U] where he proved that an infinite cardinal $\kappa$ which carries a non-trivial, $\kappa$-complete and $\kappa$-saturated ideal is weakly inaccessible. Here an ideal $\Pi$ on $\kappa$ is called non-trivial if $\mathbb{P}_{\kappa}(\kappa) \subseteq \Pi \subsetneq \mathbb{P}(\kappa)$. Fourty years later R.M. Soloway [S] could show that such a cardinal is much larger: $\kappa$ has to be a weakly Mahlo cardinal, i.e. $\kappa$ is weakly inaccessible and the set of regular cardinals below $\kappa$ is stationary (this implies that the set of all weakly inaccessible cardinals below $\kappa$ is stationary). Furthermore Soloway established that if $S \subseteq \kappa$ is stationary, the set of all $\nu < \kappa$ for which $S \cap \nu$ is stationary in $\nu$ is stationary. In particular there are $\kappa$ many weakly Mahlo cardinals below $\kappa$.

If the ideal on $\kappa$ has a stronger saturation property, $\kappa$ gets larger. By a result of A. Tarski [T] $\kappa$ carries a non-trivial $\kappa$-complete ideal which is $\nu$-saturated for some cardinal $\nu$ satisfying $2^{<\nu} = \sup\{2^{\alpha}: \alpha < \nu\} < \kappa$, only if $\kappa$ is measurable.

Now we come to an application of these results. M. Dugas and R. Göbel [DG] call a group G <u>strongly cotorsionfree</u>, if it is $\kappa$-reduced (i.e. $\mathrm{Hom}(\mathbb{Q}_{\kappa}, G) = 0$; cf. [W2]) for all infinite $\kappa$ less than every measuable cardinal.

<u>Theorem</u>: Let G be a cotorsionfree group which is $\kappa$-reduced for all $\kappa < \mathrm{nsr}^+(G)$. Suppose, that there is no cardinal $\kappa$, $\mathrm{nsr}^+(G) \le \kappa \le 2^{<\mathrm{nsr}^+(G)}$, which carries a non-trivial, $\kappa$-complete, $\kappa$-saturated ideal. (Such a $\kappa$ would be weakly inaccessible in the strong sense mentioned above.) Then G is strongly cotorsionfree.

<u>Proof</u>: We suppose that G is $\nu$-reduced for all $\nu$ less than some $\kappa$ and $\kappa$ is less than every measurable. Without loss of generality we may assume $\mathrm{nsr}^+(G) \le \kappa$. Let $\varphi$ be a homomorphism from $\mathbb{Z}$ into G satisfying $\varphi(\mathbb{Z}^{\kappa(\kappa)}) = 0$. By 3.1 of [W2] the ideal $\Pi = \{X \subseteq \kappa: \varphi(\mathbb{Z}^{[X]}) = 0\}$ is $\kappa$-complete and by 6. of this paper $\Pi$ is $\mathrm{nsr}^+(G)$-saturated. If $\kappa \le 2^{<\mathrm{nsr}^+(G)}$ we can use Soloway's result to show that $\Pi$ is trivial, i.e. $\Pi = \mathbb{P}(\kappa)$, and hence $\varphi = 0$. Otherwise, if $2^{<\mathrm{nsr}^+(G)} < \kappa$, Tarski's result applies and $\varphi = 0$ follows again. Therefore G is $\kappa$-reduced.

A classical result in Abelian group theory is the theorem of J. Łoś, that a slender group G (i.e. G is cotorsionfree and $nsr^+(G) \leq \omega$; cf.2) is strongly cotorsionfree, vgl. L. Fuchs [F] Theorem 94.4.

Corollary: If $2^\omega$ is not too large (for example: there is no weakly inaccessible cardinal $\leq 2^\omega$) every cotorsionfree group G satisfying $nsr^+(G) \leq \omega_1$ is strongly cotorsionfree.

8. Kunen asked, whether it is possible that a cardinal $\kappa$ carries a non-trivial, $\kappa$-complete, $\kappa^+$-saturated ideal. At first he showed, that such a cardinal is measurable in an inner model; that is a transitive $\in$-model of ZFC lying between L and V where L is the constructible universe and V the whole universe; cf. [K1]. Some years later he came to the following consistency result: If the existence of a "very strong large cardinal" is consistent with ZFC, then the existence of a non-trivial $\omega_1$-complete, $\omega_2$-saturated ideal on $\omega_1$ is consistent; cf. [K2].

Let us denote by $\neg L_\mu$ the assertion that there is no inner model with a measurable cardinal. This is much weaker as $V = L$, but many consequences of $V = L$, follow from $\neg L_\mu$.

For the last statements it is convenient to define $nsr^{(+)}(G) = nsr(G)$ if G has non-slender rank and $nsr^{(+)}(G) = nsr^+(G)$ otherwise. Thus, $nsr^{(+)}(G)$ is the supremum of all cardinals $\kappa$, which allowed an embedding of $Z^{[\kappa]}$ into G. By 1.d) and e) we can obtain $nsr^{(+)}(G) \leq r_o(G) \leq |G|$. Using the result of K. Kunen [K1] we establish with a similar proof as in 7.:

> $\neg L_\mu$ implies that a cotorsionfree group G is strongly cotorsionfree if G is $\kappa$-reduced for all $\kappa < nsr^{(+)}(G)$.

By an actual result of H.D. Donder [D] it is sufficient to deal with $\omega_1$-completeness instead of $\kappa$-completeness: A cardinal $\kappa$ which carries a non-trivial, $\omega_1$-complete, $\kappa^+$-saturated ideal is measurable in an inner model. Because in our business $\omega_1$-completeness follows directly from the assumption that the group is cotorsionfree we have the following theorem.

Theorem: $(\neg L_\mu)$ A cotorsionfree group is $\kappa$-reduced for all $\kappa \geq nsr^{(+)}(G)$.

Corollary: ($\neg L_\mu$) Let G be a cotorsionfree group and $nsr^+(G) \leq \omega_2$ (i.e. G is slender or G has non-slender rank $\omega$ or $\omega_1$) then G is strongly cotorsionfree.

References

[Bc] S. Balcerzyk, On groups of functions defined on Boolean algebras, Fund. Math. 50 (1962) 347-367.

[Bg] J.E. Baumgardner, Almost disjoint sets, the dence set problem and the partition calculus, Ann. Math. Logic 9 (1976) 401-439.

[Ch] S.U. Chase, On direct sums and products of modules, Pacific J.Math. 12 (1962), 847-854.

[D] H.D. Donder, in preparation

[DG] M. Dugas and R. Göbel, On radicals and products, to appear in Pacific J. Math.

[DZ] M. Dugas and B. Zimmermann-Huisgen, Iterated direct sums and products of moduls, in Abelian Group Theory. Proceedings, Oberwolfach 1981, Springer Lecture Notes 874 (1981) 179-173.

[F] L. Fuchs, Infinite Abelian Groups II, Academic Press, New York 1974.

[GW] R. Göbel and B. Wald, Wachstumstypen und schlanke Gruppen, Symp. Math. 23 (1979) 201-239.

[GWW] R. Göbel, B. Wald and P. Westphal, Groups of integer-valued functions, in Abelian Group Theory, Proceedings, Oberwolfach 1981, Springer Lecture Notes, 874 (1981) 161-178.

[I] A.V. Ivanov, Direct sums and complete direct sums of abelian groups (Russian), In Abelian Groups and Modules, Tomsk. Gos. Univ., (1 · 70-90, 136-137.

[J] T. Jech, Set Theory, Academic Press, New York, London (1978).

[JP] T. Jech and K. Prikry, Ideals over uncountable sets: application of almost disjoint functions and generic ultrapowers, Memoirs Amer. Math.Soc. 18 (1979) no. 214.

[K1] K. Kunen, Some application of iterated ultrapowers in set theory, Ann.Math.Logic 1 (1970) 179-227.

[K2] K. Kunen, Saturated ideals, J. Symbolic Logic 43 (1978) 65-76.

[KM] A. Kanamori and M. Magidor, The evolution of large cardinal axioms in set theory, in Higher Set Theory, Proceedings, Oberwolfach 1977, Springer Lecture Notes 669 (1978) 99-275.

[N] R.J. Nunke, Slender groups, Bull. Amer. Math. Soc. 67 (1961) 274-275; Acta Sci. Math Szeged 23 (1962) 67-73.

[S] R.M. Soloway, Real-valued measurable cardinals, in Axiomatic Set Theory (D. Scott, ed.), Proc. Symp. Pure Math. 13 I (1971) 397-428.

[T] A. Tarski, Ideale in vollständigen Mengen-Körpern I, Fund. Math. 32 (1939) 45-63

[U] S. Ulam, Zur Maßtheorie in der allgemeinen Mengenlehre, Fund. Math. 16 (1930) 140-150.

[W1] B. Wald, Martinaxiom und die Beschreibung gewisser Homomorphismen in der Theorie der $\aleph_1$-freien abelschen Gruppen, Manuscripta Math. 42 (1983) 297-309.

[W2] B. Wald, On $\kappa$-products modulo $\mu$-products, in Abelian Group Theory, Proceedings, Honolulu 1982/83, Springer Lecture Notes 1006 (1983) 362-370.

# A-PROJECTIVE GROUPS OF LARGE CARDINALITY

ULRICH ALBRECHT

MARSHALL UNIVERSITY

## 1. INTRODUCTION

Because free abelian groups have many useful homological properties, the question arises how closely do groups of the form $\oplus_I A$ for a torsion-free group A resemble free groups. It becomes apparent that it is necessary to restrict the choices for A. In [1] and [2], it has been shown that the best way of doing this is to restrict the class of rings from which the endomorphism ring E(A) = Hom (A,A) can be chosen. In these papers, A was a torsion-free, reduced abelian group whose endomorphism ring is semi-prime, right and left Noetherian, and hereditary. Such groups will be called generalized rank 1 groups. Their important properties are given in [2, Proposition 3.2]. These can be used to show that the A-projective groups which are direct summands of groups of the form $\oplus_I A$ closely resemble free abelian groups.

However, while this gives a satisfactory answer to the initial problem, there remains the question if there are conditions on a group G that will ensure that G is A-projective. In the case $A = \mathbb{Z}$, this was done by Eklof [4] and Shelah [7]. The goal of this paper is using those methods to extend their results to abelian groups A as in [1] and [2].

The results of these papers show that only abelian groups G with $S_A(G) = G$ have to be considered where $S_A(G) = \Sigma\{f(A) \mid f \in \mathrm{Hom}(A,G)\}$. If G is an abelian group of cardinality $\kappa$, then there are two cases to consider depending on the cofinality cf $(\kappa)$ of $\kappa$ where cf $(\kappa) = \inf\{\lambda \mid$ There is $\{\kappa_\alpha\}_{\alpha<\lambda}$ with $\sup\{\kappa_\alpha\} = \kappa\}$. If cf $(\kappa) = \kappa$, then the criterium will resemble Pontryagin's result and coincide with Eklof's for $A = \mathbb{Z}$ (Theorem 2.1). On the other hand, if cf $(\kappa) < \kappa$, then G is A-projective if $S_A(G) = G$ and each subgroup U of G with $S_A(U) = U$ whose cardinality is strictly less than $\kappa$ is A-projective. In both cases, one has to assume that A has a cardinality less than $\kappa$.

## 2. A-Projectives of Regular Cardinality

In this section, the case of cardinal numbers $\kappa$ with cf $(\kappa) = \kappa$ is discussed. They are called regular.

If A and G are torsion-free abelian groups, and $\kappa$ is a cardinal number, then G is $\kappa$-A-projective if each subgroup U of G with $S_A(U) = U$ and $|U| < \kappa$ is A-projective where $|U|$ denotes the cardinality of U. Moreover, G is strongly $\kappa$-A-projective if each subgroup U of G with $|U| < \kappa$ is contained in an A-projective subgroup V of G such that $|V| < \kappa$ and G/V is $\kappa$-A-projective, and G itself is $\kappa$-A-projective.

Proposition 2.1 : Let $\kappa$ be an uncountable, regular cardinal number. If A and G are torsion-free abelian groups such that $|A| < |G| = \kappa$, and A is a generalized rank 1 group, then G is strongly $\kappa$-A-projective if and only if it is the union of a smooth ascending chain $\{G_\nu\}_{\nu<\kappa}$ of A-projective subgroups of G such that $|G_\nu| < \kappa$ and $G/G_{\nu+1}$ is $\kappa$-A-projective for all $\nu < \kappa$.

Proof : If G is strongly $\kappa$-A-projective, then write $G = \{g_\nu \mid \nu < \kappa\}$ with $g_0 = 0$. Let $G_0 = \{0\}$ and suppose a smooth ascending chain $\{G_\mu\}_{\mu<\nu}$ has been defined such that $G_\mu$ is an A-projective subgroup of G, $|G_\mu| < \kappa$, $g_\mu \in G_{\mu+1}$, and $G/G_{\mu+1}$ is $\kappa$-A-projective for all $\mu < \nu$. If $\nu$ is a limit ordinal, then set $G_\nu = \cup_{\mu<\nu} G_\mu$. Clearly, $S_A(G_\nu) \supseteq G_\mu$ for all $\mu < \nu$ and $|G_\nu| < \kappa$. Thus, $G_\nu$ is A-projective since G is $\kappa$-A-projective. If $\nu = \mu + 1$, then choose an A-projective subgroup $G_\nu$ of G such that $|G_\nu| < \kappa$, $\{g_\mu\} \cup G_\mu \subseteq G_\nu$, and $G/G_\nu$ is $\kappa$-A-projective. Clearly, $\{G_\nu\}_{\nu<\kappa}$ satisfies the conditions of the proposition.

Conversely, each subgroup U of G with $|U| < \kappa$ is contained in some $G_{\mu+1}$. If $S_A(U) = U$, then U is A-projective by [2, Proposition 3,2]. Thus, G is $\kappa$-A-projective. □

In order to proceed further, some notations are needed. A function $f: \kappa \rightarrow \kappa$ is normal if it is increasing and $f(\lambda) = \sup\{f(\mu) \mid \mu < \lambda\}$ for all limit ordinals $\lambda < \kappa$. The set $\{\text{im}(f) \mid f: \kappa \rightarrow \kappa \text{ normal}\}$ is closed under finite intersections if $\kappa$ is an uncountable, regular cardinal number. Finally, a subset $X$ of $\kappa$ is stationary if $X \cap \text{im}(f) \neq \emptyset$ for all normal functions $f: \kappa \rightarrow \kappa$.

If G is strongly $\kappa$-A-projective, then choose a chain $\{G_\nu\}_{\nu<\kappa}$ as in Proposition 2.1 and let $\lambda(G) = \{\lambda<\kappa \mid G/G_\lambda$ is not $\kappa$-A-projective$\}$. It is easy to see that the fact that $\lambda(G)$ is stationary or not is independent of the choice of the $\{G_\nu\}_{\nu<\kappa}$. With this, one obtains

Theorem 2.2: Let $\kappa$ be an uncountable, regular cardinal number. If G and A are torsion-free abelian group such that $|A| < |G| = \kappa$, and A is a generalized rank 1 group, then G is A-projective if and only if G is strongly $\kappa$-A-projective and $\lambda(G)$ is not stationary in $\kappa$.

Proof: If G is $\kappa$-A-projective and $\lambda(G)$ is not stationary in $\kappa$, then choose a chain $\{G_\nu\}$ as in Proposition 2.1 and a normal function $f:\kappa\longrightarrow\kappa$ with $\lambda(G) \cap \text{im}(f) = \emptyset$. Let $B_\nu = G_{f(\nu)}$. Since f is normal, $\{B_\nu\}_{\nu<\kappa}$ is a smooth ascending chain whose union is G.

Since $B_{\nu+1}$ is A-projective, $S_A(B_{\nu+1}/B_\nu) = B_{\nu+1}/B_\nu$. Because $f(\nu) \notin \lambda(G)$, the group $G/B_\nu$ is $\kappa$-A-projective. Thus, $B_{\nu+1}/B_\nu$ is A-projective. By [2, Proposition 3.2], $B_\nu$ is a direct summand of $B_{\nu+1}$, say $B_{\nu+1} = B_\nu \oplus C_{\nu+1}$. With $C_0 = B_0$, one has $G = \oplus_{\nu<\kappa} C_\nu$. Thus, G is A-projective.

Conversely, if G is A-projective, then $\text{Hom}(A,G) = \oplus_{i\in I} P_i$ where $P_i$ is isomorphic to a right ideal of $E(A)$. Tensoring with A over $E(A)$ proves the theorem. □

## 3. A-Projectives of Singular Cardinality

In contrast to the last section, one is now concerned with cardinal numbers $\kappa$ such that $cf(\kappa) < \kappa$. They will be called singular.

Lemma 3.1 : Let $\kappa$ be a singular cardinal number. Suppose A is a generalized rank 1 group, and G is $\kappa$ - A-projective such that $|A| = \lambda < \kappa = |G|$. Every pure, A-projective subgroup C of G with $\lambda \leq \mu = |C| < \kappa$ is contained in a pure A-projective subgroup C* of G such that $|C^*| = \mu$ and $C'/C^*$ is A- projective for all pure, A-projective subgroups C' of G with $C' \supseteq C^*$ and $|C'| = \mu$.

Proof : Since $\kappa$ is singular, the successor cardinal number $\mu^+$ of $\mu$ is less than $\kappa$. Suppose there is a subgroup C of G for which does not exist such a C*.

Let $C_0 = C$, and suppose one has constructed a smooth ascending chain $\{C_\eta\}_{\eta<\nu}$ for some $\nu<\mu^+$ such that $C_\eta$ is a pure A - projective subgroup of G, $|C_\eta| = \mu$, and $C_{\eta+1}/C_\eta$ is not A-projective for all $\eta<\nu$. If $\nu$ is a limit ordinal, then let $C_\nu = \bigcup_{\eta<\nu} C_\eta$. Since $S_A(C_\nu) \supseteq S_A(C_\eta) = C_\eta$ and $|C_\nu| \leq \mu^+$, $C_\nu$ is A-projective and pure in G. If $\nu = \eta + 1$, then choose an A-projective, pure subgroup $C_\nu$ of G containing $C_\eta$ such that $|C_\nu| = \mu$ and $C_\nu/C_\eta$ is not A-projective.

If this has been done for each $\nu < \mu^+$, then $H = \bigcup_{\nu<\mu^+} C_\nu$ has cardinality $\mu^+$. Since G is $\kappa$ - A-projective, $S_A(H) = H$, H is A-projective. Then, $H = \bigcup_{\nu<\mu^+} D_\nu$ such that $D_\nu$ is a direct summand of H and $|D_\nu| \leq \mu$. There is a normal function $f : \mu^+ \to \mu^+$ such that $C_{f(\nu)} = D_{f(\nu)}$. But then, $C_{f(\nu)}$ is a direct summand of $C_{f(\nu)+1}$, which contradicts the fact that $C_{f(\nu)+1}/C_{f(\nu)}$ is not A-projective. □

Lemma 3.2 : Let $\kappa > \aleph_0$ be a cardinal number. If A and G are torsion-free abelian groups with $|A| = \lambda < \kappa = |G|$ and $S_A(G) = G$, then every subgroup $U$ of $G$ with $\lambda \le |U| < \kappa$ is contained in a pure subgroup $C$ of $G$ with $S_A(C) = C$ and $|C| = |U|$.

Proof: Since $S_A(G) = G$ and $|A| < \kappa$, there is an index set $I$ with $|I| = \kappa$ such that there is an epimorphism $\pi: \oplus_I A \to G$. Let $U$ be any subgroup of $G$ with $\lambda \le |U| < \kappa$. There is a subset $J_1$ of $I$ with $|J_1| = |U|$ and $\pi^{-1}(U) \le \oplus_{J_1} A$. If $C_1 = \pi(A^{(J_1)})$, then $S_A(C_1) = C_1$ and $|U| \le |C_1| \le \lambda |J_1| = |U|$. Let $D_1$ be the purification of $C_1$ in $G$. Inductively, one obtains chains $U \subseteq C_1 \subseteq D_1 \subseteq \ldots \subseteq C_n \subseteq D_n \subseteq C_{n+1} \subseteq \ldots$ in $G$ with $S_A(C_n) = C_n$, $D_n$ is pure in $G$, and $|C_n| = |D_n| = |U|$. Let $C = \bigcup_{n<\omega} C_n = \bigcup_{n<\omega} D_n$. Then, $S_A(C) = C$ and $C$ is pure in $G$. Thus, $C$ is $A$-projective and $|C| \le \aleph_0 |U| = |U|$. $\square$

Theorem 3.3[5, Theorem 3.3]: Let $\kappa$ be a singular cardinal number, and $\{\kappa_i \mid i < cf(\kappa)\}$ be a smooth, increasing chain of cardinal numbers with $\kappa_0 = 0$, $cf(\kappa) \le \kappa_1$, and $\kappa = \sup\{\kappa_i \mid i < cf(\kappa)\}$. If $G$ is a group of cardinality $\kappa$, then let $S_i$ be the set of all subgroups of $G$ of cardinality $\kappa_i$, and let $S_i' = \{0\} \cup S_i$. Suppose $\mathcal{F}$ is a class of pairs $(C_2, C_1)$ of subgroups of $G$ which satisfies the following two properties:

H1) For every $i < cf(\kappa)$, there is a function $g_i: S_i' \times S_i \to S_i$ such that whenever $A_1 \subsetneq A_2$ are in $S_i'$ and $A_1 \in \{0\} \cup (\text{range}(g_i))$, then $A_2 \subseteq g_i(A_1, A_2)$ and $(g_i(A_1,A_2), A_1) \in \mathcal{F}$.

H2) For every $i < cf(\kappa)$ and every $A_1 \subsetneq A_2$ in $S_{i+1}'$, if $(A_2, A_1) \in \mathcal{F}$, then player II has a winning strategy in the following game: in the $n^{th}$ move ($n < \omega$), player I chooses $B_n \in S_i'$ such that

$C_{n-1} \subseteq B_n$ (where $C_{-1} = 0$), and then player II chooses $C_n \in S_i'$ such that $B_n \subseteq C_n$. Player II wins if $(A_2 + \bigcup_{n<\omega} C_n, A_1 + \bigcup_{n<\omega} C_n) \in \mathcal{F}$.

Then G is the union of a smooth ascending chain $\{G_\nu | \nu < \omega\, cf(\kappa)\}$ in G of subgroups of G such that $G_0 = 0$ and $(G_{\nu+1}, G_\nu) \in \mathcal{F}$ for all $\nu$. □

Theorem 3.4 : Let $\kappa$ be a singular cardinal number. Let A and G be torsion-free abelian groups such that A is a generalized rank 1 group and $|A| = \lambda < \kappa = |G|$. G is A-projective if and only if $S_A(G) = G$ and G is $\kappa$-A-projective.

Proof : The necessity of these conditions is obvious. For the converse, Theorem 3.3 is used. In a series of lemmas, it is shown that it can be applied.

In the first step, choose an ascending smooth chain $\{\kappa_i | i < cf(\kappa)\}$ of regular cardinal numbers such that $\kappa_0 = 0$, $\kappa_1 \geq \max\{\lambda, cf(\kappa)\}$, $\kappa_{i+1} > \kappa_i$ for each $i < cf(\kappa)$, and $\kappa = \sup\{\kappa_i | i < cf(\kappa)\}$.

In the second step, $\mathcal{F}$ is defined. Let $\mathcal{F} = \{(C_2, C_1) | C_2$ is A-projective and $C_1$ is a direct summand of $C_2\}$

Lemma 3.5 : $\mathcal{F}$ satisfies H1.

Proof : Since the case $i = 0$ is trivial, assume $i > 0$. Let $(A_1, A_2)$ be in $S_i' \times S_i$. There is a pure subgroup C of G containing $A_2$ with $S_A(C) = C$ and $|A_2| = |C|$. Since G is $\kappa$-A-projective, C is A-projective. Choose a subgroup $C^*$ for C as in Lemma 3.1. Then $A_2 \subseteq C \subseteq C^*$. Let $g_i(A_1, A_2) = C^*$. Clearly, $g_i(A_1, A_2) \in S_i'$.

If $A_1 \subsetneq A_2$ and $A_1 \in \{0\} \cup \text{range}\{g_i\}$, then one has $(g_i(0, A_2), 0) = (C^*, 0) \in \mathcal{F}$. Otherwise, if $A_1 \neq 0$, then $|A_1| = \kappa_i$, and there is a pure

A-projective subgroup $D \in S_i$ such that $A_1 = D^*$ and $D^*$ is constructed according to Lemma 3.1. Consequently, $A_1$ is A-projective, and $H/A_1$ is A-projective for every A-projective, pure subgroup H of G with $H \supseteq A_1$ and $|H| = \kappa_i$. This holds in particular for $H = C^* \supseteq A_2 \supseteq A_1 = D^*$. Thus, $(C^*, D^*) \in \mathcal{F}$, and $\mathcal{F}$ satisfies H1. □

Lemma 3.6 : $\mathcal{F}$ satisfies H2.

Proof : Let $i < cf(\kappa)$ and choose $A_1 \subsetneq A_2 \in S'_{i+1}$ with $(A_2, A_1) \in \mathcal{F}$. Since the case $i = 0$ is trivial, assume $i > 0$.

Because of $(A_2, A_1) \in \mathcal{F}$, one has $A_2 = A_1 \oplus C$. Since $A_1$ and C are direct sums of groups which are subgroups of A, there is an index-set I such that $A_2 = \oplus_{i \in I} L_i$, $|L_i| \leq \lambda$ and $|I| = \kappa_{i+1}$. Moreover, there is $I_0 \subsetneq I$ such that $A_1 = \oplus_{i \in I_0} L_i$.

If $\mathcal{N} = \{\oplus_{j \in J} L_j \mid J \subseteq I\}$, then $\mathcal{N}$ is closed under unions of chains. Moreover, if $H = \oplus_{j \in J} L_j$ is in $\mathcal{N}$, then $A_1 + H = \oplus_{i \in I_0 \cup J} L_j \in \mathcal{N}$ and is a direct summand of $A_2$. Consequently, $(A_2, A_1 + H) \in \mathcal{F}$.

Player II now has the following winning strategy : Suppose, player II has chosen $C_{n-1} \in S_i'$ such that $C_{n-1} \cap A_2 \in \mathcal{N}$ and $S_A(C_{n-1}) = C_{n-1}$. Player I chooses a $B_n$ containing $C_{n-1}$.

By Lemma 3.2, it is possible to choose an A-projective, pure subgroup $D_0$ of G containing $B_n$ with $|D_0| = \kappa_i$. Since $|A_2 \cap D_0| \leq \kappa_i$, there is a subset $J_0$ of I such that $|J_0| \leq \kappa_i$ and $A_2 \cap D_0 \subseteq \oplus_{j \in J_0} L_j = \tilde{D}_0$. Since $D_0 + \tilde{D}_0 = S_A(D_0) + S_A(\tilde{D}_0)$, one has $D_0 + \tilde{D}_0$ is A-projective. Now $D_0 + \tilde{D}_0$ is contained in an A-projective, pure subgroup $D_1$ of G with $|D_1| = \kappa_i$. Continuing inductively, there are chains of

A-projective groups $D_0 \subseteq \ldots \subseteq D_m \subseteq \ldots \in S_i$ and $\tilde{D}_0 \subseteq \ldots \subseteq \tilde{D}_m \subseteq \ldots$ such that $\tilde{D}_{i-1} \subseteq D_i$, $A_2 \cap D_i \subseteq \tilde{D}_i$, $|\tilde{D}i| = |D_i| = \kappa_i$ and $\tilde{D}_i = \oplus_{j \in J} L_j$.

Let $M_n = \bigcup_{m < \omega} J$ and $C_n = \bigcup_{m < \omega} D_m$. Then $|C_n| = \kappa_i$ and $C_n$ is A-projective, since $S_A(C_n) = C_n$. Moreover, $A_2 \cap C_n = \oplus_{i \in M_n} L_i$.

Since $\mathfrak{N}$ is closed under unions of chains, $\bigcup_{n < \omega} A_2 \cap C_n \in \mathfrak{N}$. By the first part of this proof, $(A_2, A_1 + \bigcup_{n < \omega} A_2 \cap C_n)$ is in $\mathcal{F}$ and hence

$$(A_2 + \bigcup_{n < \omega} C_n)/(A_1 + \bigcup_{n < \omega} C_n) \cong A_2/(A_2 \cap (A_1 + \bigcup_{n < \omega} C_n))$$
$$= A_2/(A_1 + \bigcup_{n < \omega} (A_2 \cap C_n))$$

is A-projective. Since $S_A(A_2 + \bigcup_{n < \omega} C_n) = A_2 + \bigcup_{n < \omega} C_n$ and $|A_2 + \bigcup_{n < \omega} C_n| \leq \kappa_{i+1} < \kappa$ imply $A_2 + \bigcup_{n < \omega} C_n$ is A-projective, and $A_1 + \bigcup_{n < \omega} C_n$ is a direct summand of $A_2 + \bigcup_{n < \omega} C_n$. This means $(A_2 + \bigcup_{n < \omega} C_n, A_1 + \bigcup_{n < \omega} C_n) \in \mathcal{F}$. $\square$

<u>Conclusion of the proof of Theorem 3.4 :</u>

Since $\mathcal{F}$ satisfies H1 and H2, $G = \bigcup G_\nu$ such that $\{G_\nu \mid \nu < \omega \operatorname{cf}(\kappa)\}$ is a smooth ascending chain of A-projective subgroups of G with $G_0 = 0$ and $(G_{\nu+1}, G_\nu) \in \mathcal{F}$. Then, $G_{\nu+1} = G_\nu \oplus C_\nu$ and $G = \oplus_{\nu < \omega \operatorname{cf}(\kappa)} C_\nu$. Thus, G is A-projective. $\square$

References :


1. Albrecht, U.; Endomorphism rings and A-projective torsion-free abelian groups, in Abelian Group Theory, Proceedings Honolulu 1982/83, Springer LNM 1006, New York, 1983, 209-227.
2. Albrecht, U.; A note on locally A-projective abelian groups, to appear.
3. Chase, S.; On group extensions and a problem of J. H. C. Whitehead, in Topics in Abelian Groups, Scott - Foresman, Glenview, 1963, 173-193.
4. Eklof, P.; On the existence of $\kappa$-free abelian groups, Proc. Amer. Math. Soc. 47, 1975, 65-72.
5. Eklof, P. and Huber, M.; Abelian group extensions and the axiom of constructibility, Comment. Math. Helvetii, 54, 1979, 440 - 457.
6. Kaplansky, I.; Projective Modules, Annals of Mathematics, 68, No. 2, 1958, 372 - 376.
7. Shelah, S.; A compactness theorem for singular cardinals, free algebras, Whitehead problem and transversals, Israel G. Math., 21, 1975, 319 - 349.

## ADDITIVE GROUPS OF EXISTENTIALLY CLOSED RINGS

P.C. EKLOF and H.C. MEZ

UNIVERSITY OF CALIFORNIA, IRVINE

This is a report on work in progress on the structure of existentially closed rings and of their underlying groups. For simplicity of exposition we shall confine ourselves here to the 'local case', i.e., the case of $\mathbb{Q}_p$-algebras, for a fixed prime $p$ ; but most of the results have analogs for the general case of rings i.e., $\mathbb{Z}$-algebras.

Of course, $\mathbb{Q}_p$ is the ring of all rationals whose denominators are prime to $p$ . Here, a "$\mathbb{Q}_p$-algebra" will mean an associative $\mathbb{Q}_p$-algebra which is not necessarily commutative, and does not necessarily contain a multiplicative identity.

If $X$ is a set of new symbols, let $\mathbb{Q}_p\langle X\rangle$ denote the $\mathbb{Q}_p$-algebra of all polynomials <u>without constant terms</u> in the non-commuting variables of $X$ . Let $A(X) := A*\mathbb{Q}_p\langle X\rangle$ , the free product of the $\mathbb{Q}_p$-algebras $A$ and $\mathbb{Q}_p\langle X\rangle$ ; this is just the extension of $A$ obtained by adjoining the non-commuting indeterminates of $X$ . (See P. Cohn[1](p. 187, Ex. 8)). The elements of $A(X)$ are finite sums of monimals $\lambda c_1 \cdots c_r$ where $r \in \mathbb{N}, \lambda \in \mathbb{Q}_p - \{0\}$, $c_i \in A \cup X$, and $c_i \in A$ implies $c_{i+1} \in X$ for all $i < r$. Note also that any set function $v: X \to A$ extends to a $\mathbb{Q}_p$-algebra mapping $\Psi_v: A(X) \to A$ ('evaluation at $v$').

An equation over $A$ (resp. inequation over $A$) is an expression of the form $f = 0$ (resp. $f \neq 0$) where $f \in A(X)$. A set of equations and inequations over $A$ has a solution in $A$ if there is a map $v: X \to A$ such that every equation and inequation in the set becomes true in $A$ under $\Psi_v$.

If $A$ is a $\mathbb{Q}_p$-subalgebra of $B$, we say $A$ is existentially closed in $B$ if every finite set of equations and inequations over $A$ which has a solution in $B$ also has a solution in $A$. We say $A$ is existentially closed - abbreviated e.c. - if $A$ is existentially closed in every $\mathbb{Q}_p$-algebra extension.

A standard union-of-chains construction (see, for example Hirschfield-Wheeler[2]) shows that

LEMMA 1. Every $\mathbb{Q}_p$-algebra $B$ can be embedded as a subalgebra of an e.c. $\mathbb{Q}_p$-algebra $A$ such that $|A| \leq |B| + \aleph_0$. □

(Here $|A|$ denotes the cardinality of $A$).

The paper Eklof-Mez[3] discusses the ideals of an existentially closed algebra. Here we shall concentrate on some consequences of that work which relate to Problems 93 and 94 of Fuchs[4]. Problem 93 asks about the groups on which a ring can be defined whose only ideals are "absolute" ideals, i.e., subgroups which are ideals in every ring structure on the group. It is noted in Fuchs[4] (§117) that a sufficient condition for a subgroup of a group $A$ to be an absolute ideal is that it be a fully invariant subgroup. (An apparently weaker necessary condition is given in Theorem 117.2 of Fuchs[4]).

Problem 94 asks about the "absolute" annihilator and the "absolute" Jacobson radical of a group. Define the absolute annihilator, $\overline{\mathrm{Ann}}(M)$, of $M$ to be the subgroup of $M$ consisting of the elements of $M$ which belong to the annihilator, $\mathrm{Ann}(A)$, of every ring $A = (M,\cdot)$ on $M$. $[\mathrm{Ann}(A) = \{a\in A : \forall x\in A(a\cdot x = 0 = x\cdot a)\}]$. Similarly, the absolute Jacobson radical, $\bar{J}(M)$, of $M$ consists of the elements of $M$ which belong to the Jacobson radical, $J(A)$, of every ring $A$ on $M$.

Jackett[5] (Cor. 3.5) proves that if $M$ is a reduced algebraically compact $Q_p$-module, then $\bar{J}(M) = pM$. As a consequence, we can prove the following (where $D(M)$ denotes the divisible part of $M$):

PROPOSITION 2. If $M$ is an algebraically compact $Q_p$-module such that $D(M)$ is torsion, then $\bar{J}(M) = pM$.

<u>Proof</u>: By Theorem 3.4 of Jackett[5], $\bar{J}(M) \subseteq pM$. We claim that $\bar{J}(M/D(M)) \subseteq \bar{J}(M)/D(M)$. If true, then since

$\bar{J}(M/D(M)) = p(M/D(M)) = pM/D(M)$ (by Jackett's result) we have $pM/D(M) \subseteq \bar{J}(M)/D(M)$, so $pM \subseteq \bar{J}(M)$. Thus it remains to prove the claim. Suppose $\bar{a}(=a+D(M))$ belongs to $\bar{J}(M/D(M))$. Let $A = (M,\cdot)$ be an algebra structure on $M$; we must show that $\bar{a} \in J(A)/D(M)$. Since $D(M) \subseteq pT(M)$, $D(M) \subseteq J(A)$, (cf. Fuchs[4], Exercise 9 (section 120)). Therefore $J(A)/D(M) = J(A/D(M))$. Hence, because $\bar{a} \in \bar{J}(M/D(M)), \bar{a}$ belongs to $J(A)/D(M)$. □

We shall see below that this result does not hold if $D(M)$ is not torsion.

If $M$ is a $\mathbb{Q}_p$-module and $a \in M$, let $H(a)$ denote the height sequence, or Ulm sequence, of $a$, i.e., $H(a) = (h_p^*(a), h_p^*(pa), h_p^*(p^2a), \cdots)$ [cf. Fuchs[4], §37]. Define a $\mathbb{Q}_p$-algebra $A$ to be a fully transitive algebra if whenever $a,b \in A$ such that $H(b) \geq H(a)$, there exist $u,v \in A$ such that $b = uav$. Note that the torsion part, $T(A^+)$, of the underlying module, $A^+$, of a fully transitive algebra $A$, is a fully transitive module in the sense of Kaplansky[6] (p. 58).

THEOREM 3. If $A$ is an e.c. $\mathbb{Q}_p$-algebra, then $A$ is a fully transitive algebra, and contains a subalgebra isomorphic to $\mathbb{Z}/p\mathbb{Z} \oplus \mathbb{Q}$ (the direct sum, as rings, of the canonical rings).

Proof: Here we give only a brief sketch; details can be found in Eklof-Mez[3]. If $H(b) \geq H(a)$, then the hard work goes into proving that there is an algebra extension B of A containing elements $\tilde{u},\tilde{v}$ such that $b = \tilde{u}a\tilde{v}$ ; this is done by combining a model-theoretic with a ring-theoretic construction. Then since A is e.c., the equation $xay-b = 0$ has a solution $- x = u, y = v -$ in A .

By considering the extension $A \oplus \mathbb{Z}/p\mathbb{Z} \oplus \mathbb{Q}$ of A one can show that the following system of equations and in-equations has a solution in $A$: $x = pz = (pz)^2 \neq 0$ ; $y^2 = y \neq 0$ ; $py = 0$ . From this it follows that A contains a subalgebra isomorphic to $\mathbb{Z}/p\mathbb{Z} \oplus \mathbb{Q}$ . [Send $rpz^n$ to $r \cdot \frac{1}{p^{n-1}} \in \mathbb{Q}$ $(r \in \mathbb{Q}_p, n \geq 1)$; and send $ky$ to $k(1+p\mathbb{Z}) \in \mathbb{Z}/p\mathbb{Z}$ $(k \in \{0,1,\cdots,p-1\})$] . □

Now we give some consequences relating to Problems 93 and 94.

THEOREM 4. Let A be a fully transitive $\mathbb{Q}_p$-algebra which contains a subalgebra isomorphic to $\mathbb{Z}/p\mathbb{Z} \oplus \mathbb{Q}$ . Then

(1) Every ideal of A is a fully invariant sub-module (and hence an absolute ideal) of $A^+$ ;

(2) $\mathrm{Ann}(A) = \{0\} = \overline{\mathrm{Ann}}(A^+)$ ;

(3) $J(A) = pT(A^+) = \bar{J}(A^+)$ .

<u>Proof</u>: (1) Let I be an ideal of A , and let $\gamma : A^+ \to A^+$ be an endomorphism of $A^+$ . If $a \in I$ , then $H(\gamma(a)) \geq H(a)$ , so there exist $u,v \in A$ such that $\gamma(a) = uav \in I$ .

(2) Since $H(a) \geq H(a)$ , there exist $u,v \in A$ such that $uav = a$ . Thus if $a \neq 0$ , $ua \neq 0$ , so $a \notin Ann(A)$ .

(3) By Exercise 9 (section 120) of Fuchs[4] , $pT(A^+) \subseteq \bar{J}(A^+) \subseteq J(A)$ . Let $b = 1 + p\mathbb{Z} \in \mathbb{Z}/p\mathbb{Z} \subseteq A$ . Then $b^2 = b \neq 0$ , so $b \notin J(A)$ . If $a \in T(A^+) - pT(A^+)$ , then $H(b) \geq H(a)$ , so $b = uav$ for some $u,v \in A$ ; thus $a \notin J(A)$ . Let $c = 1 \in \mathbb{Q} \subseteq A$ . Then $c^2 = c \neq 0$ , so $c \notin J(A)$ . If $a \in A - T(A^+)$ , then $H(c) \geq H(a)$ , so $c = xay$ for some $x,y \in A$ ; thus $a \notin J(A)$ . □

Theorems 3 and 4 naturally raise the question of what we can say about the underlying module, $A^+$ , of an e.c. $\mathbb{Q}_p$-algebra $A$ . We already see from Theorem 3 that $A^+$ cannot be reduced or torsion-free. In fact, we can show even more:

THEOREM 5. Let $A$ be an existentially closed $\mathbb{Q}_p$-algebra. Then

(1) $p^\omega A^+ = D(A^+)$ ;

(2) $D(A^+) \cong Z(p^\infty)^{(\omega)} \oplus \mathbb{Q}^{(\omega)}$ ;

(3) For all $n \in \omega$ , $\dim p^n A^+[p]/p^{n+1}A^+[p] \geq \omega$ , and $\dim A^+/(pA^+ + T(A^+)) \geq \omega$ .

<u>Proof</u>: The proof of (1) and (2) can be found in Eklof-Mez[3], and the proof of (3) is an easy exercise from the definition of e.c. □

It follows from Theorem 5 that the underlying modules of any two e.c. $\mathbb{Q}_p$-algebras are elementarily-equivalent (cf. Eklof-Fisher[7]). Since any module is elementarily-equivalent to an algebraically compact module, the question arises whether the underlying module of an e.c. algebra can be algebraically compact.

Recall that an algebraically compact $\mathbb{Q}_p$-module $M$ is determined up to isomorphism by certain cardinal invariants, viz: $\alpha_n(M) \stackrel{def}{=} \dim p^nM[p]/p^{n+1}M[p]$ ; $\beta(M) \stackrel{def}{=} \dim M/pM + T(M)$ ; $\gamma(M) \stackrel{def}{=} \dim D(M)[p]$ ; and $\delta(M) \stackrel{def}{=} \operatorname{rank} D(M)/T(D(M))$ . (See Chapter VII of Fuchs[4]). It may easily be shown that there are e.c. $\mathbb{Q}_p$-algebras $A$ such that $A^+$ is not algebraically compact. On the other hand:

THEOREM 6. If $\kappa$ is a cardinal $\geq 2$ s.t. $\kappa^{\aleph_0} = \kappa$, and $B$ is a $\mathbb{Q}_p$-algebra of cardinality $\leq \kappa$, then $B$ can be embedded in an e.c. $\mathbb{Q}_p$-algebra $A$ of cardinality $\kappa$ such that $A^+$ is algebraically compact and (for all $n \in \omega$) $\alpha_n(A^+) = \beta(A^+) = \gamma(A^+) = \delta(A^+) = \kappa$ .

Proof: We shall construct $A$ as the union of a continuous chain $\{A_\nu : \nu<\kappa\}$ of $\mathbb{Q}_p$-algebras of cardinality $\kappa$ such that: $B \subseteq A_0$ ; for every even $\nu$, $A_\nu$ is e.c.; and for every odd $\nu$ , $A_\nu^+$ is algebraically compact. Then $A = \bigcup_{\nu<\kappa} A_\nu$ will be e.c. since the union of a chain of e.c. algebras is e.c.; and $A^+$ will be algebraically compact since the union of a chain of algebraically compact modules is algebraically compact, provided that the length of the chain is of uncountable cofinality (by Exercise 5, section 38, of Fuchs[4]).

Suppose $A_\nu$ has been constructed. If $\nu$ is even, Lemma 1 allows us to construct $A_{\nu+1}$. If $\nu$ is odd, we let $A_{\nu+1}$ be an $\aleph_1$-saturated elementary extension of $A_\nu$ of cardinality $\kappa$ ; this is possible by Theorem 16.4 of Sacks[8] (since $\kappa^{\aleph_0}=\kappa$) . Then $A^+_{\nu+1}$ is algebraically compact (cf. Eklof-Fisher[7]).

Since $|A| = \kappa$ , the cardinal invariants of $A^+$ are $\leq \kappa$ . In order to insure that all the invariants of $A^+$ are $\geq \kappa$ , we make an appropriate choice of $A_0$ . It is not enough that the invariants of $A_0^+$ are $\geq \kappa$ since $A_0^+$ is not necessarily a pure submodule of $A^+$ ; we have to put into $A_0$ "witnesses" to the size of the cardinal invariants. We illustrate the trick with $\alpha_0$ . Let $B' = B \oplus N$ (algebra direct sum) where $N$ has the zero multiplication and $N^+$ is the direct sum of cyclic modules $\mathbb{Z}x_\nu (\nu<\kappa)$ where $px_\nu = 0$ . Let $B'' = B'(\{u\})/I$ , where $I$ is the principal ideal generated by $pu$ . Then for any $b \in B'$ , $ub = 0$ if and only if $b \in pB'$ . Finally, let $A_0$ be an e.c. algebra (of cardinality $\kappa$) containing $B''$ . Then in any algebra extension $A$ of $A_0$ , the elements $\{x_\nu : \nu<\kappa\}$ belong to $A[p]$ and are independent mod $pA[p]$ because if $\sum_{j=1}^{m} r_j x_{\nu_j} \in pA$ , then $u(\sum_{j=1}^{m} r_j x_{\nu_j}) = 0,$

so $\sum_{j=1}^{m} r_j x_{\nu_j} \in pB'$ and hence $p|r_j$ for all $j = 1,\cdots,m$ . The other invariants are handled similarly. □

COROLLARY 7. If $M$ is an algebraically compact $\mathbb{Q}_p$-module such that (for some cardinal $\lambda \geq 2$ , for all $n \in \omega)\alpha_n(M) = \gamma(M) = \delta(M) = \lambda^{\aleph_0}$ , then $M$ admits an algebra structure $(M,\cdot)$ such that every ideal of $(M,\cdot)$ is an absolute ideal, and $J((M,\cdot)) = \bar{J}(M) = pT(M)$ . □

We do not know exactly which algebraically compact modules satisfy the conclusions of Corollary 7. However, we can calculate the absolute Jacobson radical of every algebraically compact module $M$ , using Proposition 2 in case $D(M)$ is torsion, and the following otherwise.

LEMMA 8. For any $\mathbb{Q}_p$-module $M$ , $\bar{J}(M \oplus \mathbb{Q}) \subseteq T(M)$ .

<u>Proof</u>: If $b \in M \oplus \mathbb{Q}$ is of infinite order and $b \notin \mathbb{Q}$ , choose a maximal linearly independent set $\{a_i : i \in I\}$ of torsion-free elements of $M \oplus \mathbb{Q}$ such that $a_0 = 1 \in \mathbb{Q}$ and $a_1 = b$ . Define an associative multiplication on the free submodule, $U$ , on $\{a_i : i \in I\}$ by the rule: $a_i a_j = a_0 = 1 \in \mathbb{Q}$ for all $i,j \in I$ . Then because $U^2 \subseteq \mathbb{Q}$ and $M \oplus \mathbb{Q}/U$ is torsion, this multiplication extends to one on $M \oplus \mathbb{Q}$ (cf. the proof of Theorem 119.1 of Fuchs[4]). But $b$ does not belong to the Jacobson radical of this ring because $b^2 = 1$ (cf. proof of Theorem 4(3)). □

PROPOSITION 9. If $M$ is an algebraically compact $\mathbb{Q}_p$-module such that $D(M)$ is not torsion, then $\bar{J}(M) = pT(M)$.

Proof: This follows immediately from Lemma 8, Theorem 3.4 of Jackett[5] and Exercise 9 (section 120) of Fuchs[4]. □

We do not know if an algebraically compact module always admits an algebra structure which realizes the absolute Jacobson radical. In particular, is $J_p \oplus \mathbb{Q}$ a semisimple ring group?

## REFERENCES


1. Cohn, P.M., Universal Algebra, revised edition, D. Reidel, Dordrecht, 1981.

2. Hirschfeld, J. and Wheeler, H.W., Forcing, Arithmetic, and Division Rings, Lecture Notes in Math. No. 454, Springer-Verlag, Berlin, 1975.

3. Eklof, P.C. and Mez, H.C., The ideal structure of existentially closed algebras, preprint.

4. Fuchs, L., Infinite Abelian Groups, vols I and II, Academic Press, New York, 1970 and 1973.

5. Jackett, D.R., Rings on certain mixed abelian groups, Pacific Journal of Math., 98, 365, 1982.

6. Kaplansky, I., Infinite Abelian Groups, revised edition, Univ. of Michigan Press, Ann Arbor, 1969.

7. Eklof, P.C. and Fisher, E.R., The elementary theory of abelian groups, Annals of Math Logic, 4, 115, 1972.

8. Sacks, G., Saturated Model Theory, W.A. Benjamin, Reading, 1972.

# CLASSIFYING ENDOMORPHISM RINGS OF RANK ONE MIXED GROUPS

E. TOUBASSI and W. MAY

UNIVERSITY OF ARIZONA

We are going to prove two theorems on endomorphism rings of rank one abelian groups with simply presented torsion. In Theorem 1, we describe when the endomorphism rings of two groups are isomorphic, and in Theorem 2, we tell when the endomorphism ring of a group possesses only inner automorphisms. In the proofs we utilize two results: the first reduces the global problem to endomorphism rings of local groups; the second, a local theorem, classifies isomorphisms of endomorphism rings of local groups. The bulk of the proofs are then devoted to relating such isomorphisms to p-indicators.

Let $G$ be an abelian group, $x \in G$, and $p$ a prime number. We must define some subsets of $E(G)$. Let $I(G) = \mathrm{Hom}(G,T(G))$, $I_p(G) = \mathrm{Hom}(G,T(G)_p)$, $I(G,x) = \{\alpha \in I(G) \mid \alpha(x) = 0\}$, and $I_p(G,x) = \{\alpha \in I_p(G) \mid \alpha(x) = 0\}$. Then $I(G)$ and $I_p(G)$ are ideals of $E(G)$, while $I(G,x)$ and $I_p(G,x)$ are right ideals.

Let $Z_{(p)}$ denote the p-adic rational numbers. If $G$ is an abelian group, put $G_{(p)} = Z_{(p)} \otimes G$. In case $G$ has torsion-free rank 1, then $G_{(p)}$ is a $Z_{(p)}$-module of torsion-free rank 1. We let $\gamma_p : G \to G_{(p)}$ be the homomorphism given by $\gamma_p(g) = 1 \otimes g$ for $g \in G$. It is clear that the kernel of $\gamma_p$ is $\bigoplus_{q \neq p} T(G)_q$, and that $\gamma_p$ restricts to an isomorphism of $T(G)_p$ with $T(G_{(p)})$. We note that the cokernel of $\gamma_p$ will be a torsion group with trivial p-component. Since the same is true of the kernel, we see that if $x$ is a torsion-free element of $G$, then the p-indicator of $x$ in $G$ is the same as the p-indicator of $\gamma_p(x)$ in $G_{(p)}$. We shall use the notation $x_p$ for $\gamma_p(x)$.

Now let $\alpha \in I(G_{(p)})$. Then $\gamma_p|^{-1}_{T(G_{(p)})} \circ \alpha \circ \gamma_p \in I_p(G)$, hence $\gamma_p$ induces a ring homomorphism $I(G_{(p)}) \to I_p(G)$. Since the kernel and cokernel of $\gamma_p$ are torsion groups with trivial p-components, this homomorphism is bijective. Thus we have the

__Lemma.__ The map $\gamma_p$ induces a ring isomorphism $I(G_{(p)}) \to I_p(G)$, which takes $I(G_{(p)},x_p)$ to $I_p(G,x)$ for every $x \in G$.

We now prepare to state our first theorem. Let $(\sigma_{pi})$ and

$(\tau_{pi})$ be p-indicators of torsion-free elements $x$ and $y$ respectively, and let $T$ be a torsion group with p-length $\lambda_p$. We say that the p-indicator $(\sigma_{pi})$ stabilizes at $\sigma_{pk}$ if $k$ is minimal with the property that there are no gaps at or above $\sigma_{pk}$. We say that the p-indicators $(\sigma_{pi})$ and $(\tau_{pi})$ are $T_p$-equivalent if either they are equal, or else they stabilize at $\sigma_{pk}$ and $\tau_{pk}$ respectively, where $\lambda_p + k \leq \min(\sigma_{pk}, \tau_{pk}) < \infty$, $0 < |\sigma_{pk} - \tau_{pk}| < \omega$, and $\sigma_{pi} = \tau_{pi}$ $(0 \leq i < k)$, in case $k > 0$. Now let $G$ and $H$ be groups of torsion-free rank 1. We say that $G$ and $H$ are T-equivalent if $T(G) \cong T(H) \cong T$, and if there exist torsion-free elements $x \in G$ and $y \in H$ such that the p-indicators of $x$ and $y$ are $T_p$-equivalent for every $p$.

Theorem 1. Let $G$ be an abelian group of torsion-free rank one with $T = T(G)$ simply presented. If $H$ is another abelian group of torsion-free rank one, then there exists an isomorphism $\Phi : E(G) \to E(H)$ if and only if $G$ and $H$ are T-equivalent.

Proof. First assume that $G$ and $H$ are T-equivalent for torsion-free elements $x \in G$ and $y \in H$. Suppose that $x$ and $y$ have equal p-indicators. Since $T_p$ is simply presented, by Wallace[1] we may choose an isomorphism $\phi_p : G_{(p)} \to H_{(p)}$ with $\phi_p(x_p) = y_p$. This induces an isomorphism $I(G_{(p)}) \to I(H_{(p)})$ taking $I(G_{(p)}, x_p)$ to $I(H_{(p)}, y_p)$, which in turn induces an isomorphism $\Phi_p : I_p(G) \to I_p(H)$ with $\Phi_p(I_p(G,x)) = I_p(H,y)$. Now suppose that the p-indicators of $x$ and $y$ satisfy the stabilization alternative in the definition of $T_p$-equivalent. We may choose $z \in G_{(p)}$ and $z' \in H_{(p)}$ such that $p^k z = p^k x_p$, $p^k z' = p^k y_p$, $h_p(z) \geq \lambda_p$, and $h_p(z') \geq \lambda_p$. Put $t = x_p - z$ and $t' = y_p - z'$. Choose torsion-free elements $u \in G_{(p)}$ and $u' \in H_{(p)}$ such that $h_p(u) = h_p(u') = \lambda_p$. Since $t$ is a torsion element, it follows that $h_p(au + bt) = \min (h_p(au), h_p(bt))$ $(a, b \in Z)$, and similarly for $u'$ and $t'$. Thus, there exists a height-preserving isomorphism $\langle u,t \rangle \to \langle u',t' \rangle$ taking $u$ to $u'$ and $t$ to $t'$. This extends to an isomorphism $\phi_p : G_{(p)} \to H_{(p)}$, which, by the Lemma, induces $\Phi_p : I_p(G) \to I_p(H)$. If $\alpha \in I(G_{(p)})$, then $\alpha(z) = 0$ by height considerations. Therefore $\alpha(x_p) = 0$ if and only if $\alpha(t) = 0$. Similarly for $y_p$ and $t'$. Since $t$ is taken to $t'$, we conclude that $\Phi_p(I_p(G,x)) = I_p(H,y)$. Thus we can apply May and Toubassi[2] (note that $P(G) = P(H)$ is clear by assumption), to obtain the desired isomorphism $\Phi$.

Now assume that we are given an isomorphism $\Phi : E(G) \to E(H)$. The argument here will be more substantial. Choose torsion-free elements $x \in G$ and $y \in H$. By the above reference, $\Phi$ is equivalent to a family $\{\Phi_p | p \text{ prime}\}$, where, by the Lemma, we may assume that $\Phi_p : I(G_{(p)}) \to I(H_{(p)})$ is an isomorphism for every $p$, and $\Phi_p(I(G_{(p)}, x_p)) = I(H_{(p)}, y_p)$ for almost every $p$. By the same reference, each $\Phi_p$ is induced by an isomorphism $E(G_{(p)}) \to E(H_{(p)})$. Hence by May and Toubassi[3], each $\Phi_p$ is induced by an isomorphism $\phi_p : G_{(p)} \to H_{(p)}$. (Note that we now see that

$T(G) \cong T(H)$.) For every $p$, we have a relation $mp^{\ell}\phi_p(x_p) = np^{k}y_p$ for some $m,n$, $\ell$ and $k$ with $p \nmid mn$. Note that $I(G_{(q)},\gamma_q(ax)) = I(G_{(q)},\gamma_q(x))$ and $I(H_{(q)},\gamma_q(by)) = I(H_{(q)},\gamma_q(y))$ for every prime $q$ with $q \nmid ab$. It follows that after replacing $x$ and $y$ by appropriate multiples, we may assume that $\Phi_p(I(G_{(p)},x_p)) = I(H_{(p)},y_p)$ for every $p$. We still have relations as above for every $p$. Replacing $\phi_p$ by $mn^{-1}\phi_p$ does not affect $\Phi_p$, hence we may assume for every $p$ that

$\Phi_p(I(G_{(p)},x_p) = I(H_{(p)},y_p)$, and there is a relation of (*)
form $p^{\ell}\phi_p(x_p) = p^{k}y_p$.

Fix a prime $p$. We have $I(H_{(p)},\phi_p(x_p)) = I(H_{(p)},y_p)$, and $p^{\ell}\phi_p(x_p) = p^{k}y_p$. We must show that the p-indicators of $\phi_p(x_p)$ and $y_p$ are $T_p$-equivalent. We consider three cases.

(1) Suppose $\lambda_p = \infty$. Then $\langle \phi_p(x_p) \rangle$ is characterized as the intersection of the kernels of all the homomorphisms in $I(H_{(p)},\phi_p(x_p))$. Similarly for $\langle y_p \rangle$. Thus $\langle \phi_p(x_p) \rangle = \langle y_p \rangle$, and the p-indicators of $\phi_p(x_p)$ and $y_p$ are equal.

In the remaining cases, we may assume that $\lambda_p < \infty$. We may further assume that the relation $p^{\ell}\phi_p(x_p) = p^{k}y_p$ has minimal exponents, and that $\ell \geq k$. Put $t = p^{\ell-k}\phi_p(x_p) - y_p$. Then $t$ has order $p^k$. We claim there exists an integer $m$ such that

$h_p(t + m\phi_p(x_p)) \geq \lambda_p$. Let $\alpha \in I(H_{(p)}, \phi_p(x_p))$. Then $\alpha(y_p) = 0$ by our earlier assumptions, hence $\alpha(t) = 0$. Since $H_{(p)}/\langle \phi_p(x_p) \rangle$ is simply presented, we may conclude that $h_p(t + \langle \phi_p(x_p) \rangle) \geq \lambda_p$. Since $\langle \phi_p(x_p) \rangle$ is nice, the integer $m$ exists as claimed.

(2) Suppose $\ell = k$ and $\lambda_p < \infty$. The inequality $h_p(t + m\phi_p(x_p)) \geq \lambda_p$ implies $h_p(\phi_p(x_p)) \leq h_p(m\phi_p(x_p)) \leq h_p(t)$. But $t = \phi_p(x_p) - y_p$ implies $h_p(\phi_p(x_p)) \leq h_p(y_p)$. By symmetry ($\ell = k$), we also have the reverse inequality, therefore $h_p(\phi_p(x_p)) = h_p(y_p)$. But case (2) applies to $p^i\phi_p(x_p)$ and $p^iy_p$ since $I(H_{(p)}, p^i\phi_p(x_p)) = I(H_{(p)}, p^iy_p)$ for all $i$, and $\ell - i = k - i$. Therefore the p-indicators of $\phi_p(x_p)$ and $y_p$ are equal.

(3) Suppose $\ell > k$ and $\lambda_p < \infty$. First consider $k = 0$. Then $y_p = p^\ell\phi_p(x_p)$, $\ell > 0$, thus $h_p(\phi_p(x_p)) = h_p(\phi_p(x_p) + \langle y_p \rangle)$. If $\alpha \in I(H_{(p)}, y_p)$, then $\alpha(\phi_p(x_p)) = 0$, thus we conclude $h_p(\phi_p(x_p) + \langle y_p \rangle) \geq \lambda_p$ since $H_{(p)}/\langle y_p \rangle$ is simply presented. Hence $h_p(\phi_p(x_p)) \geq \lambda_p$, and therefore the p-indicator of $\phi_p(x_p)$ stabilizes at $h_p(\phi_p(x_p))$, and $h_p(p^iy_p) = h_p(p^{i+\ell}\phi_p(x_p))$ for every $i$. If $h_p(\phi_p(x_p)) = \infty$, then the two p-indicators are equal. If $h_p(\phi_p(x_p)) < \infty$, then they are $T_p$-equivalent with $k = 0$.

Now suppose $k > 0$. First we claim that $h_p(p^k\phi_p(x_p)) \geq \lambda_p$. Since $p^ky_p = p^{\ell-k} \cdot p^k\phi_p(x_p)$, we see that $h_p(p^k\phi_p(x_p)) =$

$h_p(p^k\phi_p(x_p) + \langle p^k y_p \rangle)$. If $\alpha \in I(H_{(p)}, p^k y_p)$, then $\alpha(p^k\varphi_p(x_p)) = 0$; hence since $H_{(p)}/\langle p^k y_p \rangle$ is simply presented, we conclude $h_p(p^k\phi_p(x_p)) \geq \lambda_p$, proving the claim. Recall that we have shown there exists an integer $m$ such that $h_p(t + m\phi_p(x_p)) \geq \lambda_p$. We claim that $p \nmid m$. Suppose to the contrary, that $p|m$. By the previous claim, $h_p(p^{k-1}m\phi_p(x_p)) \geq \lambda_p$. But $h_p(p^{k-1}t + p^{k-1}m\phi_p(x_p)) \geq \lambda_p$, hence $h_p(p^{k-1}t) \geq \lambda_p$. But this contradicts the facts that $p^{k-1}t \neq 0$ and $\lambda_p < \infty$. This proves our claim. Now let $0 \leq i < k$. Then $h_p(p^i t) < \lambda_p$ and $h_p(p^i t + p^i m\phi_p(x_p)) \geq \lambda_p$ imply that $h_p(p^i t) = h_p(p^i m\phi_p(x_p)) = h_p(p^i\phi_p(x_p))$. Hence $h_p(p^i y_p)$ $= h_p(p^{\ell-k+i}\phi_p(x_p) - p^i t) = h_p(p^i t) = h_p(p^i\phi_p(x_p))$. Therefore $h_p(p^i y_p) = h_p(p^i\phi_p(x_p))$ for $0 \leq i < k$. The p-indicator of $\phi_p(x_p)$ can have no gaps at or above $h_p(p^k\phi_p(x_p))$ since $h_p(p^k\phi_p(x_p)) \geq \lambda_p$. If $h_p(p^k\phi_p(x_p)) = \infty$, then the two p-indicators are equal. If $h_p(p^k\phi_p(x_p)) < \infty$, then there exists $n$, $0 < n < \omega$, such that $h_p(p^i y_p) = h_p(p^i\phi_p(x_p)) + n$ for all $i \geq k$. Moreover, $h_p(t + m\,\phi_p(x_p)) \geq \lambda_p$ implies that $h_p(p^k\phi_p(x_p)) =$ $h_p(p^k m\phi_p(x_p)) \geq \lambda_p + k$. Since $h_p(p^{k-1}\phi_p(x_p)) = h_p(p^{k-1}t)$, we see that the p-indicator of $\phi_p(x_p)$ stabilizes at $h_p(p^k\phi_p(x_p))$. It is also clear that the p-indicator of $y_p$ stabilizes at $h_p(p^k y_p)$, thus the two p-indicators are $T_p$-equivalent, finishing the proof of case (3) and the theorem.

<u>Theorem 2.</u> Let $G$ be a mixed group of torsion-free rank one with $T = T(G)$ simply presented and $x \in G$ a torsion-free element. Then every automorphism of $E(G)$ is inner if and only if there do not exist infinitely many primes $p$ such that the p-indicator of $x$ stabilizes at $\sigma_{ps}$, where $s \geq 1$ and $\lambda_p < \sigma_{ps} < \infty$.

<u>Proof.</u> First suppose that $\Phi$ is an automorphism of $E(G)$ and that there exist only finitely many primes $p$ as in the theorem. Our starting point will be the setting (*) of the previous theorem with $H = G$. Then for every $p$, $\phi_p : G_{(p)} \to G_{(p)}$ is a local automorphism and $p^{\ell}\phi_p(x_p) = p^k y_p$ ($k$ and $\ell$ depend on $p$). We now describe adjustments in either the elements $x$ and $y$ or the automorphisms $\phi_p$ which modify (*) so that $\phi_p(x_p) = y_p$ for all $p$.

For a $p$ as in the theorem we replace $x$ and $y$ by $p^s x$ and $p^s y$. This does not alter the setting (*). By doing this for the finitely many primes $p$ in the hypothesis we may assume that there are no such $p$. Next we consider cases based on the previous theorem. For $p$ as in case 3 and $p \notin P(G) = \{p | \infty$ appears in the p-indicator$\}$, the p-indicators of $\phi_p(x_p)$ and $y_p$ are different. Since $x, y \in G$ there are only finitely many of these primes. For such a $p$ we replace $x$ by $p^{\ell}x$, $y$ by $p^k y$, and for $q \neq p$ we replace $\phi_q$ by $p^{\ell-k}\phi_q$. Arguing as in the discussion prior to (*) one obtains $\Phi_p(I(G_{(p)}), x_p) = I(G_{(p)}, y_p)$ and $\phi_p(x_p) = y_p$ for all $p$. Let $p$ be as in case 2 and $p \notin P(G)$. Then $y_p = \phi_p(x_p) - t$

and there exists an $m$ such that $h_p(t + mx_p) \geq \lambda_p$. If $t \neq 0$ then $m \neq 0$ and it would follow that $p$ would be a prime as in the hypothesis, a contradiction. Thus $t = 0$ and $\phi_p(x_p) = y_p$. For $p$ with $\lambda_p = \infty$, case 1 allows us to conclude that $\phi_p(x_p) = y_p$.

Finally we consider the case $p \in P(G)$ and $p$ not in case 1. We may decompose $G_{(p)} = F \oplus T_p$ where $F$ is a torsion-free divisible $Z_p$-module. Write $\phi_p(x_p) = f_1 + t_1$ and $y_p = f_2 + t_2$ where $f_i \in F$, $t_i \in T_p$, $i = 1,2$. Now $I(G_{(p)},t_1)$ $= I(G_{(p)},\phi_p(x_p)) = I(G_{(p)},y_p) = I(G_{(p)},t_2)$ and thus $< t_1 > = < t_2 >$. Therefore there exists an automorphism $\alpha$ of $G_{(p)}$ such that $\alpha(f_1) = f_2$, $\alpha(t_1) = t_2$, and which induces unit multiplication on $F$ and $T_p$. Hence $\alpha$ is central and $\alpha\phi_p(x_p) = y_p$. On replacing $\phi_p$ by $\alpha\phi_p$ we have $\phi_p(x_p) = y_p$. By Warfield[4], the family of automorphisms $\{\phi_p | p \text{ prime}\}$, satisfying $\phi_p(x_p) = y_p$ for all $p$, extends to an automorphism $\phi$ of $G$. Thus $\phi$ induces the family $\{\phi_p | p\}$ and by May and Toubassi[2] $\phi$ induces $\Phi$.

For the converse we suppose that there are infinitely many $p$ such that the p-indicator stabilizes at $\sigma_{ps}$, $s \geq 1$, $\lambda_p < \sigma_{ps} < \infty$ and we construct an automorphism of $E(G)$ which is not inner. For $p$ as in the hypothesis and $p$ odd, choose $z \in G$ with $h_p(z) \geq \lambda_p$ and $pz = p^s x$. Set $t = p^{s-1}x - z$ and $x(p) = x + t$. It is clear that for $\alpha \in I_p(G)$, $\alpha(t) = \alpha(p^{s-1}x)$ and hence $\alpha(x(p)) =$ $\alpha(x) + \alpha(p^{s-1}x) = \alpha((1 + p^{s-1})x)$. Since $p$ is odd and $s \geq 1$, $(p, 1 + p^{s-1}) = 1$, and therefore $I_p(G,x(p)) = I_p(G,x)$. Now

$x(p) = x + t = (1 + p^{s-1})x - z$, $h_p(z) > h_p(x)$ and $(p, 1 + p^{s-1}) = 1$ implies $h_p(x(p)) = h_p(x)$. Thus for $p$ as in the hypothesis the height preserving map $x_p \to \gamma_p(x(p))$ extends to an automorphism $\phi_p : G_{(p)} \to G_{(p)}$. For the remaining $p$ let $\phi_p = 1$.

The family $\{\phi_p : G_{(p)} \to G_{(p)} | p\}$ induces a family $\{\Phi_p : I(G_{(p)}) \to I(G_{(p)}) | p\}$ such that $\Phi_p(I(G_{(p)}, x_p)) = I_p(G_{(p)}, x_p)$ for all $p$. Thus by May and Toubassi[2] we obtain an automorphism $\Phi$ of $E(G)$. Suppose by way of contradiction that $\theta : G \to G$ induces $\Phi$. Then $\theta(mx) = nx$ for some nonzero integers $m$ and $n$. Choose $p$ as in the hypothesis, $p$ odd and $p \nmid mn$. Let $\{\theta_p : G_{(p)} \to G_{(p)} | p\}$ be the induced family of maps associated with $\theta$. It follows that $\phi_p$ and $\frac{m}{n}\theta_p$ induce $\Phi_p$ and hence there exists a central automorphism $\alpha$ of $G_{(p)}$ such that $\alpha(x_p) = \gamma_p(x(p))$. Put $\sigma = \alpha - 1$. Then $\sigma$ is central in $I(G_{(p)})$ and $\sigma(x_p) = t$. We now show that such a $\sigma$ does not exist, thus completing the proof of the theorem.

Choose a root $w$ of $z$ subject to: (1) $\lambda_p$ appears in the p-indicator of $w$ and (2) $h_p(p^i w) = s$ for $0 \leq i \leq s$. This can be done as follows. Since $\lambda_p \leq h_p(z) < \lambda_p + \omega$ we may choose a root of $z$ of p-height exactly $\lambda_p$. We then take a root of this element which is sufficiently large, adding a generator of a cyclic summand of exponent $\geq s$, if necessary. Note that $p^k w = p^s x$ for some $k > s$. Put $t' = x - p^{k-s}w$. If $\sigma(w) = 0$ then $t = \sigma(x) = \sigma(t')$. By (2) there exists a homomorphism $\tau$ such that $\tau(w) = t'$.

Consequently $\sigma\tau(w) = t$ yet $\tau\sigma(w) = 0$, a contradiction. Thus we may suppose $\sigma(w) = t'' \neq 0$. If $h_p(t'' + \langle w \rangle) \geq \lambda_p$ then from the observation in the first theorem $\lambda_p$ is not in the p-indicator of w, contradicting (1). Therefore $h_p(t'' + \langle w \rangle) < \lambda_p$ and we may choose $\tau$ with $\tau(w) = 0$ and $\sigma\tau(t'') \neq 0$. This however leads to another contradiction since $\sigma\tau(w) = 0$ and $\tau\sigma(w) \neq 0$.

## References

1. Wallace, K., On mixed groups of torsion-free rank one, *J. Algebra*, 17, 482, 1971.

2. May, W. and Toubassi, E., Isomorphisms of endomorphism rings of rank one mixed groups, *J. Algebra*, 71, 508, 1981.

3. May, W. and Toubassi, E., Endomorphisms of rank one mixed modules over discrete valuation rings, *Pacific J. Math.*, 108, 155, 1983.

4. Warfield, Robert W., Jr., The structure of mixed abelian groups, in *Abelian Group Theory, 2nd New Mexico State University Conference, 1976*, 616, Arnold, D., Hunter, R. and Walker, E., Springer-Verlag, New York, 1977.

# NOTES ON MIXED GROUPS I

PHILLIP SCHULTZ

UNIVERSITY OF WESTERN AUSTRALIA

## 1. INTRODUCTION

There is a flourishing theory of mixed groups considered as extensions of nice full torsion-free subgroups by simply presented torsion groups, and an older somewhat neglected theory of extensions of torsion groups by torsion-free groups. The purpose of these notes is to introduce an alternative approach by considering the torsion subgroup and a full torsion-free subgroup as of equal status, and likewise for a torsion-free and a torsion factor group.

Consequently certain intrinsic difficulties of the other theories are avoided; for example, the "torsion-free-by-torsion" theory is successful only for groups whose torsion-free and torsion parts have a complete system of cardinal invariants, or at least a well understood structure theory [10], whereas the "torsion-by-torsion-free" theory has already encountered the hurdle of intransigent problems of set theory [1]. In these notes, we accept all torsion and torsion-free groups as known, and we are concerned only with how they can be combined. Furthermore, we have enough extra data to make the splitting problem trivial.

Section 2 is a general theory of pullbacks of a torsion and a torsion-free group, which yields a local-global connection.

The theory of pullbacks is applied in Section 3 to classify rank 1 groups and in Section 4 to classify mixed groups with a full Murley subgroup.

In Section 5, mixed groups are characterized as pullbacks of a cotorsion group and a torsion-free group. This leads to a definition of a class of groups which can be classified by linear transformations of rational vector spaces, and some structure theorems for such groups. Sections 4 and 5 will appear separately elsewhere under the title "Notes on Mixed Groups II".

## 2. MIXED GROUPS AS PULLBACKS

Let $G$ be a mixed group with torsion subgroup $t$, let $W = G/t$, and let $\rho : G \to W$ be the natural homomorphism. Let $A$ be any full torsion-free subgroup of $G$, let $U = G/A$ and let $\pi : G \to U$ be the natural homomorphism. Since $t \cap A = 0$, the inclusions of $A$ and $T$ in $G$ induce monomorphisms $\kappa : t \to U$ and $\sigma : A \to W$ such that $U/\kappa(t) \cong W/\sigma(A) \cong G/\langle t,A\rangle$ and there is an exact commutative diagram:

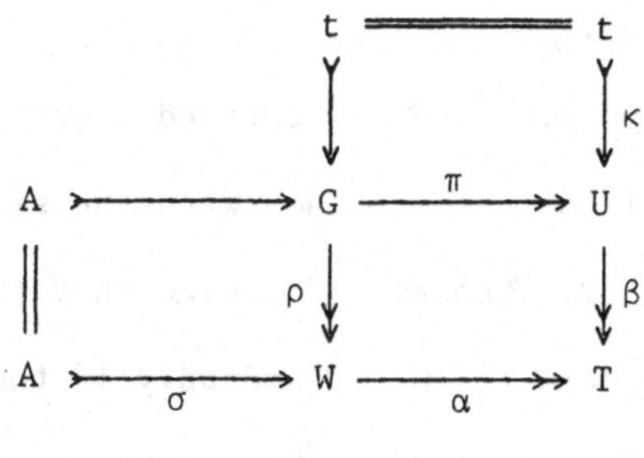


Figure 1

In this diagram, $T \cong U/\kappa(t)$ is a torsion group and $\alpha$ and $\beta$ are chosen to make the lower right square commute. Such a diagram is called a *pullback diagram for* $G$ ; we say $G$ is *the pullback of* $\alpha$ *and* $\beta$ , or $G$ is *a pullback of* $W$ *and* $U$ *with kernels* $A$ *and* $t$ . Of course $G$ has many pullback diagrams, depending on the choice of $A$ , $\alpha$ and $\beta$ .

Conversely, let

$$A \overset{\sigma}{\rightarrowtail} W \overset{\alpha}{\twoheadrightarrow} T \quad \text{and} \quad t \overset{\kappa}{\rightarrowtail} U \overset{\beta}{\twoheadrightarrow} T$$

be short exact sequences, where $A$ and $W$ are torsion-free, and $T, t$ and $U$ are torsion groups. Let

$$G = \{(w,u) \in W \oplus U : \alpha(w) = \beta(u)\} \text{ , and define}$$

$$\rho : G \to W \text{ by } (w,u) \mapsto w \text{ and } \pi : G \to U \text{ by } (w,u) \mapsto u \text{ .}$$

Then with the obvious identifications of $A$ and $t$ with subgroups of $G$ ,

Figure 1 is a pullback diagram for $G$ . This construction is essentially due to Fuchs [2, Theorem 6].

Suppose now that with $W, U$ and $T$ defined above $\alpha' : W \to T$ and $\beta' : U \to T$ are another pair of epimorphisms with the same kernels $A$ and $t$ as $\alpha$ and $\beta$ . Fuchs [2, Theorem 11] proved that the pullback $G'$ of $\alpha'$ and $\beta'$ is isomorphic to $G$ if and only if there exist interlacing automorphisms of $T$ induced by automorphisms of $W$ and $U$ . More precisely, let $\bar{\alpha}, \bar{\alpha}' : W/\sigma(A) \to T$ and $\bar{\beta}, \bar{\beta}' : U/\kappa(t) \to T$ be the isomorphisms induced by $\alpha, \alpha', \beta$ and $\beta'$ respectively. Let $\theta = \bar{\beta}\bar{\alpha}^{-1}$ and $\theta' = \bar{\beta}'\bar{\alpha}'^{-1}$ be the induced automorphisms of $T$ . Then $G \cong G'$ if and only if there exist $\xi, \eta \in \mathrm{Aut}(T)$ such that $\xi\alpha = \alpha\bar{\xi}$ for some $\bar{\xi} \in \mathrm{Aut}(W)$ , $\eta\beta = \beta\bar{\eta}$ for some $\bar{\eta} \in \mathrm{Aut}(U)$ , and $\theta\xi = \eta\theta'$ .

It follows immediately [4, Theorem 1] that if every automorphism of $T$ lifts to an automorphism of $W$ or to an automorphism of $U$ , then all pullbacks of $W$ and $U$ with kernels $A$ and $t$ are isomorphic. This fact will be used in Sections 3 and 4 to classify certain mixed groups.

Our object now is to show how a mixed group represented by the pullback diagram Figure 1 can in some sense be decomposed into its "primary components" which are p-mixed groups. This decomposition is just the Universal Algebra concept of direct sum with amalgamated subgroup [9, Theorem 3.4].

Theorem 2.1. Let $G$ be a mixed group with pullback diagram as in Figure 1. Let $R$ be the set of primes relevant for $U$ . For each $p \in R$ , let:

$$G^p = \pi^{-1}(U_p) \; ; \; W^p = \rho(G^p) \; ; \; \pi_p = \pi \restriction G^p \; ; \; \rho_p = \rho \restriction G^p \; ;$$

$$\kappa_p = \kappa \restriction t_p \quad ; \; \beta_p = \beta \restriction U_p \quad ; \; \alpha_p = \alpha \restriction W^p \; .$$

Then $G^p$ has pullback diagram:

$$\begin{array}{ccccc}
 & & t_p & = & t_p \\
 & & \downarrow & & \downarrow{\scriptstyle \kappa_p} \\
A & \rightarrowtail & G^p & \xrightarrow{\pi_p} & U_p \\
\| & & \downarrow{\scriptstyle \rho_p} & & \downarrow{\scriptstyle \beta_p} \\
A & \rightarrowtail & W^p & \xrightarrow[\alpha_p]{} & T_p
\end{array}$$

Figure 2

Conversely, let $A$ be a torsion-free group and $R$ a non-empty set of primes. For each $p \in R$ , let

$$A^p \overset{\sigma_p}{\rightarrowtail} W^p \overset{\alpha_p}{\twoheadrightarrow} T_p \quad \text{and} \quad t_p \overset{\kappa_p}{\rightarrowtail} U_p \overset{\beta_p}{\twoheadrightarrow} T_p$$

be short exact sequences in which $t_p, U_p$ and $T_p$ are p-groups, $W^p$ is torsion-free and $A^p \cong A$ . Let $G^p$ be the pullback of $\alpha_p$ and $\beta_p$ , so $G^p$ has Figure 2 as pullback diagram, and let $G$ be the direct sum of the $G^p$ with amalgamated subgroup $A$ , and $W$ the direct sum of the $W^p$ with amalgamated subgroup $A$ .

Let $\kappa = \oplus\kappa_p$ , $\beta = \oplus\beta_p$ , $t = \oplus t_p$ , $U = \oplus U_p$ , $T = \oplus T_p$ .

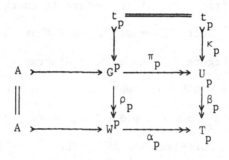

Let $\pi, \rho$ and $\alpha$ be the natural homomorphisms induced by $\oplus\pi_p$ , $\oplus\rho_p$ and $\oplus\alpha_p$ .

Then $G$ is the pullback of $\alpha$ and $\beta$ and has Figure 1 as pullback diagram. The relation so defined between pullbacks of $\alpha$ and $\beta$ , and families of pullbacks of $\alpha_p$ and $\beta_p$ , $p \in R$ , is a $1 - 1$ correspondence.

Proof. For the first part, it is routine to check that $t_p$ is the torsion subgroup of $G^p$ with cokernel $W^p$ and that $A$ is the kernel of $\pi_p$ and of $\alpha_p\rho_p$ , so Figure 2 is a pullback diagram.

Conversely, let $B = \oplus G^p$ , $C = \oplus W^p$ and let $D$ be the subgroup of $B$ generated by $\{a_p - a_q : a \in A , p,q \in R\}$ , where for all $p \in R$ , $a \mapsto a_p$ is the given isomorphism $A \to A^p$ . Then $G = B/D$ , and $D$ can be identified with a subgroup of $C$ such that $W = C/D$ .

The mappings $\pi, \rho$ and $\alpha$ are defined by:

$$\pi : \Sigma g^p + D \mapsto \Sigma\pi_p(g^p)$$

$$\rho : \Sigma g^p + D \mapsto \Sigma\rho_p(g^p)$$

and $\alpha : \Sigma w^p + D \mapsto \Sigma\alpha_p(w^p)$ .

It is routine to check that they are well-defined epimorphisms and that $\beta\pi = \alpha\rho$ .

To complete the proof that the correspondence is $1 - 1$ , it remains to show that if the $G^p$ are constructed from $G$ , then $G$ is the direct sum of the $G^p$ with amalgamated subgroup $A$ , and that if $G$ is constructed from the $G^p$ , then $G^p = \pi^{-1}(U_p)$ . The first statement follows from the fact that if $g \in G$ , then $\pi(g)$ is a finite sum $\Sigma u_p$ , with $u_p \in U_p$, and the second from the fact that the natural homomorphism of $G^p$ into $G$ has kernel $0$ and image $\pi^{-1}(U_p)$.

## 3. CLASSIFICATION OF RANK 1 GROUPS

Let $t$ be a torsion group and $A$ the infinite cyclic group. Let $\mathcal{G}$ be the set of all rank 1 mixed groups with torsion subgroup $t$. By the results of the previous section, for each $G \in \mathcal{G}$ there is a rank 1 torsion-free group $W$ and a torsion group $U$ such that $G$ is a pullback of $W$ and $U$ with kernels $A$ and $t$. In this section, we shall determine for each $G \in \mathcal{G}$, all the pullback diagrams which represent $G$.

Let $K(t)$ be the set of all monomorphisms $\kappa : t \to U$ such that $U$ is a torsion group and $U/\kappa(t)$ is locally cyclic. Note that if $G \in \mathcal{G}$ has Figure 1 as pullback diagram, then $T$ is locally cyclic since $W$ is a rank 1 group. Thus the embedding of $A$ in $G$ determines a unique $\kappa \in K(t)$.

Conversely, each $\kappa \in K(t)$ gives rise to a $G \in \mathcal{G}$ and an embedding $A \leq G$ as follows: let $T = U/\kappa(t)$ and let $\beta : U \to T$ be any epimorphism with kernel $t$. Let $W$ be the unique subgroup of $\mathbb{Q}$ containing $\mathbb{Z}$ such that $W/\mathbb{Z} \cong T$, and let $\alpha : W \to T$ be any epimorphism with kernel $\mathbb{Z}$. Then the pullback $G$ of $\alpha$ and $\beta$ is in $\mathcal{G}$; with $A = \ker \pi$, Figure 1 is a pullback diagram for $G$.

This correspondence between $\mathcal{G}$ and $K(t)$ involves choices of $\alpha$ and $\beta$ and a particular embedding of $A$ in $G$. In order to obtain a satisfactory classification it is necessary to eliminate these arbitrary choices. We achieve this by defining on $K(t)$ an equivalence relation which identifies those monomorphisms $\kappa$ which yield isomorphic groups $G$.

Let $\kappa : t \to U$ and $\kappa' : t \to U'$ be in $K(t)$; we say $\kappa$ <u>dominates</u> $\kappa'$ if there is an epimorphism of $U$ onto $U'$ which fixes $t$ and has a cyclic kernel. In terms of pullbacks, $\kappa$ dominates $\kappa'$ if there is a pullback diagram

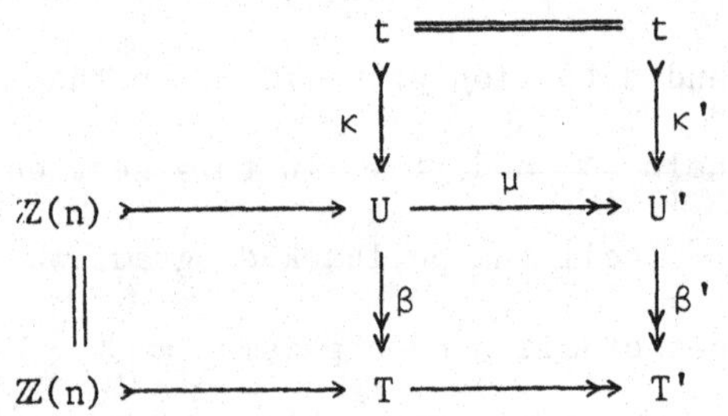


<u>Figure 3</u>

where $T$ and $T'$ are locally cyclic.

If $\kappa : t \to U$ and $\kappa' : t \to U'$ are elements of $K(t)$, we say that $\kappa$ is <u>equivalent</u> to $\kappa'$, and $U$ is <u>t-equivalent</u> to $U'$ if there exists $\kappa^* \in K(t)$ such that $\kappa^*$ dominates both $\kappa$ and $\kappa'$.

Clearly t-equivalence is reflexive and symmetric, and is preserved by isomorphisms of $U$ which fix $t$. The following theorem shows that it is also transitive.

<u>Theorem 3.1.</u> Let $t$ be a torsion group $A$ the infinite cyclic group and $\kappa, \kappa' \in K(t)$. Let $\kappa : t \to U$ determine $G \in \mathcal{G}$ and an embedding

of $A$ in $G$, and let $\kappa' : t \to U'$ determine $G' \in \mathcal{G}$ and an embedding of $A$ in $G'$ as described above.

Then $G \cong G'$ if and only if $\kappa$ is equivalent to $\kappa'$.

Proof. Suppose $\phi : G' \to G$ is an isomorphism. Let $b$ be a generator of $A$ in $G'$, and $a$ a generator of $A$ in $G$. Since $G$ has rank 1, $a$ and $\phi(b)$ have a common integral multiple $c = na = m\phi(b)$ in $G$. Let $C$ be the cyclic subgroup of $G$ generated by $c$, and let $\kappa^* : t \to U^*$ be the element of $K(t)$ determined by the embedding of $C$ in $G$.

Now $A/C \cong \mathbb{Z}(n)$ is a subgroup of $U^* = G/C$ which misses $\kappa^*(t)$, and the natural homomorphism of $U^*$ onto $U$ fixes $t$, so $\kappa^*$ dominates $\kappa$; similarly $\phi^{-1}$ induces a homomorphism of $U^*$ onto $U'$ fixing $t$, whose kernel is isomorphic to $\mathbb{Z}(m)$, so $\kappa^*$ dominates $\kappa'$. Hence $\kappa$ and $\kappa'$ are equivalent.

Conversely, suppose $\kappa$ and $\kappa'$ are equivalent. Without loss of generality we may assume $\kappa$ dominates $\kappa'$, and that Figure 1 is a pullback diagram for $G$, and Figure 3 is a pullback diagram for $U$. Let

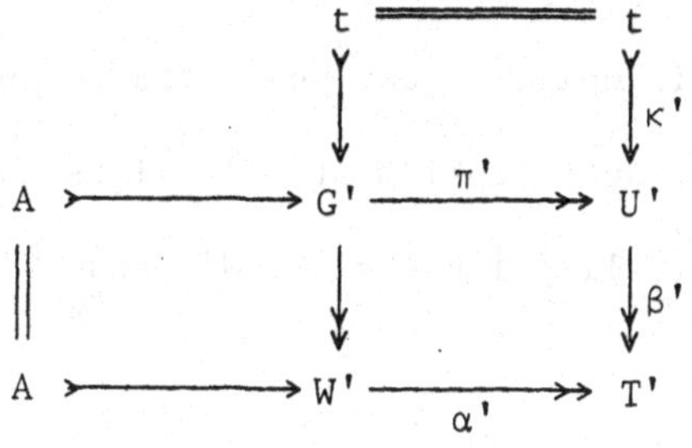


Figure 4

be a pullback diagram for $G'$.

Since $T$ and $T'$ are locally cyclic, Figure 3 implies that they differ only by a finite summand, so $W$ and $W'$ are rank 1 torsion-free groups with the same type. Hence there exists an isomorphism $\theta : W \to W'$.

Now the composite mappings $\beta' \circ \mu \circ \pi$ and $\alpha' \circ \theta \circ \rho$ are epimorphisms of $G$ onto $T'$, so there is an automorphism $\psi$ of $T'$ such that $\psi \circ \beta' \circ \mu \circ \pi = \alpha' \circ \theta \circ \rho$. Hence $G$ has a pullback diagram

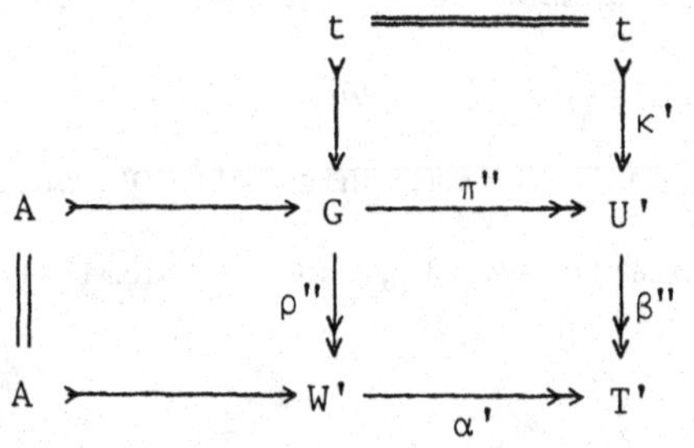


Figure 5

in which $\pi'' = \mu \circ \pi$, $\beta'' = \psi \circ \beta'$ and $\rho'' = \theta \circ \rho$.

For each relevant prime $p$, every automorphism of $T'_p$ is a multiplication by a p-adic unit, so can be lifted to an automorphism of $U'_p$. Consequently, every automorphism of $T'$ lifts to an automorphism of $U'$ so by [4, Theorem 1], all pullbacks of $W'$ and $U'$ with kernels $A$ and $t$ are isomorphic.

In particular, $G \cong G'$.

It is important to see how this classification can be used to recover the important invariants of rank 1 groups, such as the height matrices of elements of infinite order. The following theorem shows how to compute the height matrix of $a \in G$ in terms of heights in the torsion group $U$. We use $(\sigma_n)$ to denote the p-indicator of an element of infinite order in $G$, and to avoid consideration of special cases, we put $\sigma_{-1} = -1$ ; $h_A^p(x)$ denotes the generalized p-height of an element $x$ of a group $A$.

Theorem 3.2. Let $G$ be a pullback of a rank 1 torsion-free group $W$ and a torsion group $U$ with kernels $A$, a cyclic group, and $t$. Let $\beta : U \to T$ be the epimorphism of Figure 1 and let $a$ be a generator of $A$. Let $p$ be a prime and let $(\sigma_n)$ be the p-indicator of $a$; for all $0 \le n < \omega$, let

$$U(n) = \{u \in U : u \text{ has order } p^{n+1} \text{ and } \beta(u) \text{ has order } p\}, \text{ and let}$$

$$\tau(n) = \max\{\sup\{h_U^p(p^n u) + 1 : u \in U(n)\}, \sigma_{n-1} + 1\}.$$

Then for all $0 \le n < \omega$, $\sigma_n = \tau(n)$.

Proof. With $W$ and $\alpha$ as in Figure 1,

$$G = \{(w,u) \in W \oplus U : \alpha(w) = \beta(u)\}$$

If $\sigma_n > \sigma_{n-1} + 1$, then $p^n a = p(w',u')$ for some $(w',u') \in G$ of p-height $\ge \sigma_{n-1} + 1 \ge n$, so $p^n a = p^{n+1}(w,u)$ for some $(w,u) \in G$ such that $p^n w = w'$ and $p^n u = u'$. Thus $w = p^{-1}a$ and $p^{n+1}u = 0$. If $p^n u = u' = 0$, then $(w',u') = p^{n-1}a$, so $(w',u')$ has p-height $\sigma_{n-1}$, a contradiction. Hence $u$ has order $p^{n+1}$. Since $w = p^{-1}a$, $\beta(u) = \alpha(w)$ has order $p$, so $u \in U(n)$. It follows by the contrapositive that if $U(n)$ is empty, then $\sigma_n = \sigma_{n-1} + 1$, and the formula for $\sigma_n$ is correct in this case.

If $U(n)$ contains an element $u$ such that $h_U^p(p^n u) > \sigma_{n-1}$, there exists $w \in W$ such that $(w,u) \in G$. Since $A$ is p-nice in $G$ [3, Lemma 104.1], we can even choose $w$ so that $h_G^p(p^n w, p^n u) = h_U^p(p^n u)$. Since $\alpha(w) = \beta(u)$ has order $p$, $w = p^{-1} sa$ for some integer $s$ prime to $p$. Then

$$\sigma_n = h_G^p(p^n sa) = h_G^p(p(p^n w, p^n u)) \geq 1 + h_U^p(p^n u) \text{ ; hence } \sigma_n \geq \tau(n) .$$

Suppose then that some $\sigma_n > \tau(n)$. Thus $p^n a = p(w',u')$ for some $(w',u') \in G$ with p-height $\geq \tau(n) \geq n$, and there exists $(w,u)$ in $G$ with $(w',u') = p^n(w,u)$. Hence $p^n a = p^{n+1}(w,u)$ so $w = p^{-1}a$, $p^{n+1}u = 0$, and $p^n u = u' \neq 0$ (else $(w',u') = p^{n-1}a$ has p-height $\sigma_{n-1} < \tau(n)$), so $u \in U(n)$. But $h_U^p(p^n u) = h_U^p(u') \geq h_G^p(w',u') \geq \tau(n)$, a contradiction. Hence $\sigma_n = \tau(n)$ for all $n$.

Corollary. If $U_p = 0$, then $(\sigma_n) = (0,1,2,\ldots)$. In particular in the localization (Section 2) of a rank 1 group, the q-indicator of a generator $a$ of $A \leq G^p$ is $(0,1,2,\ldots)$.

Problem. Given $t$ and an admissible height matrix $M$, construct $U$ containing $t$ so that the corresponding rank 1 mixed group $G$ has an element of infinite order realizing $M$.

REFERENCES


1. Eklof, P., Applications of logic to the problem of splitting abelian groups, in *Logic Colloquium 76*, Gandy, R. and Hyland, M., Eds., North-Holland, Amsterdam, 1977, 287.
2. Fuchs, L., On subdirect unions I, *Acta Math. Acad. Sci. Hung.* 3, 103, 1952.
3. Fuchs, L., *Infinite Abelian Groups*, Vols. I and II, Academic Press, New York, 1970 and 1973.
4. Fuchs, L., and Loonstra, F., On the cancellation of modules in direct sums over Dedekind domains, *Koninkl. Nederl. Akad. Van Wetenschappen*, 74, 163, 1971.
5. Harrison, D.K., Infinite abelian groups and homological methods, *Annals of Math.*, 69, 366, 1959.
6. Griffith, P., On direct sums of p-mixed groups, *Archiv der Math.*, 19, 359, 1968.
7. Murley, C., The classification of certain classes of torsion-free Abelian groups, *Pacific J. of Math.* 40, 647, 1972.
8. Oppelt, J., A decomposition of mixed abelian groups, *Trans. Amer. Math. Soc.* 127, 341, 1967.
9. Pierce, R.S., *Introduction to the theory of abstract algebras*, Holt, Rinehart and Winston, New York, 1968.
10. Richman, F., Mixed groups, *Abelian Group Theory*, Lecture Notes in Maths. 1006, Springer-Verlag, Berlin, 1983, 445.

11. Rotman, J., A completion functor on modules and algebras, *J. of Algebra,* 9, 369, 1968.

12. Stratton, A.E., A note on Ext-completions, *J. of Algebra,* 17, 110, 1971.

13. Mader, A., Extensions of abelian groups, in *Studies on abelian groups,* Charles, B., Ed., Dunod, Paris, 1968, 259.

# COUNTABLE $\aleph_0$-INDECOMPOSABLE MIXED ABELIAN GROUPS OF TORSION-FREE RANK 1

ALEXANDER SOIFER

UNIVERSITY OF COLORADO AT COLORADO SPRINGS

## 1. INTRODUCTION

In [2] the author found all countable reduced torsion abelian groups T such that there exists an $\aleph_0$-indecomposable extension G of T by a rank 1 torsion-free group. This work enabled the author to come up with the following conjecture (Problem Session, The Honolulu Conference on Abelian Group Theory): let $0 \to T \to G \to R \to 0$ be a short exact sequence, with T a countable reduced torsion group and R a rank one torsion-free group. If $t(R) < \{r_p(T)\}_{p \in P}$ (where $t(R)$ is the type of R, $r_p(T)$ is the p-rank of T, and P is the set of primes), then G is $\aleph_0$-decomposable (i.e. G decomposes into a direct sum of $\aleph_0$ non-zero summands).

This conjecture is proven here. It allows to find all pairs (T,R) with countable reduced torsion group T, and a rank one torsion-free group R, such that there exists an $\aleph_0$-indecomposable extension G of T by R. Such an extension G exists if and only if T, R satisfy the following conditions:

1) a basic subgroup B of T can be presented as a direct sum $B = F \oplus H$, where F is a finite group and every non-zero p-component $H_p$ of H has the structure

$$H_p = \oplus_{i \in N_p} Z(p^{k_i})$$

where $N_p$ is the set of natural numbers N or any beginning subset of N, and every $i \in N_p$ such that $i + 1 \in N_p$ satisfies the inequality $k_{i+1} - k_i \geqslant 2$.

2) $$\{r_p(T)\}_{p \in P} \leqslant t(R)$$

Section 4 contains number of conjectures.

The author acknowledges the support of this research by the University of Colorado at Colorado Springs.

## 2. PRELIMINARIES

All groups in this paper are abelian. We will mostly use the notations and terminology of [1]. In addition, N will stand for the set of natural numbers;

$N^* = \{0\} \cup N$ ; P -- the set of prime numbers; (m,n) --the greatest common divisor of m, $n \in Z$ ; $h_p(g)_G$ -- the p-height of g in G, ($g \in G$) ; $f_k^p(G)$ -- the Ulm-Kaplansky invariants, ($p \in P$, $k \in N^*$) ; $B_n[p^n]$ --the direct sum of a set of cyclic groups $Z(p^n)$ of order $p^n$ ; if T is a subgroup of G and a, $b \in G$, then $a \equiv b \pmod T$ means exactly $a - b \in T$.

Let $\{G_i\}_{i \in \Delta}$ be a set of groups. The subgroup $S_{i \in \Delta} G_i$ of the direct product $\prod_{i \in \Delta} G_i$ consisting of all sequences $\{g_i\}_{i \in \Delta}$, ($g_i \in G_i$), such that for any natural n almost all components of $\{g_i\}_{i \in \Delta}$ satisfy $g_i \in nG_i$, is called the regular direct sum of the groups $G_i$, $i \in \Delta$.

The group G is said to be fully reduced if $\bigcap_{n \in N} nG = 0$.

LEMMA 1. ([3], theorem 4.3) Let G be a fully reduced group and

$$B = \oplus_{p \in P} \oplus_{n \in N} B_n[p^n]$$

an arbitrarily fixed basic subgroup of the torsion part tG of G, ($\delta: B \to G$ is the corresponding embedding).

Then there exists a group $\hat{G}$ and a monomorphism $\sigma: G \to \hat{G}$ such that

$$\hat{G} = \prod_{p \in P} S_{n \in N} B_n[p^n] \oplus G'$$

where the group G' is the regular direct sum of some set of additive groups of rings of p-adic integers (with distinct or identical p), the subgroup $\sigma G$ is pure in

$\hat{G}$, and the composition $\sigma\,\delta$ is the canonical embedding of the direct sum into the regular direct sum.

LEMMA 2. Let B be a basic subgroup of a reduced countable torsion group T. There exists an $\aleph_0$-indecomposable extension of T by a finite rank torsion-free group R if and only if there exists an $\aleph_0$-indecomposable extension of B by R.

THEOREM 1. ( [2] , theorem 2) Let T be a countable reduced torsion group. There exists an $\aleph_0$-indecomposable extension of T by a torison-free group of the torsion-free rank 1 if and only if a basic subgroup B of T can be presented as a direct sum $B = F \oplus H$, where F is a finite group and every non-zero p-component $H_p$ of H has the structure

$$H_p = \oplus_{i \in N_p} Z(p^{k_i}) \qquad (0)$$

where $N_p$ is the set of natural numbers N or any beginning subset of N, and every $i \in N_p$ such that $i + 1 \in N_p$ satisfies the inequality $k_{i+1} - k_i \geqslant 2$.

REMARK. Rank 1 torsion-free group R constructed in [2] in the proof of the sufficient condition of theorem 2, in fact satisfies the following condition:

$$t(R) = \{r_p(T)\}_{p \in P}.$$

## 3. COUNTABLE COUNTABLY-INDECOMPOSABLE GROUPS OF THE TORSION-FREE RANK 1, n-DECOMPOSABLE FOR ANY FINITE n.

THEOREM 2. Given a countable reduced torsion group T and a torsion-free group R of rank 1. There exists an extension of T by R which is an $\aleph_0$-indecomposable group if and only if T, R satisfy the following conditions:

1) $T = F \oplus H$, where F is a finite group and for every prime p and non-negative integer k

$$f_k^p(H) + f_{k+1}^p(H) \leqslant 1 \qquad (1)$$

2) $$\{r_p(T)\}_{p \in P} \leqslant \mathbf{t}(R) \qquad (2)$$

PROOF. The sufficient condition follows from theorem 1 and remark below it.

NECESSARY CONDITION. Let G be an $\aleph_0$-indecomposable group with the torsion part T(G) = T and G T $\tilde{=}$ R ($\tau : G \to R$ corresponding epimorphism).

Due to theorem 1, T satisfies the condition (0), or equivalently, condition (1).

Assume the groups T, R do not satisfy the condition (2), then one of the following two cases takes place.

CASE 1. For some prime p $r_p(T) = \aleph_0$, and p-component of $\mathbf{t}(R)$ is finite. In this case G has a countable direct summand which is a p-group, therefore G is $\aleph_0$-decomposable.

CASE 2. Let $\bar{g} \in R$. There exists an infinite set of primes $P_1$ such that for every $p \in P_1$

$$r_p(T) > h_p(\bar{g})_R \tag{3}$$

Due to lemma 2, we can assume without losing generality that the group T is a direct sum of cyclics. Also, F is a direct summand of G, so we can assume F=0 without losing generality. Decomposition (0) can be written as follows:

$$T = \oplus_{p \in P_1} \oplus_{i \in N_p} Z(p^{k_i}) \oplus T_1 \tag{4}$$

where $(\forall p \in P_1)(\forall i \in N_p) \ni (i+1 \in N_p): k_{i+1} - k_i \geqslant 2$ and $T_1$ is a direct sum of p-components $T_p$ of T for $p \in P \setminus P_1$.

According to lemma 1, we can present G as a pure subgroup of

$$\hat{G} = \prod_{p \in P_1} S_{i \in N_p} Z(p^{k_i}) \oplus G_1 \tag{5}$$

with the direct sum T (decomposition (4)) canonically embedded into G and $T_1 \leqslant G_1$. Choose an inverse image $g \in G$ of $\bar{g} \in R$ at the epinorphism $\tau$. In accordance with the decomposition (5), g can be written as the sequence of its components in $Z(p^{k_i})$ and $G_1$:

$$g = \{\{ [g]_i^p \mid p \in P_1, i \in N_p\} ; g_1\}$$

If for infinitely many $p \in P_1$ and $i, i+1 \in N_p$

$$o([g]_{i+1}^{p}) < o([g]_{i}^{p}) \tag{6}$$

or for infinitely many $p \in P_1$ and $i, i+1 \in N_p$

$$h_p([g]_{i+1}^{p}) < h_p([g]_{i}^{p}) \tag{7}$$

then by changing generators of $Z(p^{k_i})$ (if (6) takes place) or $Z(p^{k_{i+1}})$ (if (7) takes place), we can create $\aleph_0$ zero components $[g]_{i}^{p}$ of g. Then by appropriate divisions of g in G by $h_p(\bar{g})_R$, $(p \in P_1)$ modulo T , we can prove the existence of a countable direct summand of G, which is a direct sum of cyclics, and thus show that G is $\aleph_0$-decomposable.

This contradiction proves that inequalities (6), (7) do not hold for almost all $p \in P_1$ ; $i, i+1 \in N_p$. It is easy to show that without losing generality we can omit the word "almost" in the previous sentence, i.e., we have:

$$(\forall p \in P_1)(\forall i \in N_p) \ni (i+1 \in N_p): \; o([g]_{i+1}^{p}) > o([g]_{i}^{p}) \tag{8}$$

and

$$h_p([g]_{i+1}^{p}) > h_p([g]_{i}^{p}) \tag{9}$$

If for infinitely many $p \in P_1$ $[g]_{1}^{p} = 0$, it can be easily shown that G is $\aleph_0$ -decomposable. Therefore $[g]_{1}^{p} \neq 0$ for almost all $p \in P_1$, and we can assume without losing generality, that

$$(\forall p \in P_1): [g]_{1}^{p} \neq 0 \tag{10}$$

Let $p \in P_1$. Define w(p) by $p^{w(p)} = h_p(\bar{g})_R$. Due to (3), (8), (9), (10) there exists $m(p) \in N_p$ such that $[g]^{p}_{m(p)} \neq 0$ and $p^{w(p)} \leqslant h_p([g]^{p}_{i})$ for $i \geqslant m(p)$, $i \in N_p$. Define $g_p$ by

$$g \equiv p^{w(p)} g_p \pmod{T'_p} \tag{11}$$

where $T'_p = \oplus_{\substack{i < m(p) \\ i \in N_p}} Z(p^{k_i})$.

Denote $G' = \langle T, \{g_p\}_{p \in P_1} \rangle$. We are ready now to prove that the following direct decomposition takes place:

$$G' = \oplus_{p \in P_1} \oplus_{\substack{i \geqslant m(p) \\ i \in N_p}} Z(p^{k_i}) \oplus \langle T', \{g_p\}_{p \in P_1} \rangle, \tag{12}$$

where $T' = \oplus_{p \in P_1} T'_p \oplus T_1$.

Let

$$h = \sum_{p \in \Delta} n_p g_p \tag{13}$$

where $\Delta$ is a finite subset of $P_1$; for every $p \in \Delta, n_p \in Z$; and $o(h) < \infty$.

Denote

$$L = \sum_{p \in \Delta} (n_p \prod_{q \in \Delta \setminus \{p\}} q^{w(q)}) \tag{14}$$

Since $o(h) < \infty$, $[h]^{p}_{m(p)} = 0$ for almost all $p \in P_1$. Therefore there exists $s \in P_1$ such that

$$s > \max\{\,|L|, \{p\}_{p \in \Delta}\} \tag{15}$$

and

$$[h]^{s}_{m(s)} = 0 \tag{16}$$

Statements (11), (13), (16) imply that

$$\frac{L}{M}[g]^{s}_{m(s)} = 0 \tag{17}$$

where $M = \prod_{p \in \Delta} p^{w(p)}$. Due to (10) and (9), $[g]^{s}_{m(s)} \neq 0$; also $(s, M) = 1$, therefore (17), (15) imply $L = 0$. Due to (14) it means that

$$(\forall p \in \Delta):\quad n_p \equiv 0 \pmod{p^{w(p)}} \tag{18}$$

Congruences (11) and (18) allow us to interpret (13) as

$$h \equiv ug \pmod{T'}$$

for some $u \in Z$. Due to the fact that $o(h) < \infty$ and $o(g) = \infty$, it implies $h \equiv 0 \pmod{T'}$. Decomposition (12) is proven.

To generate G we have to add to G' only generators $g_{p,i}$; $p \in P \setminus P_1$, $p^i \leqslant h_p(\bar{g})_R$ satisfying congruences $g \equiv p^i g_{p,i} \pmod{T}$. While doing that, we can easily show (please see (12)), that $T_0 = \oplus_{p \in P_1} \oplus_{i \geqslant m(p),\ i \in N_p} Z(^{k_i})$ is not only a direct summand of G', but of G as well. G is $\aleph_0$-decomposable.

In both cases 1, 2 we showed that the group G is $\aleph_0$-decomposable. This contradiction proves that the groups T, R satisfy the condition (2).

The proof of theorem 2 with some modification of construction from [2] allow us to classify all the countable $\aleph_0$-indecomposable mixed groups of torsion-free rank 1:

THEOREM 3. Countable fully reduced mixed abelian group G of torsion-free rank 1 is $\aleph_0$-indecomposable if and only if its torsion part T and factor G/T satisfy the inequalities (1), (2), and the components of an element of infinite order $g \in G$ in cyclic factors of n-adic completion $\hat{G}$ (please see (5)) satisfy the inequalities (8), (9) for almost all p and i.

## 4. CONJECTURES AND PROBLEMS

Another problem raised by the author on the Problem Session in Honolulu was: find all countable reduced torsion groups T such that there exists an $\aleph_0$-indecomposable extension G of T by a completely decomposable finite rank torsion-free group.

The author proposes the following conjecture:

CONJECTURE 1. Let T be a countable reduced torsion group. There exists an $\aleph_0$-indecomposable extension G of T by a completely decomposable torsion-free group of rank n if and only if T can be presented as a direct sum $T = F \oplus H$, where F is a finite group and

$$(\forall\, p \in P)(\forall\, k \in N^*): \qquad f_k^p(H) + f_{k+1}^p(H) \leqslant n$$

This criterion does not change if to omit the words "completely decomposable".

The proof of the necessary condition has just been obtained by the author.

PROBLEM 2. Given a countable reduced torsion group T and a torsion free group R. Find the necessary and sufficient condition for the existence of an $\aleph_0$-indecomposable extension of T by R.

PROBLEM 3. Is it true that for any countable reduced torsion group T without infinite bounded summands, there exists an $\aleph_0$-indecomposable extension of T by a countable torsion-free group H?

## REFERENCES


1 L. Fuchs. Infinite Abelian Groups, Vols I and II, Academic Press, New York, 1970 and 1973.

2 A. Soifer. Countable Countably-Indecomposable Abelian Groups, n-Decomposable For Any Finite n, Proceedings of the Honolulu Conference on Abelian Group Theory, Springer-Verlag, 1983, 425-435.

3 A. Soifer, Criteria for the Existence of Irreducible Systems of Generators for Abelian Groups, Siberian Math. J. 15 (1974), 86-108.

# REALIZING GALOIS FIELDS

R. S. PIERCE
UNIVERSITY OF ARIZONA

1. INTRODUCTION. If $F$ is an algebraic number field of degree $n$ over $Q$ and $p$ is a prime, then $F$ is p-realizable if there is a torsionfree abelian group $A$ of rank $n$ such that $qA = A$ for all primes $q \neq p$ and $F$ is isomorphic to the quasi-endomorphism algebra of $A$. The question "for which $F$ and $p$ is $F$ p-realizable?" was the subject of the paper by Pierce and Vinsonhaler.[1] The results in that work were incomplete. In particular, the characterization of the fields that are p-realizable for all $p$ was not considered. This note provides fragmentary information on that question in the case that $F/Q$ is a Galois extension.

Throughout this paper $F$ denotes a finite Galois extension of $Q$ with the Galois group $G$. For a prime $p$, let $C_p$ denote the decomposition group of an extension to $F$ of the p-adic valuation $v_p$ of $Q$. Thus, $C_p$ is a certain subgroup of $G$ that is determined up to conjugation by the prime $p$. The ambiguity of the choice of $C_p$ is unimportant for our purposes. The following criterion for p-realizability was proved in Section 2 of the paper.[1]

1.1 THEOREM. $F$ is p-realizable if and only if there is a union $X$ of right cosets of $C_p$ such that $\sigma X \neq X$ for all $\sigma \in G - \{1\}$.

A subgroup $C$ of $G$ will be called abnormal (in $G$) if it satisfies the criterion of this theorem, that is, there exists $X \subseteq G$ such that $XC = X$ and $\sigma X \neq X$ for all $\sigma \neq 1$. It is clear that if $C$ is abnormal, then so are its conjugates. Our problem is to characterize the fields $F$ such that all decomposition groups of $F$ are abnormal.

2. REALIZABILITY ALMOST EVERYWHERE. The question "which fields are p-realizable for almost all p?" was studied at length in Pierce and Vinsonhaler.[1] It is a simpler problem than everywhere realizability because of two results from algebraic number theory: $C_p$ is cyclic for almost all $p$; every conjugate class of cyclic subgroups of $G$ is represented by $C_p$ for infinitely many primes. Thus, $F$ is p-realizable for almost all $p$ if and only if every cyclic subgroup of $G$ is abnormal.

It is clear that a non-trivial normal subgroup $C$ of $G$ cannot be abnormal. Indeed, if $C \triangleleft G$ and $XC = X$, then $\sigma X = X$ for all $\sigma \in C$. Therefore, a necessary condition for $F$ to be p-realizable for almost all $p$ is that there are no non-trivial, normal, cyclic subgroups of $G$. It is somewhat surprising that this condition is nearly sufficient.

2.1 THEOREM. If $G$ has no non-trivial, normal, cyclic subgroups, and if $G$ is not isomorphic to the symmetric group $S_4$ or the alternating group $A_4$, then every cyclic subgroup of G is abnormal. In this case, F is p-realizable for almost all primes p.

The proof of this theorem comes from a long, elementary group theoretical argument; it can be found in Pierce.[2]

The statement of Theorem 2.1 leaves an unanswered question: which finite groups have no non-trivial, normal, cyclic subgroups? Plainly, if the center $Z(G)$ is not trivial, then there are normal, cyclic subgroups of prime order in G. Thus, it is reasonable to limit our attention to groups with a trivial center.

For any finite group H, denote the set of prime divisors of the order of H by $\pi(H)$. As usual, $G'$ designates the commutator subgroup of the group G. We will write $G'G^k$ for $\cup_{\sigma\varepsilon G} G'\sigma^k$, where k is any natural number. Since $G/G'$ is abelian, $G'G^k$ is a normal subgroup of G.

2.2 PROPOSITION. For a finite group G with $Z(G) = 1$ and for each $p \varepsilon \pi(G)$, the following conditions are equivalent.

(a) There is a normal, cyclic subgroup C of G such that $p \varepsilon \pi(C)$.

(b) There is a normal, cyclic subgroup C of G with $C \subseteq Z(G')$ such that $|C| = p$.

(c) There is a normal subgroup N of G such that $p \varepsilon \pi(Z(N))$, $G/N$ is cyclic of order m, and m divides $p - 1$.

(d) $p \varepsilon \pi(Z(G'G^{p-1}))$.

(e) There is a normal subgroup M of G with $p \varepsilon \pi(Z(M))$, such that $G/M$ is an abelian group whose exponent divides $p - 1$.

Proof. Clearly, (b) implies (a), and (e) is a special case of both

(c) and (d). If (a) is satisfied, then $C[p]$ is a non-trivial, characteristic subgroup of the normal, cyclic subgroup $C$. Thus, $C[p]$ is a normal, cyclic subgroup of order $p$ in $G$, say $C[p] = \langle \sigma \rangle$. If $N = C_G(\sigma)$, then $N \triangleleft N_G(C[p]) = G$; and $G/N$ is isomorphic to a subgroup of $\mathrm{Aut}\langle \sigma \rangle$, a cyclic group of order $p - 1$. It follows that $G' \subseteq N$, and there exists $\tau \in G$ such that $G = \langle N,\tau \rangle$. Hence, $\tau\sigma\tau^{-1} = \sigma^k$ with $1 \leq k < p$. The equality $k = 1$ is excluded by the assumption that $Z(G) = 1$. Thus, $\sigma \in \langle \sigma^{k-1} \rangle = \langle \tau\sigma\tau^{-1}\sigma^{-1} \rangle \subseteq G' \subseteq G'G^{p-1} \subseteq C_G(\sigma)$, $\langle \sigma \rangle \subseteq Z(G')$, and $p \in \pi(Z(G'G^{p-1}))$. This argument shows that (a) implies (b), (c), and (d). Assume that (e) holds true. Denote $A = G/M$. In additive notation, the hypothesis (e) implies that $V = Z(M)[p]$ is a non-trivial, $\mathbf{F}_pA$-module, with the action of $A$ induced by conjugation. Since $A$ has exponent less than $p$, it follows from Maschke's Theorem that $V = V_1 \oplus \ldots \oplus V_r$, where each $V_i$ is a simple $\mathbf{F}_pA$-module. If $A$ is decomposed as a product $A_1 \times \ldots \times A_s$ of cyclic groups with $|A_j| = m(j)$, then $\mathbf{F}_pA \cong \mathbf{F}_pA_1 \otimes \ldots \otimes \mathbf{F}_pA_s$ and $\mathbf{F}_pA_j \cong \mathbf{F}_p[x]/(x^{m(j)} - 1) \cong \mathbf{F}_p[x]/(x-a_{j1}) \oplus \ldots \oplus \mathbf{F}_p[x]/(x-a_{jm(j)})$ as $\mathbf{F}_pA_j$-modules. Indeed, $x^{m(j)} - 1$ splits into a product of distinct linear factors in $\mathbf{F}_p[x]$ because $m(j)$ divides $p - 1$. Thus, the simple modules $V_i$ are one dimensional, say $V_i = \mathbf{F}_p\sigma_i$. When this observation is translated into multiplicative form, it yields the conclusion that $\langle \sigma_i \rangle$ is a normal subgroup of order $p$ in $G$.

Part (d) of the proposition provides an algorithm to determine whether $G$ has non-trivial, normal, cyclic subgroups. Note that by (b) it is only necessary to test the primes in $\pi(Z(G'))$. In particular, if

$Z(G) = Z(G') = 1$, then $G$ has no non-trivial, normal, cyclic subgroups.

3. REALIZABILITY EVERYWHERE. This section illustrates the methods used to find the exact set of primes at which a given field is realizable. We construct Galois extensions of $\mathbb{Q}$ that are p-realizable for all $p$, and other extensions with the same Galois group that are not realizable everywhere.

For any subset $X$ of a finite group $G$, let $G_X = \{\sigma \in G: \sigma X = X\}$. By definition, a subgroup $C$ of $G$ is abnormal if $G_X = 1$ for some union $X$ of left cosets of $C$. If $X = \tau_1 C \cup \ldots \cup \tau_m C$, then each $\sigma \in G_X$ induces a permutation of the cosets $\tau_1 C, \ldots, \tau_m C$. This permutation is the identity map if and only if $\sigma \in \cap_{i=1}^{m} \tau_i C \tau_i^{-1}$. Consequently, $\cap_{i=1}^{m} \tau_i C \tau_i^{-1} = 1$ is a necessary condition for $G_X = 1$. The triviality of this intersection is generally not sufficient for $G_X = 1$, but it does guarantee that the permutation action of $G_X$ on the set $\{\tau_i C: 1 \le i \le m\}$ is faithful.

The following technical result collects the group theoretical facts that are needed in the rest of the paper. The group $G$ that we consider is the symmetric group $S_n$ whose elements are the permutations of the set $\{0, 1, 2, \ldots, n-1\}$.

3.1 LEMMA. Let $G = S_n$ with $n \ge 5$. For $0 \le k_1 < \ldots < k_m \le n - 1$, denote the subgroup of $G$ consisting of permutations of $\{k_1, \ldots, k_m\}$ by $S(k_1, \ldots, k_m)$.

(a) Every cyclic subgroup of $G$ is abnormal.

(b) If $n \ge 6$ and $C_1$ is a cyclic subgroup of $S(2, \ldots, n-1)$, or if

$n = 5$ and $|C_1| \leq 2$, then $C = \langle (0,1) \rangle C_1$ is abnormal in $G$.

(c) If $n$ is prime, $n \geq 11$, and $C$ is the normalizer in $G$ of a Sylow $n$-subgroup of $G$, then $C$ is abnormal in $G$.

(d) If $n = 5$, and if $C$ and $N$ are subgroups of $G$ such that $|C| = 10$ and $|N| = 20$, then $C$ is abnormal, and $N$ is not abnormal.

Proof. The result (a) is a special case of Theorem 2.1. To prove (b), let $H = \langle (0,1) \rangle S(2,\ldots,n-1)$. By Lemma 7.5 of the paper,[1] $[H:C] = [S(2,\ldots,n-1):C_1] \geq n-2$. Since $H \cap (02)H(02) \cap (03)H(03) \cap \ldots \cap (0n-2)H(0n-2) = 1$, it follows from Proposition 6.5 of Pierce and Vinsonhaler,[1] that $C$ is abnormal in $G$. Assume that $n$ is a prime and $n \geq 11$. It is convenient to view $G$ as the group of all permutations of the finite field $\mathbf{F}_n$. A typical Sylow $n$-subgroup of $G$ is the cyclic group $\langle \gamma \rangle$, where $\gamma = (012\ldots n-1)$. Our aim is to show that $C = N_G(\langle \gamma \rangle)$ is abnormal. Let $W = \{\pi \in G: C \cap \pi C \pi^{-1} \neq 1\}$ and $U = \{\pi \in G: \pi C$ contains an involution$\}$. It will suffice to prove that $|W| + |U| < |G|$. In this case, there exists $\pi \in G$ with $\pi \notin W$ and $\pi \notin U$, so that $C \cap \pi C \pi^{-1} = 1$, and there is no involution in $\pi C$. These properties imply that if $X = C \cup \pi C$, then $G_X = 1$. To estimate $|W|$ and $|U|$, it is useful to identify $C$. If $\phi \in C$ satisfies $\phi(0) = 0$ and $\phi(1) = 1$, then $\phi = 1$. Indeed, $\phi \gamma \phi^{-1} = \gamma^k$ implies $(0,k,2k,\ldots) = \gamma^k = (\phi(0),\phi(1),\phi(2),\ldots) = (0,1,\phi(2),\ldots)$, so that $k = 1$ and $\phi(a) = a$ for all $a \in \mathbf{F}_n$. A calculation shows that if $m \in \mathbf{F}_n^*$ and $r \in \mathbf{F}_n$, then $\phi_{mr}(a) = ma + r$ defines an element of $N_G(\langle \gamma \rangle) = C$. The collection $\{\phi_{mr}: m \in \mathbf{F}_n^*,\ r \in \mathbf{F}_n\}$ is a doubly transitive subgroup of $C$. Our previous remark then implies that every element of $C$ has the form $\phi_{mr}$.

In particular, $|C| = n(n-1)$. Since every element of $U$ is the product of an element of $C$ with an involution, and there are $n!/((n-1)/2)!\,2^{(n-1)/2}$ involutions in $G$, it follows that

$$|U|/|G| \leq (n)(n-1)/((n-1)/2)!2^{(n-1)/2}. \tag{1}$$

If $m = 1$ and $r \neq 0$, then $\phi_{mr} = \gamma^r$ has order $n$. On the other hand, $m \neq 1$ implies $\phi_{mr}(a) = m(a - s) + s$, where $s = r(1 - m)^{-1}$. Thus, $\phi_{mr}^k(a) = m^k(a - s) + s$. It follows that the order of $\phi_{mr}$ is the order of $m$ in the group $\mathbf{F}_n^*$, hence a divisor of $n - 1$. Moreover, $\phi_{mr}$ has the unique fixed point $s$. Since $\mathbf{F}_n^*$ is cyclic, it is clear that if $\phi, \psi \in C - \langle\gamma\rangle$ have the same order and fixed point, then $\langle\phi\rangle = \langle\psi\rangle$. These remarks lead to the following description of $W$:

$$W \subseteq C \cup \bigcup\langle\gamma\rangle N_G(\langle\phi\rangle), \tag{2}$$

where the union ranges over the cyclic subgroups $\langle\phi\rangle$ of $C$ that have prime order dividing $n - 1$. In fact, if $\pi \in W$, then there exists $\phi \in C$ of prime order such that $\pi\phi\pi^{-1} \in C$. If the order of $\phi$ is $n$, then $\pi \in C$. Otherwise $\pi\phi\pi^{-1} = \phi_{mr}$ for some $m \neq 1$ in $\mathbf{F}_n^*$ and $r \in \mathbf{F}_n$. If $\pi\phi\pi^{-1}$ fixes $s$ and $\phi$ fixes $t$, then $\gamma^{t-s}\pi\phi\pi^{-1}\gamma^{s-t}$ also fixes $t$. Thus, $\pi \in \langle\gamma\rangle N_G(\langle\phi\rangle)$, which completes the proof of (2). If $p$ is a prime divisor of $n - 1$, then the elements of order $p$ in $C - \langle\gamma\rangle$ are of the form $\phi_{mr}$, where $m$ has order $p$ in $\mathbf{F}_n^*$ and $r \in \mathbf{F}_n$ is arbitrary. Since $\mathbf{F}_n^*$ is cyclic, it follows that there are exactly $n$ cyclic groups $\langle\phi\rangle$, where $\phi \in C - \langle\gamma\rangle$ has order $p$. Any such $\phi$ is a product of $(n-1)/p$ disjoint p-cycles; hence, $|C_G(\phi)| = ((n-1)/p)!p^{(n-1)/p}$ and $|N_G(\phi)| = (p - 1)|C_G(\phi)|$. Using these computations and (2) gives the estimate

$$|W| \le n(n-1) + \Sigma n^2(p-1)((n-1)/p)!p^{(n-1)/p}, \tag{3}$$

summed over the prime divisors $p$ of $n-1$. An easy computation based on (1), (3), Stirling's inequalities for the factorial, and the estimate $\Sigma_{p|n-1}(p-1) \le n-1$ yields the desired inequality $|W|/|G| + |U|/|G| < 1$ for $n \ge 11$. For the proof of (d) we can assume that $\langle\gamma\rangle \subset C \subset N$, where $\gamma$ is a 5-cycle, say $(01234)$, and $N = N_G(\langle\gamma\rangle)$. The proof of (c) shows that $N = \langle\gamma,\tau\rangle$, where $\tau$ is a 4-cycle; and $C = \langle\gamma,\sigma\rangle$, where $\sigma = \tau^2$. Thus, $\sigma$ is the product of two disjoint 2-cycles and $\phi\gamma\phi^{-1} = \gamma^{-1}$. Indeed, every element of $C - \langle\gamma\rangle$ has these properties. If $\pi \in G$ normalizes $C$, then necessarily $\pi\langle\gamma\rangle\pi^{-1} = \langle\gamma\rangle$, so that $\pi \in N$. Conversely, since $\sigma = \tau^2$, the elements of $N$ normalize $C$. Thus, $N_G(C) = N$. Let $\pi \in G - N$ be an odd permutation. Then $\pi \notin N_G(C)$, so that $C \cap \pi C\pi^{-1} \subset C$. Moreover, $\langle\gamma\rangle \cap \pi C\pi^{-1} = 1$ because $\langle\gamma\rangle$ has prime order. Hence, $|C \cap \pi C\pi^{-1}| \le 2$. If $\gamma^i\pi C = \gamma^j\pi C$, then $\gamma^{i-j} \in \pi C\pi^{-1} \cap \langle\gamma\rangle = 1$. It follows that the cosets $\pi C$, $\gamma\pi C, \ldots, \gamma^4\pi C$ are distinct. If $1 \ne \phi \in C \cap \pi C\pi^{-1}$ and $1 \le i < 5$, then $\phi \notin \gamma^i\pi C\pi^{-1}\gamma^{-i}$; otherwise, $\phi\gamma^{2i} = \gamma^{-i}\phi\gamma^i \in C \cap \pi C\pi^{-1} = \{1,\phi\}$, which is impossible. Thus, $C \cap \pi C\pi^{-1} \cap (\gamma^i\pi)C(\gamma^i\pi)^{-1} = 1$. With these preparations made, define $X = C \cup \pi C \cup \gamma\pi C \cup \gamma^2\pi C$. Suppose that $\phi X = X$. The above remark implies that $\phi$ is faithfully represented by its action on the cosets $\{C, \pi C, \gamma\pi C, \gamma^2\pi C\}$. The elements of $C$ are even permutations, and the elements in the other cosets are odd. It follows that $\phi \in C$. Hence, $\phi$ must be an involution, so that $\phi$ must fix at least two cosets. We can suppose that $\phi\pi C = \pi C$. Consequently, $\phi \in C \cap \pi C\pi^{-1}$. If $\phi \ne 1$, then $\phi\gamma\pi C = \gamma^2\pi C$ and $\gamma^{-2}\phi\gamma \in \pi C\pi^{-1}$. This relation gives the contradiction $\phi\gamma^3 \in C \cap \pi C\pi^{-1} = \{1,\phi\}$. Therefore, $G_X = 1$, so that $C$ is abnormal in $G$. Turning our attention to $N$, we note that the distinct left cosets of $N$ in $G$ are $N$, $(01)N$, $(12)N$, $(23)N$, $(34)N$, and $(40)N$. Suppose that $XC = X$ and $G_X = 1$. By a counting argument, no two conjugates of $M$ have trivial intersection, so that $X$ must be a union of three cosets of $N$. It can be assumed that $N$ and $(01)N$ are among these three cosets because $G_{G-X} = G_X$, $G_{\phi X} = \phi G_X \phi^{-1}$, and $N$ is a doubly transitive subgroup of $G$. Thus, $X$ can be one of the following sets: (1) $N \cup (01)N \cup (12)N$; (2) $N \cup (01)N \cup (23)N$; (3) $N \cup (01)N \cup (34)N$; or (4) $N \cup (01)N \cup (40)N$. A calculation shows that $(02)(34) \in G_X$ in case 1, $(03)(12) \in G_X$ in case 2, $(05)(13) \in G_X$ in case 3, and $(14)(23) \in G_X$ in case 4. Thus, $C$ is not abnormal in $G$.

It seems likely that the normalizers of Sylow 7-subgroups of $S_7$ are abnormal, but the crude estimate in the proof of (c) doesn't settle this question.

Our discussion of realizability will involve some standard results from algebraic number theory. For convenience, these facts are collected in the next lemma.

3.2 LEMMA. Let $\Phi(x) = x^n + a_1x^{n-1} + \ldots + a_n$ be an irreducible polynomial with integral coefficients. Denote the splitting field of $\Phi$ over $Q$ by $F$, and write $G$ for $\mathrm{Gal}(F/Q)$. For each rational prime $p$, let $\hat{F}_p$ be the splitting field of $\Phi$ over the field $\hat{Q}_p$ of p-adic numbers, $\hat{G}_p = \mathrm{Gal}(\hat{F}_p/\hat{Q}_p)$, and denote the unique extension of $v_p$ to $\hat{F}_p$ by $\hat{v}_p$.

(a) If $\phi: F \to \hat{F}_p$ is an embedding, then $\tau \to \phi^{-1}\tau\phi$ maps $\hat{G}_p$ isomorphically to the decomposition group $C_p$ of the extension $\hat{v}_p\phi$ of $v_p$.

(b) If $\hat{v}_p$ is unramified over $v_p$, then $\hat{G}_p$ is cyclic.

(c) If $\hat{v}_p/v_p$ is ramified, then $p$ divides the discriminant $D$ of the polynomial $\Phi$.

(d) If $p$ divides $a_i$ for $1 \leq i \leq n$, $p$ does not divide $n$, and $p^2$ does not divide $a_n$, then $\hat{G}_p$ is a split extension of a cyclic group of order $n$ by a cyclic group of order $f$, where $f$ is the multiplicative order of $p$ modulo $n$.

Proof. Results that are equivalent to (a) and (b) can be found in Section 4-10 of Weiss's book.[3] To obtain (c), note that $\hat{v}_p/v_p$ is ramified if and only if $p$ divides the discriminant $\Delta$ of $\hat{F}_p/\hat{Q}_p$. Since $\hat{F}_p$ can be constructed by a chain of extensions $\hat{Q}_p = K_0 \subset K_1 \subset K_2 \subset \ldots \subset K_r = \hat{F}_p$, where each $K_{j+1}$ is a simple extension of $K_j$ by the root of a monic divisor of $\Phi$ in $K_j[x]$, it follows that $\Delta$ divides a power of $D$. Thus, if $\hat{v}_p/v_p$ is ramified, then $p$ divides $D$. More details of this sketch can be found in Section 3.7 of Weiss.[3] To prove (d), let $z$ be a root of $\Phi$ in $\hat{F}_p$, and denote $E = \hat{Q}_p(z)$. The extension $E/\hat{Q}_p$ is totally ramified of degree $n$ by Weiss,[3] p. 86; and since $p$ does not divide $n$, the ramification is tame. Hence, there is a unit $u \in \hat{Z}_p$ such that $E$ is the splitting field of $X_n - pu$ over $\hat{Q}_p$, as is shown in Weiss,[3] p. 89. By Kummer Theory, $\hat{F} = \hat{Q}_p(w,v)$, where $w^n = pu$ and $v$ is a primitive n'th root of unity over $\hat{Q}_p$. Thus, if $L = \hat{Q}_p(v)$, then $L/\hat{Q}_p$ is an unramified, cyclic Galois extension of degree $f$, where $f$ is the order of $p$ modulo $n$. See Serre's book,[4] p. 77.

Since $F/\hat{Q}_p$ is totally ramified and $L/\hat{Q}_p$ is unramified, it follows that $E \cap L = \hat{Q}_p$. Thus, $\hat{G}_p$ is a split extension of $\mathrm{Gal}(\hat{F}_p/L)$ by $\mathrm{Gal}(\hat{F}_p/E)$, and $\mathrm{Gal}(\hat{F}_p/L)$ is cyclic of order $n$ by Kummer Theory.

We can now prove a result that leads to many interesting examples.

3.3 PROPOSITION. Let $\Phi(x) = x^n - ax + b$ be an irreducible polynomial in $\mathbf{Z}[x]$, where $n$ is a prime greater than 3. The discriminant of $\Phi$ is $D = (-1)^{n(n-1)/2}(n^n b^{n-1} - (n-1)^{n-1}a^n)$. Let $F$ be the splitting field of $\Phi$ over $\mathbf{Q}$ and $G = \mathrm{Gal}(F/\mathbf{Q})$. If $p$ is a prime divisor of $D$ such that $p^2$ does not divide $D$, then $F$ is p-realizable. Moreover, if such a prime $p$ exists, then $G \cong S_n$.

Proof. The derivative of $\Phi$ is $\Phi' = nx^{n-1} - a$. Thus,

$$n\Phi - x\Phi' = nb - (n-1)ax. \tag{1}$$

The discriminant of $\Phi$ is given as usual by

$$D = \Pi_{i<j}(r_j - r_i)^2 = (-1)^{n(n-1)/2}\Pi_{j=1}^{n}\Phi'(r_j), \tag{2}$$

where $r_1, \ldots, r_n$ are the roots of $\Phi$ in some extension of $\mathbf{Q}$. Substitute $r_j$ for $x$ in (1), and take the product over all $j$ to get

$(-1)^{n(n-1)/2}bD = \Pi_{j=1}^{n}(-r_j)\Phi'(r_j) = \Pi_{j=1}^{n}(nb - (n-1)ar_j) =$

$(n-1)^n a^n \Pi_{j=1}^{n}(nb/(n-1)a - r_j) = (n-1)^n a^n \Phi(nb/(n-1)a) = n^n b^n - (n-1)^{n-1}a^n b.$

Consequently, $D = (-1)^{n(n-1)/2}(n^n b^{n-1} - (n-1)^{n-1}a^n)$, as we claimed. Assume that $p$ is a prime divisor of $D$ and $p^2$ does not divide $D$. Then $p$ does not divide $nb$ or $(n-1)a$. In particular, $p$ is odd. Since $D$ is the discriminant of $\Phi$, it follows that $\Phi$ and $\Phi'$ have a common factor modulo $p$. By (1), this factor is linear. Specifically, $\Phi(x) \equiv (x-c)^2\Psi(x) \pmod p$, where $(n-1)ac \equiv nb \pmod p$, $\Psi \in \mathbf{Z}[x]$ is monic of degree $n - 2$, $(x - c, \Psi(x)) \equiv 1 \pmod p$, and $\Psi$ has no factor in common with its derivative (so that $p$ does not divide the discriminant of $\Psi$). By Hensel's Lemma, $\Phi(x) = \Omega(x)\Lambda(x)$ in $\hat{Z}_p[x]$, where $\deg \Omega(x) = 2$, $\deg \Lambda(x) = n-2$, $(\Omega(x), \Lambda(x)) = 1$, $\Omega(x) = x^2 + ex + f \equiv (x-c)^2 \pmod p$, and $\Lambda(x) \equiv \Psi(x) \pmod p$. Note that the discriminant $e^2 - 4f = pd$ of $\Omega$ divides the discriminant $D$ of $\Phi$ in $\hat{Z}_p$, so that $d \notin p\hat{Z}_p$ because $p^2$ does not divide $D$. Let $\hat{F}_p$ be the splitting field of $\Phi$ over $\hat{Q}_p$ and $\hat{G}_p = \mathrm{Gal}(\hat{F}_p/\hat{Q}_p)$. Then $\hat{F}_p = KL$, where $K$ is the splitting field of $\Omega$ over $\hat{Q}_p$ and $L$ is the splitting field of $\Lambda$ over $\hat{Q}_p$. Since $(\Lambda, \Lambda') \equiv (\Psi, \Psi') \equiv 1 \pmod p$, it follows that $L/\hat{Q}_p$ is unramified. Thus $L/\hat{Q}_p$ is a cyclic Galois extension by Serre,[4] p. 54. On the other hand, $K = \hat{Q}_p(s)$, where $s^2 = pd$, so that $K$ is a totally ramified extension of $\hat{Q}_p$. Hence, $K \cap L = 1$, and $\hat{G}_p = \mathrm{Gal}(\hat{F}_p/K) \times \mathrm{Gal}(\hat{F}_p/L)$. If we view $\hat{G}_p$ as a permutation group on the roots of $\Phi$ and list the roots $r_0, r_1, \ldots, r_{n-1}$ so that $r_0$ and $r_1$ are the solutions of $\Omega(x) = 0$, then $\hat{G}_p = \langle (01) \rangle C_1$, where $C_1$ is a cyclic subgroup of $S(2,\ldots,n-1)$. With the analogous interpretation of $G$ as a group of permutations of the roots of $\Phi$ in the same order, it follows from Lemma 3.2(a) that the decomposition group $C_p = \langle (01) \rangle C_1$. In particular, $G$ includes the 2-cycle $(01)$. Moreover, since $\Phi$ is irreducible over $\mathbb{Q}$, $n$ divides the order of $G$. The hypothesis that $n$ is prime then implies that there is an $n$-cycle in $G$, so that $G = S_n$. If $n > 5$, or $n = 5$ and $|C_1| \leq 2$, then $C_p$ is abnormal in $G$ by Lemma 3.1(b). If $n = 5$ and $|C_1| = 3$, then $C_p$ is cyclic, hence abnormal in $G$ by Lemma 3.1(a). In both cases, $F$ is $p$-realizable by Theorem 1.1.

3.4 COROLLARY. Suppose that $n$ and $q$ are distinct primes with $n \geq 11$, $c$ and $d$ are integers with $d$ not divisible by $q$, and $n^n d^{n-1} - (n-1)^{n-1} qc^n$ is square free and not $\pm 1$. If $F$ is the splitting field over $\mathbb{Q}$ of $\Phi(x) = x^n - qcx + qd$, then $F$ is p-realizable for all primes $p$.

<u>Proof</u>. By Eisenstein's criterion, $\Phi$ is irreducible. The discriminant of $\Phi$ is $\pm q^{n-1}(n^n d^{n-1} - (n-1)^{n-1} qc^n)$. Hence, if $p$ is a prime that is different from $q$, then $F$ is p-realizable by Proposition 3.3. Moreover, the Galois group of $F/\mathbb{Q}$ is isomorphic to $S_n$. By Lemma 3.2, the decomposition group $C_q$ is a split extension of a cyclic group of order $n$. Thus, $C_q$ is a subgroup of the normalizer of a Sylow n-subgroup of $S_n$. It follows from Lemma 3.1 that $C_q$ is abnormal in $S_n$. Consequently, $F$ is q-realizable by Lemma 1.1.

3.5 COROLLARY. Let $q$ be a prime that is different from 5. Suppose that $c$ and $d$ are integers with $d$ not divisible by $q$, such that $5^5 d^4 - 4^4 qc^5$ is square free and not $\pm 1$. If $F$ is the splitting field over $\mathbb{Q}$ of $x^5 - qcx + qd$, then the Galois group of $F/\mathbb{Q}$ is $S_5$, $F$ is p-realizable for all primes $p \neq q$, and $F$ is q-realizable if and only if $q \equiv \pm 1 \pmod 5$.

<u>Proof</u>. Except for the last statement, the assertions of this corollary come from the proof of Corollary 3.4. By Lemma 3.2, $C_q$ is a split extension of a cyclic group of order 5 by a cyclic group of order $f$, where $f$ is the multiplicative order of $q$ modulo 5. If $f = 1$ or $f = 2$, that is $q \equiv \pm 1 \pmod 5$, then $C_q$ is abnormal by Lemma 3.1. If $f = 4$, then $C_q$ is not abnormal. Therefore, the last assertion of the

corollary follows from Theorem 1.1.

3.6. EXAMPLES. (a) Let $\Phi(x) = x^5 - 7x + 7$. The discriminant of $\Phi$ is $7^4(5^5 - 4^4\cdot 7) = 7^4\cdot 31\cdot 43$. If $F$ is the splitting field of $\Phi$, then by Corollary 3.5, $\mathrm{Gal}(F/Q) = S_5$ and $F$ is p-realizable for all primes except $p = 7$.

(b) Let $\Phi(x) = x^5 - 62x - 93$. The discriminant of $\Phi$ is $-31^4\cdot 827$. If $F$ is the splitting field of $\Phi$, then by Corollary 3.5, $\mathrm{Gal}(F/Q) = S_5$ and $F$ is realizable everywhere.

## References


1. Pierce, R.S. and Vinsonhaler, C.I., Realizing algebraic number fields, *Honolulu Symposium on Abelian Groups*, Lecture Notes in Mathematics, vol. 1006, Springer-Verlag, New York, 1983.

2. Pierce, R.S., Permutation representations with trivial set stabilizers, *Jour. of Alg.* (to appear).

3. Weiss, E., *Algebraic Number Theory*, McGraw-Hill, New York, 1963.

4. Serre, J.-P., *Local Fields*, Graduate Texts in Mathematics, vol. 67, Springer-Verlag, New York, 1979.

# ON THE RADICAL OF THE ENDOMORPHISM RING OF A PRIMARY ABELIAN GROUP

A. D. SANDS
Department of Mathematical Sciences
The University, Dundee DD1 4HN

Since it was shown by Baer and Kaplansky that any isomorphism of the endomorphism rings of primary abelian groups is induced by an isomorphism of the groups attention has been given to the study of these rings. Pierce[1] raised the question of describing the radical of an endomorphism ring in terms of its action on the group and solved this problem in the torsion-complete case. Liebert[2] has solved this problem for direct sums of cyclic groups; Hausen[3] has solved it for totally projective groups; Hausen and Johnson[4] have solved it for sufficiently projective groups.

In this note we shall restrict our attention to separable primary abelian groups, apart from Theorem 1. The fundamental notations and definitions may be found in Fuchs.[5,6] We give a lower bound for the radical and show that this lower bound is attained when a certain condition is satisfied. This condition is satisfied by torsion-complete groups, by direct sums of cyclic groups and by separable sufficiently projective groups. We use the usual embedding of $G$ into $\overline{B}$, where $\overline{B}$ is the torsion-completion of a basic subgroup $B$ of $G$. From a given group $G$ we construct a group $\hat{G}$ satisfying the condition with $G \subseteq \hat{G} \subseteq \overline{B}$. Pierce[1] has constructed a group for which $G \neq \hat{G}$, but we do not know of any example of a separable primary abelian group where the radical differs from this lower bound.

Let $G$ be a separable abelian group, primary with respect to the prime $p$. As usual we use the subgroups $p^nG$ as a basis of neighbourhoods of $0$ to define the p-adic topology on $G$. A sequence $\{g_i\}$ of elements of $G$ is said to be Cauchy if, for each $n$, there exists $N(n)$ such that $g_i - g_j \in p^nG$ for all $i,j \geq N(n)$. A sequence $\{g_i\}$ is convergent to $g \in G$ if, for each $n$, there exists $N(n)$ such that $g - g_i \in p^nG$ for all $i \geq N(n)$. Clearly convergent sequences are Cauchy sequences. The group $G$ is said to be torsion-complete if every Cauchy sequence of elements of bounded order in $G$ is convergent in $G$. We shall say that $G$ satisfies condition (C) if given any Cauchy sequence $\{g_i\}$ in $G[p]$ which is not convergent in $G$ there exists a group $H$, which is a direct sum of cyclic groups, and a homomorphism $\alpha : G \to H$ such that $\{g_i\alpha\}$ is Cauchy but not convergent in $H$. It is clear that torsion-complete groups and direct sums of cyclic groups satisfy condition (C). A separable primary abelian group $G$ is sufficiently projective if every countable subset of $G$ is contained in a direct summand $H$ of $G$ such that $H$ is a direct sum of cyclic groups. Hill[7] has given an example of such a group which is not itself a direct sum of cyclic groups. If $\{g_i\}$ is Cauchy but not convergent in such a group $G$ it is clear that the projection mapping $\pi$ to this subgroup $H$ containing the elements $g_i$ is such that $\{g_i\pi\}$ is not convergent in $H$. Thus separable sufficiently projective groups also satisfy (C).

We denote the height of an element $a$ in a group $G$ by $h_G(a)$, or simply $h(a)$ if the meaning is clear. We denote by $E(G)$ the endomorphism ring of a group $G$, writing mappings on the right. $J(E(G))$ denotes the Jacobson radical of $E(G)$. Following Pierce[1] we define the ideal $H(G)$ of the endomorphism ring of a separable group by

$$H(G) = \{\alpha \in E(G) \mid h(a) < h(a\alpha) \text{ for all non-zero } a \in G[p]\}.$$

Pierce[1] has shown that $J(E(G)) \subseteq H(G)$. We define $C(G)$ by

$$C(G) = \{\alpha \in E(G) \mid g_i \in G[p] \text{ and } \{g_i\} \text{ a Cauchy sequence implies that } \{g_i\alpha\} \text{ is a convergent sequence in } G\}.$$

It is clear that $C(G)$ is an ideal of $E(G)$. We shall show that $C(G) \cap H(G) \subseteq J(E(G))$.

Hausen and Johnson[4] remark that in the known cases the Jacobson radical of $E(G)$ has been given in terms of its action on $G[p]$. In our first theorem we present this as a formal result.

THEOREM 1. *Let* $G$ *be a primary abelian group. Let* $\alpha, \beta$ *in* $E(G)$ *be such that their restrictions to* $G[p]$ *are equal. Then* $\alpha \in J(E(G))$ *if and only if* $\beta \in J(E(G))$ .

*Proof.* Assume $\alpha \in J(E(G))$ . Then $1-\alpha$ is bijective. Hence $\ker(1-\beta) \cap G[p] = \ker(1-\alpha) \cap G[p] = 0$ . Therefore $\ker(1-\beta) = 0$ and $1-\beta$ is injective. We show that $G(1-\beta) = G$ by induction on the orders of elements. Assume that all elements of $G$ of orders less than $p^m$ belong to $G(1-\beta)$ . Let $a \in G$ and let $a$ have order $p^m$ . Since $1-\alpha$ is surjective there exists $b \in G$ with $b(1-\alpha) = a$ . Then $0 = p^m a = p^m b(1-\alpha)$ and so $p^m b = 0$ . Thus $p^{m-1} b \in G[p]$ and $p^{m-1} b(\beta-\alpha) = 0$ , since $\alpha$ and $\beta$ agree on $G[p]$ . By the inductive assumption there exists $c \in G$ with $c(1-\beta) = b(\beta-\alpha)$ . Then $(b+c)(1-\beta) = b(1-\alpha) = a$ . It follows by induction that $G(1-\beta) = G$ . Thus $1-\beta$ is surjective and hence bijective. Now, for any $\gamma \in E(G)$ we have $\alpha\gamma \in J(E(G))$ and also $\alpha\gamma$ and $\beta\gamma$ agree on $G[p]$ . From above it follows that $1-\beta\gamma$ is bijective. Therefore the right ideal of $E(G)$ generated by $\beta$ is quasi-regular. It follows that $\beta \in J(E(G))$ .

We have defined the ideal $C(G)$ of $E(G)$ in terms of sequences. There are equivalent definitions in terms of series.

THEOREM 2. *Let* $G$ *be a separable primary abelian group. Let*

$$C_1(G) = \{\alpha \in E(G) \mid \{a_i\} \text{ in } G[p],\ h(a_i)$$

*tends to infinity, implies* $\Sigma\, a_i\alpha$ *converges in* $G\}$ ,

$$C_2(G) = \{\alpha \in E(G) \mid \{a_i\} \text{ in } G[p],\ h(a_i)$$

$< h(a_{i+1})$, *for all* $i$, *implies* $\Sigma\, a_i\alpha$ *converges in* $G\}$ .

*Then* $C(G) = C_1(G) = C_2(G)$ .

*Proof.* It is clear that $C(G) \subseteq C_1(G) \subseteq C_2(G)$ . Let $\alpha \in C_2(G)$ . Let $\{g_j\}$ be a Cauchy sequence in $G[p]$ . Let $B$ be a basic subgroup of $G$ with $B = \oplus B_n$ , where $B_n$ is a direct sum of cyclic groups of order $p^n$ . Then we may assume $G \subseteq \overline{B}$ , where $\overline{B}$ is the torsian subgroup of $\Pi B_n$ . Then the Cauchy sequence $\{g_j\}$ tends to a limit $a$ in $\overline{B}$ . If $a \in G$ then $\{g_j\alpha\}$ converges to $a\alpha \in G$ . If $a \notin G$ let $a = \Sigma b_{n_i}$ where $0 \neq b_{n_i} \in B_{n_i}$ , and $n_i < n_{i+1}$ for each $i$ . Then $h_G(b_{n_i}) < h_G(b_{n_{i+1}})$ and, since $\alpha \in C_2(G)$ , $\Sigma b_{n_i}\alpha$ converges to $g \in G$ . It follows that $\{g_j\alpha\}$ converges to $g \in G$ . Therefore $\alpha \in C(G)$ and we have $C_2(G) \subseteq C(G)$ .

We now show that $C(G) \cap H(G)$ is a lower bound for the Jacobson radical.

THEOREM 3. *Let* $G$ *be a separable primary abelian group. Then* $C(G) \cap H(G) \subseteq J(E(G))$ .

*Proof.* Since $C(G) \cap H(G)$ is an ideal of $E(G)$ it is sufficient to show that if $\alpha \in C(G) \cap H(G)$ then $1-\alpha$ is bijective. Let $a \in G[p]$ with $a(1-\alpha) = 0$ . If $a \neq 0$ then, $\alpha \in H(G)$, implies $h(a) < h(a\alpha)$ , which contradicts $a = a\alpha$ . Therefore $a = 0$ and so $\ker(1-\alpha) = 0$ .

Now let $g \in G[p]$. Since $\alpha \in H(G)$ we have $h(g\alpha^m) < h(g\alpha^{m+1})$ or $g\alpha^m = 0$ . Since $\alpha \in C(G) = C_2(G)$ it follows that $\sum_{m=0}^{\infty} g\alpha^m$ converges to $a$ in $G$ . Let $g_n = \sum_{m=0}^{n} g\alpha^m$ . Then $g_n(1-\alpha) = g - g\alpha^{n+1}$ . Now $\{g_n(1-\alpha)\}$ converges to $a(1-\alpha)$ and $\{g-g\alpha^{n+1}\}$ converges to $g$ , as $n$ tends to infinity. Therefore $a(1-\alpha) = g$ and $g \in G(1-\alpha)$ .

Now assume that all elements of $G$ of order less than $p^n$ belong to $G(1-\alpha)$. Let $c \in G$ and let $c$ have order $p^n$. Then $p^{n-1}c \in G[p]$. From above $\sum_{m=0}^{\infty} p^{n-1}c\alpha^m$ converges to $d \in G[p]$ such that $d(1-\alpha) = p^{n-1}c$. Also there exists $N$ such that $d - \sum_{m=0}^{N} p^{n-1}c\alpha^m \in p^{n-1}G$. Hence $d \in p^{n-1}G$. Let $d = p^{n-1}e$. Then $p^{n-1}(c-e(1-\alpha)) = 0$. From the inductive assumption it follows that $c - e(1-\alpha) \in G(1-\alpha)$. Therefore $c \in G(1-\alpha)$. It follows by induction that $G(1-\alpha) = G$.

Therefore $C(G) \cap H(G)$ is a quasi-regular ideal and so $C(G) \cap H(G) \subseteq J(E(G))$.

For torsion-complete groups. Pierce[1] has shown that $H(G) = J(E(G))$. In this case it is clear that $C(G) = E(G)$ and so $J(E(G)) = H(G) \cap C(G)$. For direct sums of cyclic groups Liebert[2] has shown that $J(E(G)) = \{\alpha \in E(G) | \alpha \in H(G)$ and there exists $k$ such that $(p^kG)[p]\alpha = 0\}$. We could show that this implies $J(E(G)) = H(G) \cap C(G)$ but it is as convenient to show directly that, in this case, $J(E(G)) \subseteq C(G)$. The result then follows from previous results.

THEOREM 4. *Let the primary group* $G$ *be a direct sum of cyclic groups. Then* $J(E(G)) \subseteq C(G)$.

*Proof.* If every Cauchy sequence in $G[p]$ is convergent then $C(G) = E(G)$. Otherwise let $\underline{g} = \{g_i\}$ be a Cauchy sequence in $G[p]$ which is not convergent. Let $A(\underline{g}) = \{\alpha \in E(G) | \{g_i\alpha\}$ is convergent in $G\}$. Clearly $A(\underline{g})$ is a right ideal of $E(G)$ and $C(G) = \cap A(\underline{g})$. We show that $J(E(G)) \subseteq C(G)$ by showing that $A(\underline{g})$ is a maximal right ideal of $E(G)$. Let $\beta \notin A(\underline{g})$. Then it suffices to show that $\gamma$ exists in $E(G)$ such that $1-\beta\gamma \in A(\underline{g})$. Let $G = \oplus B_n$, where each

$B_n$ is a direct sum of cyclic groups of order $p^n$. Let $\overline{G}$ be the torsion subgroup of $\Pi B_n$. Let $g_i\beta = h_i$. Then $\{h_i\}$ is not convergent in $G$, but $\{g_i\}$ and $\{h_i\}$ converge, to $a$ and $b$ say, in $\overline{G}$. Let $a = \Sigma a_n$ and $b = \Sigma b_n$, where $a_n, b_n \in B_n[p]$. Since $b \notin G$ there exists an infinite sequence $n_1 < n_2 < n_3 < \ldots$ such that $b_{n_i} \neq 0$. Let $c_i = \sum_{n=n_i}^{n_{i+1}-1} a_n$. Then there exists $\gamma_i \in \text{Hom}(B_{n_i}, G)$ such that $b_{n_i}\gamma_i = c_i$. Define $\gamma \in E(G)$ by $\gamma$ restricted to $B_{n_i}$ is equal to $\gamma_i$, for each $i$, and $\gamma$ restricted to $B_n$ is the zero mapping if $n \neq n_i$, for any $i$. Let $j$ satisfy $n_i \leq j < n_{i+1}$. Let $c = \sum_{n=1}^{n_1-1} a_n$, where $c = 0$ if $n_1 = 1$. Then $\sum_{n=1}^{j}(b_n\gamma - a_n) = c + d_j$ where $h(d_j) \geq n_i$. It follows that, as $j$ tends to infinity, $\Sigma(b_n\gamma - a_n)$ converges to $c$. Hence $\{g_j(1-\beta\gamma)\}$ converges to $-c \in G$. Therefore $1 - \beta\gamma \in A(\underline{g})$ and the right ideal $A(\underline{g})$ is maximal, as required.

COROLLARY. *If* $G$ *is a direct sum of cyclic groups then* $J(E(G)) = H(G) \cap C(G)$.

Let $G$, $H$ be abelian groups and let $R = E(G)$, $V = \text{Hom}(G,H)$, $W = \text{Hom}(H,G)$ and $S = E(H)$. Then $(R,V,W,S)$ forms a Morita context, where composition of mappings is used to define the multiplications. It has been shown by Amitsur[8] that, for any Morita context, one has $VJ(S)W \subseteq J(R)$.

THEOREM 5. *Let the separable primary abelian group* $G$ *satisfy condition* (C) . *Then* $J(E(G)) = C(G) \cap H(G)$ .

*Proof.* It suffices to show that $J(E(G)) \subseteq C(G)$ . Let $\alpha \in J(E(G))$ . If possible suppose that $\alpha \notin C(G)$ . Then there exists a Cauchy sequence $\{g_i\}$ in $G[p]$ such that $\{g_i\alpha\}$ does not converge in $G$ . By Fuchs,[5] p.143, there is a basic subgroup $B$ of $G$ with $g_i \in B$ . By condition (C) there is a group $H$ which is a direct sum of cyclic groups and a homomorphism $\beta : G \to H$ such that $\{g_i\alpha\beta\}$ does not converge in $H$ . Let $L = B \oplus H$ . Then $L$ is a direct sum of cyclic groups. Define $\gamma \in \mathrm{Hom}(L,G)$ by letting $\gamma$ be the inclusion mapping on $B$ and the zero mapping on $H$ . Define $\delta \in \mathrm{Hom}(G,L)$ as the composition of $\beta$ and the injection of $H$ into $L$ . By the result of Amitsur[3] from $\alpha \in J(E(G))$ , we deduce that $\gamma\alpha\delta \in J(E(L))$ . Since $B$ is pure in $G$ , $h_B(g_i) = h_G(g_i)$ and so $\{(g_i,0)\}$ is a Cauchy sequence in $L[p]$ . It follows by the Corollary to Theorem 4 that $\{(g_i,0)\gamma\alpha\delta\}$ is convergent in $L$ ; however $(g_i,0)\gamma\alpha\delta = (0,g_i\alpha\beta)$ and so this sequence is not convergent in $L$ . It follows that $\alpha \in C(G)$ . Therefore $J(E(G)) = C(G) \cap H(G)$ .

THEOREM 6. *Let* $G$ *be a separable primary abelian group with basic subgroup* $B$ *and* $B \subseteq G \subseteq \overline{B}$ . *Let* $\hat{G} = \{a \in \overline{B} \mid \forall\alpha \in E(\overline{B})$ , $G\alpha \subseteq B$ *implies* $a\alpha \in B\}$ . *Then* $\hat{G}$ *is a separable primary abelian group satisfying condition* (C) .

*Proof.* It is clear that $\hat{G}$ is a subgroup of $\overline{B}$ and that $G \subseteq \hat{G}$ . Let $\{a_i\}$ be a Cauchy sequence in $\hat{G}[p]$ which is not convergent in $\hat{G}$ . Then $\{a_i\}$ converges to $a \in \overline{B}$ , where $a \notin \hat{G}$ . Since $a \notin \hat{G}$ , there exists $\alpha \in E(\overline{B})$ such that $G\alpha \subseteq B$ and $a\alpha \notin B$ . $G\alpha \subseteq B$ implies

$\hat{G}\alpha \subseteq B$ . By restricting domain and codomain we have a homomorphism $\alpha : \hat{G} \to B$ such that $\{a_i\alpha\}$ does not converge in $B$ . Since $h_B(a_i\alpha) \geq h_{\hat{G}}(a_i)$ it follows that $\{a_i\alpha\}$ is a Cauchy sequence in $B$ . Since $B$ is a direct sum of cyclic groups it follows that $\hat{G}$ satisfies condition (C) .

Pierce[1] has considered the ideal $E_s(G)$ of $E(G)$ consisting of the small endomorphisms of $G$ and has shown that $J(E_s(G)) = H_s(G) = H(G) \cap E_s(G)$ . Since the Jacobson radical is hereditary we have $J(E_s(G)) = J(E(G)) \cap E_s(G)$ . However Pierce's result is consistent with the relation $J(E(G)) = H(G) \cap C(G)$ . For if $\alpha \in E_s(G)$ , then $\ker \alpha$ contains a large subgroup of $G$ and so, by 2.10 of Pierce,[1] there exists $n$ such that $(p^nG)[p]\alpha = 0$ . Hence $\alpha \in C(G)$ and so $E_s(G) \subseteq C(G)$ . In 15.4 of Pierce[1] there is constructed an example of a group and, in a private communication, he has pointed out to me that for this group $G$ one has $\hat{G} = \bar{B}$ . However $E(G) = R_p.1 + E_s(G)$ , where $R_p$ denotes the $p$-adic integers, and it is not difficult to see, using $J(R_p) = pR_p$ and the above results about $J(E_s(G))$, that for this group also one has $J(E(G)) = C(G) \cap H(G)$ . So it remains an open question whether this relation holds for the endomorphism rings of all separable primary abelian groups.

REFERENCES

1. Pierce, R.S., Homomorphisms of primary abelian groups, in *Topics in Abelian Groups*, Irwin, J.M. and Walker, E.A., Eds., Scott, Foresman and Co., Chicago, 1963.

2. Liebert, W., The Jacobson radical of some endomorphism rings, *J. Reine Angew. Math.*, 262, 166, 1973.

3. Hausen, J., Quasi-regular ideals of some endomorphism rings, *Ill. J. Math.*, 22, 845, 1977.

4. Hausen, J. and Johnson, J.A., Ideals and radicals of some endomorphism rings, *Pacific J. Math.*, 22, 845, 1977.

5. Fuchs, L., *Infinite Abelian Groups, vol 1*, Academic Press, New York and London, 1970.

6. Fuchs, L., *Infinite Abelian Groups, vol 2*, Academic Press, New York and London, 1973.

7. Hill, P., On the decomposition of groups, *Canad. J. Math.*, 21, 762, 1969.

8. Amitsur, S.A., Rings of quotients and Morita contexts, *J. Algebra*, 17, 273, 1971.

# GROUPS AND MODULES THAT ARE SLENDER AS MODULES OVER THEIR ENDOMORPHISM RINGS

Adolf Mader

University of Hawaii

## I. INTRODUCTION

It was shown in Huber [4; Theorem 3.3] that an R-module $E$ is slender as a module over its endomorphism ring $S = \mathrm{End}_R E$ if it is of the form $E = M^{(\mathbb{N})}$ for some R-module $M$. This had important consequences for questions of E-reflexivity. In Mader [6; Lemma 5.1] it was shown that the abelian group $E = \oplus_{i=1}^{\infty} Z(p^i)$ is slender over $S = \mathrm{End}\, E$ and this was a key step in proving that the completion of a group $A$ of non-measurable cardinality equipped with the direct-sum-of-cyclics topology of D'Este [1] is algebraically the group $\mathrm{Hom}_S(\mathrm{Hom}_{\mathbb{Z}}(A, E), E)$. At the Honolulu Conference on Abelian Group Theory 1982/83, Martin Huber suggested to me that [6; Lemma 5.1] should be extendable to more general direct sums. In this note we prove such a theorem and use it to characterize completely the totally projective p-groups which are slender as modules over their endomorphism rings.

The proof uses ideas of L. Lady [5]. Lady's paper contains an unfortunate error, and in Section II we first make the necessary revisions in Lady's results. Next we prove the main theorem. Section III contains the applications to abelian groups and specifically totally projective p-groups. The characterization of those separable p-groups which are slender as modules over their endomorphism rings remains an open question.

The notation in this article is standard, see Fuchs [3]. Maps are written opposite to scalars. All rings have identity and modules are unitary. The natural numbers $\mathbb{N}$ do not include 0. A "linear basis" for a linear topology on a module is a neighborhood basis at 0.

## II. SLENDERNESS AND TOPOLOGY

Lady [5] defines the "$M$-adic topology" on a module and the "strong topology" on a product of modules as follows. For a set $M$ of two-sided ideals of the ring S and a module $B_S$, the submodules of the form $Bm$ with $m$ a finite intersection of ideals in $M$ define the "$M$-adic topology" on B. The "strong topology" on the product $P = \Pi\{A_i: i \in I\}$ is the least linear topology on P which is finer than the $M$-adic topology and the product topology (for discrete $A_i$). A linear basis for the strong topology is given by the submodules $\Pi\{A_i: i \in J\}m$ where $m$ is a finite intersection of ideals of $M$ and J is a cofinite subset of I. It is then claimed that P is complete in the strong topology and each open submodule $\Pi\{A_i: i \in J\}m$ is closed in the product topology. The following counterexample to these claims originated in Essen and reached me via Ray Mines. (Ed. note: See M.Dugas,

R. Göbel; Quotients of reflexive modules, Fund. Math. 114 (1981), p. 20)

2.1 EXAMPLE. Let $k$ be a field, $S = k[X_1, X_2, \ldots]$, and $m = (X_1, X_2 \ldots)$, the ideal generated by the countably many indeterminates $X_i$. Let $M = \{m\}$, $P = \Pi\{A_i: i \in \mathbb{N}\}$ with $A_i = S$, and for each $n \in \mathbb{N}$, let $a_n = [X_1, \ldots, X_n, 0, \ldots] \in P$. Then the strong topology is metrizable with linear basis $\Pi\{A_i: i \geq j\}m$, $a_1, a_2, \ldots$ is a Cauchy sequence in $Pm$ which does not converge in the strong topology while it converges in the product topology to the point $[X_1, X_2, \ldots]$ which does not lie in $Pm$. ///

The problem is caused by the fact that $(\Pi A_i)m$ and $\Pi(A_i m)$ may differ. We therefore modify the definitions as follows.

2.2 DEFINITION. Let $M \subset S$ and $B_S$ an S-module. The $M$-<u>topology</u> of $B$ is the linear topology given by the linear basis consisting of subgroups $Bs_1 \cap Bs_2 \cap \ldots \cap Bs_t$, $s_i \in M$, $t \in \mathbb{N}$. The <u>strong topology</u> on $P = \Pi\{A_i: i \in I\}$ is the supremum of the $M$-topology and the product topology as before. ///

We can now verify the needed properties of the strong topology. We emphasize that these topologies are GROUP topologies, not module topologies as the open subgroups $Bs$, $s \in M$, need not be submodules.

2.3 LEMMA. Let $P = \Pi\{A_i: i \in I\}$.

a) A linear basis for the strong topology of $P$ is given by the subgroups

$$\Pi\{A_i s_1 \cap A_i s_2 \cap \ldots \cap A_i s_t : s_j \in M,\ t \in \mathbb{N},\ i \in J \text{ and } J \text{ is cofinite in } I\}.$$

b) $P$ is complete in the strong topology.

c) Each subgroup $U = \Pi\{A_i s_1 \cap A_i s_2 \cap \ldots \cap A_i s_t : i \in J\}$ is closed in the product topology on $P$.

PROOF: a) If $\mathcal{U}$ and $\mathcal{V}$ are linear bases for two topologies on a group $B$ then, clearly, $\{U \cap V: U \in \mathcal{U}$ and $V \in \mathcal{V}\}$ is a linear basis for the supremum of the two topologies. Hence the subgroups

$$\Pi\{A_i : i \in J\} \cap (Ps_1 \cap \ldots \cap Ps_t) = \Pi\{A_i s_1 \cap \ldots \cap A_i s_t : \ i \in J\}$$

form a linear basis for the strong topology on $P$.

b) Let $a_d$, $d \in D$, be a Cauchy net in $P$. Then, given $U = \Pi\{A_i s_1 \cap \ldots \cap A_i s_t : i \in J\}$ there is $n(U)$ such that

$$a_d - a_e \in U \text{ whenever } d, e > n(U). \tag{2.4}$$

In particular, for $V = \Pi\{A_i : i \in J\}$ and $n(J) = n(V)$ we have

$$(a_d)_i = (a_e)_i \text{ for } i \notin J \text{ and } d, e > n(J). \tag{2.5}$$

Set $a_i = (a_d)_i$ for $d > n(J)$ and $i \notin J$, and let $a = [a_i] \in P$. We claim that $a = \lim a_d$. In fact, let $U = \Pi\{A_i s_1 \cap \ldots \cap A_i s_t : i \in J\}$ be given and choose $m \geq n(U), n(J)$. We claim that $a - a_d \in U$ whenever $d > m$. Let $d > m$. Then in case $i \notin J$, $a_i - (a_d)_i = 0$ by (2.5) and the definition of $a$. If $i \in J$, choose $k > n(J - \{i\}), m$. Then, by (2.4), $a_i - (a_d)_i = [a_i - (a_k)_i] + [(a_k)_i - (a_d)_i] = 0 + [(a_k)_i - (a_d)_i] \in A_i s_1 \cap \ldots \cap A_i s_t$.

c) Suppose $a \in P - U$. If $a_i \neq 0$ for some $i \in I - J$ then

$(a + \Pi\{A_i: i \in J\}) \cap U = \emptyset$. If $a_i = 0$ for all $i \notin J$ then there is $j \in J$ such that $a_j \notin A_j s_1 \cap \ldots \cap A_j s_t$. Now $(a + \Pi\{A_i: i \in J - \{j\}\}) \cap U = \emptyset$. ///

Using Lady's Lemma 1 and 2.3 we obtain immediately

2.6 LEMMA. If $M \subset S$ then $\Pi\{A_i: i \in \mathbb{N}\}$ is not the countable union of sets which are nowhere dense for the strong topology. ///

I do not know whether Lady's Theorem 1 is true or not but Theorem 2.7 below is a correct alternative. First recall that $B_S$ is <u>slender</u> if any one of the following equivalent conditions are satisfied (see Dimitric [2; 1.3], Huber [4; p. 474]).

(1) For every family $\{A_i: i \in \mathbb{N}\}$ of right S-modules and every $f \in \mathrm{Hom}_S(\Pi\{A_i: i \in \mathbb{N}\}, B)$ there is $n \in \mathbb{N}$ such that $f(\Pi\{A_i: i \geq n\}) = 0$.

(2) For every family $\{A_i: i \in \mathbb{N}\}$ of right S-modules and every $f \in \mathrm{Hom}_S(\Pi\{A_i: i \in \mathbb{N}\}, B)$ it follows that $f|A_i = 0$ for all but finitely many $i \in \mathbb{N}$.

(3) For every $f \in \mathrm{Hom}_S(S^{\mathbb{N}}, B)$ it follows that $f(e_i) = 0$ for almost all $i$ where $e_i = [\delta_{ij}] \in S^{\mathbb{N}}$.

(4) (LOS) For any family $A_i$, $i \in I$, of right S-modules with non-measurable index set $I$ the natural map

$$\oplus\{\mathrm{Hom}_S(A_i, B): i \in I\} \to \mathrm{Hom}_S(\Pi\{A_i: i \in I\}, B)$$

is an isomorphism.

2.7 THEOREM. Let $B$ be a countable $S$-module and suppose that there exists a set $\mathcal{M}$ of commuting elements of $S$ such that each $s \in \mathcal{M}$ is injective on $B$ and $B$ is Hausdorff in the $\mathcal{M}$-topology. Then $B_S$ is slender.

PROOF. Let $f \in \mathrm{Hom}_S(P, B)$ where $P = \Pi\{A_i: i \in \mathbb{N}\}$. Then $f$ is continuous for the $\mathcal{M}$-adic topologies on $P$ and $B$ since $f(Ps) = f(P)s \subset Bs$. Now $\{f^{-1}(b): b \in B\}$ is a countable closed cover of $P$ and by 2.6 there is $b \in B$, $a \in P$ and $U = \Pi\{A_i s_1 \cap \ldots \cap A_i s_t: i \geq n\}$ such that $a + U \subset f^{-1}(b)$. Hence $f(a) + f(U) \subset \{b\}$ and this implies that $f(U) = 0$. But $\Pi\{A_i s_1 s_2 \ldots s_t: i \geq n\} \subset \Pi\{A_i s_1 \cap \ldots \cap A_i s_t: i \geq n\}$ since the $s_i$ are pairwise commuting so $0 = f(\Pi\{A_i s_1 s_2 \ldots s_t: i \geq n\}) = f(\Pi\{A_i: i \geq n\})\, s_1, s_2 \ldots s_t$. Since the $s_i$ act injectively, we have $f(\Pi\{A_i: i \geq n\}) = 0$ as desired. ///

We now come to the main theorem of this paper.

2.8 THEOREM. Let ${}_R M$ be an $R$-module, $S = \mathrm{End}_R M$ and assume that $M = \oplus\{M_k: k \in \mathbb{N}\}$ such that for all $i$, $\oplus\{M_k: k < i\}$ contains no nonzero $S$-submodule of $M$. Then $M_S$ is slender.

PROOF. Let $\pi_i: M \longrightarrow \oplus\{M_k: k \geq i\}$ be the projections and let $K_i = \mathrm{Ker}\pi_i = \oplus\{M_k: k < i\}$. Note that

$$\pi_j \pi_i = \pi_i \pi_j = \pi_j \quad \text{for } j \geq i. \tag{2.9}$$

The family $\{M\pi_i: i = 1, 2, \ldots\}$ defines a linear Hausdorff topology on $M$; each $\pi_k: M \to M$ is trivially continuous and $K_k = \{0\}\pi_k^{-1}$ is closed in $M$; furthermore $M = \cup\{K_k: k \in \mathbb{N}\}$. Let $A_k$ be a right

S-module for $k \in \mathbb{N}$, $P = \Pi\{A_k: k \in \mathbb{N}\}$ and choose $M = \{\pi_k: k \in \mathbb{N}\}$. By 2.3 b) $P$ is complete in the strong topology. Also note that $P$ has a linear basis consisting of groups $\Pi\{A_k\pi_j: k > i\}$ since $A_k\pi_j \subset A_k\pi_{j'}$ for $j \geq j'$ by (2.9). Suppose $f \in \mathrm{Hom}_S(P, M)$. Then $f$ is continuous since $f(P\pi_k) = f(P)\pi_k \subset M\pi_k$. Hence $\{f^{-1}(K_k): k \in \mathbb{N}\}$ is a countable closed cover of $P$ and by 2.6 there are $a \in P$ and integers $i, j, m$ such that $f(a + \Pi\{A_k\pi_j: k \geq i\}) \subset K_m$. It follows that $f(a) \in K_m$ and $f(\Pi\{A_k\pi_j: k \geq i\}) = f(\Pi\{A_k: k \geq i\})\pi_j \subset K_m$. Thus $f(\Pi\{A_k: k \geq i\}) \subset K_{j+m}$. But $f(\Pi\{A_k: k \geq i\})$ is an S-submodule of $B$ and so by hypothesis $f(\Pi\{A_k: k \geq i\}) = 0$. ///

## III. ABELIAN GROUPS AS MODULES OVER THEIR ENDOMORPHISM RINGS

Let us call a left R-module <u>E-slender</u> if it is slender as a right module over its endomorphism ring. We begin by exhibiting examples of groups which are not E-slender.

3.1 LEMMA. A finite non-zero elementary p-group is not E-slender.

PROOF. Let $A$ be a finite elementary p-group, i.e. a finite dimensional vector space over $\mathbb{Z}/(p)$. Then $S = \mathrm{End}A$ is a full matrix ring over $\mathbb{Z}/(p)$ and therefore semi-simple. Hence every S-module is semi-simple and the sequence $0 \to S^{(\mathbb{N})} \to S^{\mathbb{N}} \to S^{\mathbb{N}}/S^{(\mathbb{N})} \to 0$ is split-exact. Here $S^{(\mathbb{N})}$ is free on the set $e_i = [\delta_{ij}] \in S^{\mathbb{N}}$ and there is an S-homomorphism $f: S^{(\mathbb{N})} \to A$ such that $f(e_i) \neq 0$. Since $S^{(\mathbb{N})}$ is a summand of $S^{\mathbb{N}}$, $f$ extends to $S^{\mathbb{N}}$ showing that $A$ is not E-slender. ///

The following easy lemma is useful in extending the supply of non-E-slender groups. (I am indebted to the referee for improvements of 3.2 and 3.3.)

3.2 LEMMA. Let $B$ be a fully invariant $R$-submodule of ${}_RA$. If $B$ is not E-slender than neither is $A$.

PROOF. Let $r\colon \mathrm{End}A \to \mathrm{End}B$ be the restriction map. Suppose $f\colon (\mathrm{End}B)^{\mathbb{N}} \to B$ is an EndB-homomorphism. Any EndB-module $M$ becomes an EndA-module via $x \cdot t = x \cdot (tr)$ for $x \in M$, $t \in \mathrm{End}A$. In this fashion $f\colon (\mathrm{End}B)^{\mathbb{N}} \to B$ becomes an EndA-homomorphism. Preceding $f$ with the EndA-homomorphism $\bar{r}\colon (\mathrm{End}A)^{\mathbb{N}} \to (\mathrm{End}B)^{\mathbb{N}}$ induced by the restriction map $r$ we obtain an EndA-homomorphism $g = \bar{r}f\colon (\mathrm{End}A)^{\mathbb{N}} \to B \leq A$. If $f$ violates the condition for slenderness so does $g$ since
$(0\ldots0\ 1_A\ 0\ldots)g = (0r\ldots0r\ 1_Ar\ 0r\ldots)f = (0\ldots0\ 1_B\ 0\ldots)f$. ///

We are now in a position to show that many p-groups are not E-slender.

3.3 COROLLARY. If $A$ is a p-group of length $\lambda + 1$ such that $p^\lambda A$ is finite then $A$ is not E-slender.

PROOF. The group $(p^\lambda A)[p]$ is a fully invariant subgroup of $A$. An application of 3.1 and 3.2 completes the proof. ///

In particular, 3.3 describes a class of non-E-slender totally projective p-groups. We shall show that any other totally projective p-group is E-slender. This is achieved by showing that the group decomposes in such a way that 2.8 applies. A totally projective p-group of limit length is a direct sum of groups of smaller length ([3; II, p. 99]). Thus the following case remains to consider.

3.4 LEMMA. Let $A$ be a totally projective $p$-group of length $\lambda + 1$. Then $A \cong A_o \oplus (\oplus\{A_x: x \in Y\})$ with $p^\lambda A_o = 0$, $p^\lambda(\oplus\{A_x: x \in Y\}) = p^\lambda A$ and $|Y| = \dim p^\lambda A$.

PROOF. The totally projective group $A$ is simply presented, say $A = \langle X; \Sigma \rangle$ using the notation of Fuchs [3; Vol. II, p. 95]. For $0 \neq a \in p^\lambda A$ we have $a = s_1 x_1 + \ldots + s_k x_k$, $x_i \in X$, $0 < s_i < p$ by [3; II, p. 96 (b)]. By [3; II, p. 96 (c)], $x_i \in p^\lambda A$ for $1 \le i \le k$. Hence $p^\lambda A$ is generated by $Y = \{x \in X: x \in p^\lambda A\}$. Clearly $Y$ is a set of minimal elements hence ([3; II, p. 97 (d)]) $A = A_o \oplus (\oplus\{\langle x_x \rangle: x \in Y\})$ where $A_o = \oplus\{\langle x_x \rangle: x \in X, x$ is minimal and $x \notin p^\lambda A\}$. The lemma follows with $A_x = \langle x_x \rangle$, $x \in Y$. ///

The next lemma shows how the hypothesis of Theorem 2.8 can be secured.

3.5 LEMMA. Suppose $A = \oplus \{A_i: i \in I\}$ is a totally projective $p$-group such that for every finite subset $J$ of $I$ there is $k \in I \setminus J$ such that $\text{length}(A_k) \ge \text{length}(A_i)$ for $i \in J$. Then for every finite subset $J$ of $I$ the group $B = \oplus \{A_i: i \in J\}$ does not contain a non-zero fully invariant subgroup of $A$.

PROOF. It suffices to show that for every non-zero element $a \in B$ there is $f \in \text{End}A$ such that $af \notin B$. Let $h = \text{height}(a)$, and note that $h < \max\{\text{length}(A_i): i \in J\}$. So $0 \neq a + p^{h+1}B \in p^h(B/p^{h+1}B) \subset (B/p^{h+1}B)[p]$. By hypothesis there is $k \in I \setminus J$ such that $\text{length}(A_k) > h$. Hence there is $x \in A_k[p]$ with $\text{height}(x) \ge h$ and there is a homomorphism $g: p^h(B/p^{h+1}B) \to A_k$ such that $g$ does not decrease heights and $(a + p^{h+1}B)g = x$. By Fuchs [3; II, p. 74 (e)] $p^h(B/p^{h+1}B)$ is nice in $B/p^{h+1}B$ and hence $g$ extends to a homomorphism $\bar{g}: B/p^{h+1}B \to A_k$ ([3; II, p. 89 (E)]). Now let $\bar{f}$ be the composite $B \xrightarrow{\text{nat.}} B/p^{h+1}B \overset{\bar{g}}{\to} A_k$ and extend it to an endomorphism of A. ///

We can now complete the characterization of E-slender totally projective p-groups.

3.6 THEOREM. Let $A$ be a totally projective p-group of length $\lambda$.

a) If $\lambda - 1$ exists and $p^{\lambda-1}A$ is finite then $A$ is not slender as a module over its endomorphism ring.

b) If $\lambda - 1$ exists and $p^{\lambda-1}A$ is infinite then $A$ is slender as a module over its endomorphism ring.

c) If $\lambda$ is a limit ordinal then $A$ is slender as a module over its endomorphism ring.

PROOF. Part a) is Corollary 3.3. For part b) we use Lemma 3.4 to get a decomposition $A = A_o \oplus (\oplus\{A_i: i \in I\})$ with $I$ infinite, $p^{\lambda-1}A_o = 0$ and $\text{length}(A_i) = \lambda$. Clearly, we can combine summands in order to get a countable decomposition $A = A_o \oplus (\oplus\{A_i': i \in \mathbb{N}\})$ with $\text{length}(A_i') = \lambda$ for $i \in \mathbb{N}$. Now 3.5 supplies the hypothesis for an application of 2.8. In the c) we have $A = \oplus\{A_i: i \in I\}$ with $\text{length}(A_i) < \lambda = \text{length}(A)$ by [3; II, p. 97 (e)]. If $\text{cof}(\lambda) = \omega$ let $\lambda_o = 0 < \lambda_1 < \lambda_2 < \ldots$ be a sequence of ordinals converging to $\lambda$.

Define $B_i = \oplus\{A_j: \lambda_{i-1} \le \text{length}(A_j) < \lambda_i\}$. Then $A = \oplus\{B_i: i \in \mathbb{N}\}$ satisfies the hypothesis of 3.5 and again 2.8 does the rest. If $\text{cof}(\lambda) > \omega$ suppose $\lambda(\alpha)$, $\alpha \in \text{cof}(\lambda)$, is a set of lengths of groups $A_i$ converging to $\lambda$ and such that $\lambda(\alpha) > \lambda(\beta)$ if $\alpha > \beta$. For $i \in \mathbb{N}$ let $S_i = \{\lambda(\alpha + i - 1): \alpha \in \text{cof}(\lambda)$ and $\alpha$ is a limit ordinal$\}$. Then for all $i \in \mathbb{N}$, $\lim S_i = \lambda$ and setting $B_i = \oplus\{A_j : \text{length}\, A_j \in S_i\}$ and letting $B_o$ be the direct sum of all remaining $A_i$, we obtain $A = B_0 \oplus B_1 \oplus \ldots$ with $\text{length}(B_i) = \lim S_i = \lambda$ for $i \in \mathbb{N}$. Again 3.5 and 2.8 complete the proof. ///

We conclude with some miscellaneous results.

3.7 PROPOSITION. If the p-group $A$ is not reduced then $A$ is not slender as a module over its endomorphism ring.

PROOF. Justified by 3.2 we assume without loss of generality that $A$ is a non-zero divisible group. If $A[p]$ is finite it is not E-slender by 3.1, and if it is infinite it is not E-slender by 2.8. So $A$ is not E-slender by 3.2. ///

3.8 PROPOSITION. If $A$ is torsion-complete and unbounded then $A$ is not slender as a module over its endomorphism ring.

PROOF. Let $S = \text{End}A$ and let $a_1, a_2, \ldots$ be elements of $A$ with $e(i) = \text{exponent}(a_i)$ such that $0 < e(1) < e(2) < \ldots$ . Then $f: S^{\mathbb{N}} \to A: f([s_1, s_2, \ldots]) = \Sigma\{p^{e(i)-1} a_i s_i: i \in \mathbb{N}\}$ defines an S-homomorphism which shows that $A$ is not E-slender. Note that the infinite series converges since $A[p]$ is complete in the p-adic topology. ///

It is clear what happens when $A$ is a bounded group. The group is E-slender if and only if the last non-zero Ulm invariant is infinite. If a p-group $A$ without elements of infinite height is not E-slender then in any direct decomposition $A = \oplus \{A_i: i \in I\}$ there is a common bound for all but finitely many summands. This is a consequence of 2.8, and it shows that a non-E-slender p-group without elements of infinite height must be close to being torsion-complete.

## REFERENCES

[1] D'Este, G.; The $\oplus_c$-topology on abelian p-groups, Annali della Scuola Norm. Sup. di Pisa, VII, 241, 1980.

[2] Dimitric, R.; Slenderness in abelian categories, Abelian Group Theory, Proceedings, Honolulu 1982/83, Lecture Notes in Mathematics 1006, 375.

[3] Fuchs, L.; *Infinite Abelian Groups* I, II, Academic Press 1970, 1973, New York-London.

[4] Huber, M.; On reflexive modules and abelian groups, J. Algebra 82, 469, 1983.

[5] Lady, E. L.; Slender rings and modules, Pac. J. Math., 49, 397, 1973.

[6] Mader, A.; Completions via duality, Abelian Group Theory, Proceedings, Honolulu 1982/83, Lecture Notes in Mathematics 1006, 562.

# STABILIZER CLASSES DETERMINED BY SIMPLY PRESENTED MODULES

TEMPLE H. FAY
DEPARTMENT OF MATHEMATICS
UNIVERSITY OF SOUTHERN MISSISSIPPI
HATTIESBURG, MS 39406
USA

Abstract: In this note stabilizer classes (classes closed under formation of direct sums and homomorphic images) which consist of groups G for which $G/\tau G$ is divisible ($\tau G$ denotes the torsion subgroup) are examined. Particular attention is given to singly generated stabilizer classes-those determined by a single group. It is shown that simply presented torsion-free rank one groups determine stabilizer classes that depend only on the height matrix of the group. Localization techniques are employed to isolate at a prime and thereby to consider modules over the integers localized at a prime p.

AMS (MOS): 20K21, 20K12.
Key Words: socle, stabilizer class, simply presented, torsion-free rank one, Ulm sequence.

1. Preliminaries. In this note we are concerned with certain stabilizer classes of groups (all groups are abelian). Stabilizer classes are classes closed under formation of direct sums and homomorphic images. These classes are in one to one correspondence with socles (idempotent preradicals). For more information on socles and preradicals in general we refer the reader to [1], [2], and [3] of which this work can be viewed as a continuation. All groups G considered in this note will have the property that $G/\tau G$ is divisible (here $\tau G$ denotes the torsion subgroup of G). These groups form an extensive stabilizer class and have been called special by Matlis [8].

If X is a group, then the stabilizer class determined by X consists precisely of all homomorphic images of direct sums of copies of X. If $\mathcal{S}$ is a stabilizer class for which there is a group X determining $\mathcal{S}$ in this way, then $\mathcal{S}$ is called singly generated. The unique socle associated with such an $\mathcal{S}$ is denoted by $S_X$ and its value on a group A is given by $S_X A = \sum\{\mathrm{Im} f\colon\ f \in \mathrm{Hom}\,(X,A)\}$; it is clear that $\mathcal{S} = \{A\colon\ S_X A = A\}$. The socle $S_X$ is a radical precisely when, for each group A, $S_X(A/S_X A) = 0$, and this occurs exactly when $\mathcal{S}$ is closed under formation of extensions.

If S and T are preradicals, then we form the co-composition S:T by defining, for each group A, S:TA to be that subgroup of A defined by the formula $(S{:}TA)/TA = S(A/TA)$. The stabilizer class of S:T is precisely the class of all groups A for which A/TA belongs to the stabilizer class of S.

The author would like to acknowledge helpful conversations with his colleague G.L. Walls and with Professor L. Fuchs of Tulane University where, during a semester sabatical leave, this work was initiated.

Thus in particular, the stabilizer class of d: $\tau$ is precisely the class of special groups (here d denotes the divisible subgroup functor).

If S is an arbitrary socle with stabilizer class $\mathcal{S}$, then for any group A, $SA = \sum\{S_xA: x\in\mathcal{S}\}$. It is in this sense then that we say that singly generated socles $S_x$ form "building blocks" for socles in general.

2. Some Stabilizer Classes. We begin by discussing some of the more obvious stabilizer classes of special groups and show that each is singly generated. Moreover, we will identify all the stabilizer classes of special groups that are closed under formation of extensions.

The first class of groups we consider are the additive groups of $\pi$-regular rings (see [4], Vol. II). For each prime p let m(p) be a positive integer, let $[p^{m(p)}]$ denote the $p^{m(p)}$ - socle subgroup functor, and let $T = \sum_p[p^{m(p)}]$. Our intention is to show that the class of all groups G with $\tau G = TG$ and G/TG divisible is singly generated. Let $e_p$ denote the multiplicative unit of $\mathbb{Z}(p^{m(p)})$ and let $e = (\ldots, e_p, \ldots)$ in $\Pi_p \mathbb{Z}(p^{m(p)})$. We let M be that subgroup containing e satisfying $M/\oplus_p\mathbb{Z}(p^{m(p)}) \cong \mathbb{Q}$. This group M is in fact a commutative $\pi$-regular ring having the property that if G is the additive group of a $\pi$-regular ring with $\tau_pG = G[p^{m(p)}]$ for each prime p, then G is a unital M-algebra [4, Theorem 125.3]. Thus we have each such G the epimorphic image of a direct sum of copies of M.

Another way to put this is that $d{:}T = S_M$. This group M is particularly nice; it is socular in the terminology of [3]. This means M is cyclic considered as a module over it endomorphism ring and consequently, for each group A, $S_M A = \{f(e): f \in Hom(M,A)\}$. A consequence of this observation is that this stabilizer class is closed under formation of products (see [2]).

Our next result identifies those stabilizer classes of special groups closed under formation of extensions.

2.1. Theorem. Let S be a preradical satisfying $d \leq S \leq d:\tau$. Then S is a socle-radical if and only if $S = d:T$ where $T = \sum_\pi \tau_p$ for some set of primes $\pi$.

Proof. We suppose that S is a socle-radical and that $S \neq d$. Then there is a group G stabilized by S which is not divisible. Without loss of generality we may assume G to be reduced. Because $G/\tau G$ is divisible, G is not torsion-free. It follows that $pG \neq G$ for some prime p and hence G/pG is stabilized by S. This implies that $\mathbb{Z}(p)$ and consequently $\mathbb{Z}(p^m)$ for each m is stabilized by S. Let $\pi = \{p: \mathbb{Z}(p)$ is stabilized by $S\}$. It follows that $T = \sum_\pi \tau_p \leq S$ and, as S is a radical, $d: T \leq S$.

If $d: T \neq S$, then there exists a group G annihilated by d:T but for which $SG \neq 0$. If SG were torsion-free, then SG would also be reduced and thus belong to the annihilator class of S - that is $0=S(SG) = SG$ contrary to assumption. Hence $\tau SG \neq 0$. Since $TSG = 0$, SG has a non-zero q-torsion component for some prime q not belonging to $\pi$. This implies $qSG \neq SG$ and, then as above, that $\mathbb{Z}(q)$ is stabilized by S, contrary to the choice of q.

The converse follows from the observation that if T is a socle-radical, then so is d:T.

We next show that if $T = \sum_\pi \tau_p$ where $\pi$ is a set of primes, then d:T is a singly generated socle. That is, there exists a group $M_T$ with $M_T/TM_T$

divisible, so that, for any group G, G/TG is divisible if and only if G is the homomorphic image of a direct sum of copies of $M_T$. Let $B = \oplus_\pi \oplus_m \mathbb{Z}(p^m)$ and for each E belonging to Ext ($\mathbb{Q}$, B) choose a representative sequence:

$$E:\quad 0 \to B \to M_E \to \mathbb{Q} \to 0.$$

We set $M_T = \oplus_E M_E$.

2.2. Lemma. If G is countable and $G/TG = \mathbb{Q}$, then G is an epimorphic image of a direct sum of copies of $M_T$.

Proof. Pick $\{x_1, x_2, \ldots\} \subseteq G$ so that in $G/TG = \mathbb{Q}$, the image of $x_m$ is $1/m!$ and set $G^* = \langle x_1, x_2, \ldots\rangle$. Then $TG + G^* = G$ and $G^*$ is countable. There are a sequence and epimorphisms so that the following diagram commutes

$$\begin{array}{ccccccccc} 0 & \to & B & \to & H & \to & \mathbb{Q} & \to & 0 \\ & & \downarrow & & \downarrow & & \| & & \\ 0 & \to & \tau G^* & \to & G^* & \to & \mathbb{Q} & \to & 0 \end{array}$$

Because H is an epimorphic image of $M_T$, so is $G^*$. Clearly, $G = TG + G^*$ is an epimorphic image of a direct sum of copies of $M_T$.

2.3. Theorem. The socle-radical d:T is singly generated as a socle by $M_T$.

Proof. If $G/TG = \oplus_I \mathbb{Q}$, then for each $i \in I$, choose a family of generators as done in the previous lemma to produce a subgroup $G_i$ of G with $G_i/\tau G_i = \mathbb{Q}$. Then $G = \sum_I G_i + \tau G$ and as $\tau G = TG$ and each $G_i$ is the epimorphic image of a direct sum of copies of $M_T$, G is as well.

3. Simply Presented Modules. If G is a special group, then $G/\tau G = \oplus_I \mathbb{Q}$. By pulling back each factor $\mathbb{Q}_i$, we can produce a family of torsion-free rank one subgroups of G, each containing $\tau G$, and so that G may be realized

as the direct sum of the $G_i$'s with $\tau G$ amalgamated. Thus each G is a sum of special torsion-free rank one groups. Each torsion-free rank one can be obtained as the image of a direct sum of copies of a simply presented rank one group having the same height matrix as the original group. Thus the "building blocks" for stabilizer classes of special groups might be those singly generated classes determined by simply presented groups. In this section we show that such singly generated stabilizer classes determined by a simply presented rank one group depends only on its height matrix.

We begin our treatment by recalling some definitions and results primarily from the development discussed by Warfield [16]. First we fix a prime p and denote the ring of integers localized at p by $\mathbb{Z}_p$, and we localize groups by tensoring with $\mathbb{Z}_p$, thus obtaining $\mathbb{Z}_p$-modules. This is just a convenience for dealing with p-local groups, groups for which multiplication by each prime q different from p is an automorphism. We shall employ what has been called the abelian group theorist's Hasse principle [14], [5] to go from "the local case to the global case".

If G is a module (over $\mathbb{Z}_p$) and $\alpha$ is an ordinal, we define $p^\alpha G$ in the usual inductive way, and set $p^\infty G = \cap p^\alpha G$ where the intersection is taken over all ordinals. The generalized p-height of $x \in G$, $h^*_p(x)$, is defined to be that ordinal $\alpha$ for which $x \in p^\alpha G \backslash p^{\alpha+1} G$ if such an ordinal exists, otherwise it is defined to be $\infty$. The p-Ulm sequence (or more simply the Ulm sequence) of x is the sequence $U(x) = \{h_p^*(p^m x): m \geq 0\}$. An Ulm sequence is a sequence of ordinals or symbols $\infty$, $\{\alpha_0, \alpha_1, \ldots\}$ such that $\alpha_{i+1} \geq \alpha_i$ and $\alpha_{i+1} = \alpha_i$ only if $\alpha_i = \infty$.

Two Ulm sequences $\{\alpha_i\}$ and $\{\beta_i\}$ are said to be equivalent provided there are integers s and t so that $\alpha_{s+k} = \beta_{t+k}$ for all $k \geq 0$. If G is a torsion-free rank one module, then any two elements of infinite order have equivalent Ulm sequences, and thus the equivalence class of such sequences is an invariant of the group.

A simply presented module is one which can be defined in terms of generators and relations in such a way that the relations all have the form $px = 0$ or $px = y$. The most important observation about simply presented modules is that they are direct sums of simply presented torsion-free rank at most one submodules. From [16] we have the following slightly modified result.

3.1. Theorem. If G and H are simply presented torsion-free rank one modules, then G is isomorphic to H if and only if G and H have equivalent Ulm sequences and the same Ulm invariants.

This result implies that when considering singly generated socles determined by simply presented modules, we need only consider the Ulm sequences of the modules.

3.2. Theorem. Let G and H be (reduced) simply presented modules of torsion-free rank one having equivalent Ulm-sequences. Then $S_G = S_H$.

Proof. First note that from the reducedness assumption on G, $p^m G \neq G$ for every m, so $G/p^m G$ belongs to the stabilizer class of $S_G$, and hence $\mathbb{Z}(p^m)$ does as well. This implies any p-group is stabilized by $S_G$ (and similarly for $S_H$).

We can find a simply presented torsion group T with Ulm invariants so large that $G \oplus T$ and $H \oplus T$ have the same Ulm invariants. From Theorem 3.1 we have $G \oplus T \cong H \oplus T$ and thus $S_G = S_{G \oplus T} = S_{H \oplus T} = S_H$.

Simply presented torsion-free rank one modules exist in abundance as the following theorem from [16] illustrates. We sketch the proof as the details will be useful in what follows.

3.3. Theorem. If $u = \{u_i\}$ is an Ulm sequence, then there is a simply presented module $T_u$ of torsion-free rank one such that $T_u$ contains an element $x_u$ of infinite order such that $U(x_u) = u$. Moreover, $T_u / \tau T_u \cong \mathbb{Q}$.

Proof. If $u_1 = \infty$, then the module $\mathbb{Q}$ is chosen to be $T_u$. If $u_{m+1} = \infty$ and $u_m < \infty$, then set $T_u = H \oplus \mathbb{Q}$ where $H = H_{u_m+1}$ is the generalized Prüfer group and set $x_u = (y, 1)$ where y is one of the generators of H. Finally, if all the $u_i$'s are ordinals, we let $T_u$ be the module with generators consisting of finite sequences of ordinals $(\alpha_1, \alpha_2, \ldots, \alpha_m)$, for various m's, $m \geq 1$ where $\alpha_{i+1} > \alpha_i$ and $\alpha_m$ is one of the ordinals $u_i$ $(i \geq 0)$. We impose the relations

$$p(\alpha_1, \ldots, \alpha_m) = (\alpha_2, \ldots, \alpha_m)$$

and

$$p(u_i) = (u_{i+1}).$$

Clearly this description yields a module of the desired form with $x_u = (u_0)$.

The next result is of fundamental importance as it provides a criterion for extending homomorphisms, although in most applications it has been used to lift isomorphisms and to obtain classification theorems. This result is an extension of a result first proved by Hill for p-groups. A more general version was given by Walker [12] and a version for modules over a discrete valuation ring is stated in [16]. It is this latter version we state.

3.4. <u>Theorem</u>. Let G and H be modules and K a nice submodule of G with G/K torsion and totally projective (simply presented), and let f: $K \rightarrow H$ be a homomorphism satisfying for each ordinal $\alpha$, $f(K \cap p^{\alpha} G) \subseteq p^{\alpha} H$. Then f extends to a homomorphism of G into H.

This brings us to the main result of this note. In analogy with equivalence for Ulm sequences let us define $\{\alpha_i\} \geq^* \{\beta_i\}$ for two Ulm sequences $\{\alpha_i\}$ and $\{\beta_i\}$ provided there exist integers s and t so that $\alpha_{s+k} \geq \beta_{t+k}$ for all $k \geq 0$. Thus one might say that the inequality $\geq^*$ holds on tails of the $\{\alpha_i\}$ and $\{\beta_i\}$.

3.5. <u>Theorem</u>. Let G be a non-split simply presented rank one module as above, containing $(u_0)$ with $U(u_0) = \{u_i\}$. Let H be a module with $h \in H$, h of infinite order, satisfying $U(h) \geq^* U(u_0)$. Then there exists a homomorphism $f: G \rightarrow H$ with h in the image of f.

<u>Proof</u>. Let $U(h) = (\sigma_i)$. There exist integers s and t so that $\sigma_{s+k} \geq u_{t+k}$ for all $k \geq 0$.

<u>Case</u> 1. $s \leq t$.

In this case, $\sigma_{s+k} \geq u_{t+k} \geq u_{s+k}$ for all k. Pick the generator $y \in G$ where $y = (0, 1, \ldots, s-1, u_s)$ and observe that $U(y) = (0, 1, \ldots, s-1, u_s, u_{s+1}, \ldots)$ and that coordinatewise $U(y) \leq U(h)$.

<u>Case</u> 2. $t < s$ and $u_t > s$.

Observe that $y = (0, 1, \ldots, s, u_t) \in G$ and coordinatewise

$U(y) = (0,1, \ldots, s, u_t, u_{t+1}, \ldots) \leq U(h)$.

Case 3. $t < s$ and $u_t \leq s$ and there exists an $l$ so that $u_l \geq \omega$.

Clearly $l > t$ so that $u_l \leq \sigma_{s+l-t}$. Pick the generator

$y = (0, 1, \ldots, s + l - t - 1, u_l) \in G$ and observe that coordinatewise

$U(y) = (0, 1, \ldots, s + l - t - 1, u_l, u_{l+1}, \ldots) \leq U(h)$.

Case 4. $t < s$, $u_t \leq s$ and all the $u_i$'s are finite.

In this case G is countably generated and non-split, so, by a result of Megibben [9], there are an infinite number of gaps in the sequence $\{u_i\}$ (i.e., places where $u_{k+1} > u_k + 1$). Thus there exists an $l$ such that $u_l > l + (s - t) - 1$. Pick the generator $y \in G$ so that

$y = (0, 1, \ldots, s - t + l - 1, u_l)$ and observe that coordinatewise

$U(y) = (0, 1, \ldots, s - t + l - 1, u_l, u_{l+1}, \ldots) \leq U(h)$ since

$u_{l+k} = u_t + (l-t) + k \leq \sigma_{s+l-t+k}$ for $k \geq 0$.

In each of the above cases we have produced an element of infinite order, $y \in G$, so that the map between infinite cyclic subgroups $\langle y\rangle \longrightarrow \langle h\rangle$ respects valuations. Because $\langle y\rangle$ is nice in G and $G/\langle y\rangle$ is torsion and totally projective, this map extends to a homomorphism $f: G \longrightarrow H$ with h in the image of f.

3.6. Corollary. If G is a non-split simply presented rank one module and $u = U(x)$ for $x \in G$, x of infinite order, then for any module H,

$$S_G(H) = \{h \in H \mid U(h) \geq^* u\}.$$

Proof. First observe that $\tau_p \leq S_G$ as G is reduced and so $G/p^m G \neq 0$ for each m. Thus $\mathbb{Z}(p^m)$ is fixed by $S_G$ for each m and so $\tau_p \leq S_G$. Consequently,

$\tau H = \tau_p H \leq S_G(H)$. If h is of finite order, then for some m, $p^m h = 0$ which implies $h_p^*(p^k h) = \infty$ for all $k \geq m$ and so $U(h) \geq^* u$. Now the equality is obvious from the theorem.

To see how we can obtain global results from the local case, we recall the Hasse principle for abelian group theorists.

3.7. Theorem. [14,5] Let G and H be abelian groups, K a subgroup of G with G/K torsion, and let $f: K \rightarrow H$ be a homomorphism. If, for each prime p, there is a map $g_p$: $G \otimes \mathbb{Z}_p \rightarrow H \otimes \mathbb{Z}_p$ that extends $f_p = f \otimes 1_{\mathbb{Z}_p}$, then there exists a homomorphism $f^*$ from G to H extending f and such that $f^* \otimes 1_{\mathbb{Z}_p} = g_p$ for each prime p.

3.8. Corollary. If G is a simply presented rank one group, then $S_G(H) = H$ if and only if for each prime p, $H \otimes \mathbb{Z}_p = S_{G \otimes \mathbb{Z}_p}(H \otimes \mathbb{Z}_p)$.

As a closing remark, we mention that summands of simply presented modules (Warfield modules) yield no additional stabilizer classes; for if A is a Warfield module, then there is a totally projective torsion module B so that $A \oplus B$ is simply presented [5]. It follows that

$S_{A \oplus B} = S_A + S_B = S_A$.

## References

1. T.H. Fay, E.P. Oxford, and G.L. Walls, "Preradicals in abelian groups", Houston J. Math. 8 (1982) 39-52.

2. T.H. Fay, E.P. Oxford, and G.L. Walls, "Preradicals induced by homomorphisms", Abelian Group Theory, Springer Lecture Notes 1006 (1983) 660-670.

3. T.H. Fay, E.P. Oxford, and G.L. Walls, "Singly generated socles and radicals", Abelian Group Theory, Springer Lecture Notes 1006 (1983) 671-684.

4. L. Fuchs, Infinite Abelian Groups, Vols. I, II, (Academic Press, New York and London, 1970 and 1973).

5. R. Hunter and F. Richman, "Global Warfield groups", Trans. Amer. Math. Soc. 266 (1981) 555-572.

6. R. Hunter, F. Richman, and E.A. Walker, "Warfield modules", Abelian Group Theory, Springer Lecture Notes 616 (1977) 87-123.

7. R. Hunter, F. Richman, and E.A. Walker, "Existence theorems for Warfield groups", Trans. Amer. Math. Soc. 235 (1978) 345-362.

8. E. Matlis, "Cotorsion modules", Memoirs Amer. Math. Soc. 49 (1964) pp 66.

9. C.K. Megibben, "On mixed groups of torsion-free rank one", Ill. J. Math. 11 (1967) 134-144.

10. F. Richman, "Mixed groups", Abelian Group Theory, Springer Lecture Notes 1006 (1983) 445-470.

11. J. Rotman, "Torsion-free and mixed abelian groups", Ill. J. Math. 5 (1961) 131-143.

12. E.A. Walker, "Ulm's theorem for totally projective groups", Proc. Amer. Math. Soc. 37 (1973) 387-392.

13. K.D. Wallace, "On mixed groups of torsion-free rank one with totally projective primary components", J. Alg. 17 (1971) 482-488.

14. R.B. Warfield, Jr., "The structure of mixed abelian groups", Abelian Group Theory, Springer Lecture Notes 616 (1977) 1-38.

15. R.B. Warfield, Jr., "Classification theory of abelian groups. I: Balanced projectives", Trans. Amer. Math. Soc. 222 (1976) 33-63.

16. R.B. Warfield, Jr., "Classification theory of abelian groups, II: Local theory", Abelian Group Theory, Springer Lecture Notes 874 (1981) 322-349.

---

Department of Mathematics
University of Southern Mississippi
Hattiesburg, MS 39406 USA

# On divisible modules over domains

L. Fuchs
Tulane University, New Orleans, La.

This note is devoted to the study of divisible modules over arbitrary commutative domains R with 1. Whereas divisible modules over Dedekind domains can be completely characterized by numerical invariants, not much is known about their structures in the general case. Divisible modules have been studied by Matlis [6] who was the first to distinguish between divisibility in general and h-divisibility. He established an important duality between h-divisible torsion modules and complete torsion-free R-modules [7]. Matlis [6] and Hamsher [4] characterized those domains R for which all divisible R-modules are h-divisible, e.g. by the property that the field Q of quotients of R has projective dimension 1.

Recently, de la Rosa and Fuchs [2] introduced a dimension function for h-divisible torsion modules to measure their discrepancy from the direct sum of copies of Q/R. In case p.d.Q = 1, this dimension turns out to be intimately related to the projective dimension.

In this note, we investigate divisible R-modules in general. Our main tool is a divisible R-module $\partial$ which has a number of remarkable features. E.g., it has projective dimension 1 and is a generator of the subcategory of divisible R-modules. We show that, for Prüfer domains R, all divisible R-modules of projective dimension 1 are summands of a direct sum of copies of this $\partial$.

To prove our results, we need a couple of theorems on modules of projective dimension $\leq 1$. These are proved in § 2. Preliminaries are collected in § 1.

## §1. Preliminaries

Throughout, R will stand for an arbitrary commutative domain with 1, and Q for its field of quotients. R is a Prüfer domain if its finitely generated ideals are projective, and a valutation domain if its ideals form a chain under inclusion.

All R-modules will be unital. A submodule N of M is relatively divisible (RD) if, for each $r \in R$, $rN = N \cap rM$ where $rM = \{rx \mid x \in M\}$. An exact sequence $0 \longrightarrow A \xrightarrow{\alpha} B \longrightarrow C \longrightarrow 0$ is RD-exact if $\alpha A$ is RD in B. The Ulm submodule $M^1$ of M is defined as $\{rM \mid 0 \neq r \in R\}$.

An R-module D is called divisible if $rD = D$ for all $0 \neq r \in R$; it is h-divisible (Matlis [ 6 ]) if it is an epic image of an injective R-module, or, equivalently, an epic image of $\oplus Q$ for sufficiently large copies of Q. Every module M contains a unique largest divisible submodule which we denote by dM. The following is easy to prove.

Lemma A. An R-module D is divisible if and only if $\mathrm{Ext}^1_R(R/L,D) = 0$ for every projective ideal L of R.

We shall also need the following result.

<u>Lemma B.</u> (Matlis [6], Hamsher [4]). For a domain R, the following are equivalent:

(i) all divisible R-modules are h-divisible;

(ii) the torsion submodule of each divisible R-module is a summand;

(iii) p.d.Q = 1;

(iv) p.d.Q/R = 1.

Observe that, by Kaplansky [5], a valuation domain R satisfies (iii) if and only if Q is a countably generated R-module.

The next lemma is elementary.

<u>Lemma C.</u> For $0 \neq r \in R$ and R-module M,

$$\mathrm{Tor}_1^R(R/Rr,M) \cong \{x \in M \mid rx = 0\}.$$

<u>Proof.</u> Using the exact sequence $0 \longrightarrow R \xrightarrow{r} R \longrightarrow R/Rr \longrightarrow 0$, we infer that $\mathrm{Tor}_1(R/Rr,M)$ is naturally isomorphic to the kernel of the homomorphism $Rr \otimes M \to R \otimes M$. This is precisely the set of elements $r \otimes x$ ($x \in M$) such that $1 \otimes rx = 0$ in $R \otimes M$. This holds in case $rx = 0$.//

We will also require the well-known

<u>Auslander's Lemma</u> [1]. For some ordinal $\tau$, let

(1) $$0 = M_0 \leq M_1 \leq \dots \leq M_\alpha \leq \dots \quad (\alpha < \tau)$$

be a well-ordered ascending chain of submodules of a module M such that

(i) M is the union of the chain (1);

(ii) the chain is continuous, i.e. $M_\beta = \cup\{M_\alpha \mid \alpha < \beta\}$ for limit ordinals $\beta < \tau$;

(iii) there is an integer $n \geq 0$ such that $\mathrm{p.d.}M_{\alpha+1}/M_\alpha \leq n$ for all $1 \leq \alpha+1 \leq \tau$.

Then p.d.$M \leq n$.

The following lemma which generalizes Auslander's Lemma for $n = 1$ will be needed.

Lemma D. Let (1) be a well-ordered ascending chain of submodules of an R-module M satisfying (i) and (ii). Suppose that, for some R-module X,

$$\mathrm{Ext}^1_R(M_{\alpha+1}/M_\alpha, X) = 0 \qquad \text{for all } \alpha<\tau.$$

Then

$$\mathrm{Ext}^1_R(M, X) = 0.$$

Proof. Let $0 \to X \to E \to M \to 0$ be an extension of X by M. We wish to prove that it splits by constructing a complement to X in E.

Let $0 \to X \to E_\alpha \to M_\alpha \to 0$ be the exact sequence induced by the inclusion $M_\alpha \to M$. Obviously, this splits for $\alpha = 0$. Regard E as the union of the ascending chain $0 = E_0 \leq E_1 \leq \ldots \leq E_\alpha < \ldots$ $(\alpha<\tau)$, and suppose that we have found R-submodules $A_\beta$ of $E_\beta$ for each $\beta<\alpha$ such that $0 = A_0 \leq \ldots \leq A_\beta \leq \ldots$ $(\beta < \alpha)$ is a well-ordered continuous ascending chain satisfying $E_\beta = X \oplus A_\beta$ $(\beta <\alpha)$. If $\alpha$ is a limit ordinal, then set $A_\alpha = \bigcup_{\beta<\alpha} A_\beta$. This will satisfy $E_\alpha = X \oplus A_\alpha$. If $\alpha-1$ exists, then $E_\alpha/A_{\alpha-1}$ is an extension of $E_{\alpha-1}/A_{\alpha-1} \cong X$ by $E_\alpha/E_{\alpha-1} \cong M_\alpha/M_{\alpha-1}$. By hypothesis, this splits, i.e. $E_\alpha/A_{\alpha-1} = E_{\alpha-1}/A_{\alpha-1} \oplus A_\alpha/A_{\alpha-1}$ for some $A_\alpha \geq A_{\alpha-1}$. Evidently, $E_\alpha = X + A_\alpha$. On the other hand, $X \cap A_\alpha = X \cap E_{\alpha-1} \cap A_\alpha = X \cap A_{\alpha-1} = 0$, thus $E_\alpha = X \oplus A_\alpha$. Obviously, $A = \bigcup A_\alpha$ satisfies $E = X \oplus A$.//

§2. Modules of projective dimension one.

We proceed to prove results on modules of projective dimension one which are relevant for our study of divisible modules.

Proposition 1. Let R be a Prüfer domain. A finitely generated R-module M is finitely presented if and only if

$$p.d._R M \leq 1.$$

Proof. Let $0 \to H \to F \to M \to 0$ be an exact sequence with F finitely generated free. If R is a Prüfer domain, then H is finitely generated exactly if it is projective. Hence the claim is immediate.//

A less trivial result is the content of the following lemma.

Lemma 2. Let R be a Prüfer domain, F a torsion-free R-module, and H a free submodule of F. If F/H is finitely generated, then F is projective and F/H finitely presented.

Proof. Manifestly, the proof can be restricted to the case in which F/H is cyclic and torsion. Let $\{x_i \mid i \in I\}$ denote a basis of H. As F is torsion-free, we may think of F as being contained in the Q-vector space V with basis $\{x_i \mid i \in I\}$. Therefore, if F/H is generated, say, by the coset $a + H$ $(a \in F)$, then there exists a finite subset $\{1,\ldots,n\}$ of I and non-zero elements $r, r_1,\ldots,r_n \in R$ such that

$$ra = r_1x_1 + \ldots + r_nx_n. \tag{2}$$

It is obvious that any relation between a and the basis elements $x_i$ can be obtained from (2) if we multiply (2) be a suitable $s \in R$ and then divide it by some $t \in R$. Hence $F_0 = Ra + \sum_{i=1}^{n} Rx_i$ is a summand of F. As a finitely generated torsion-free R-module, $F_0$ is projective. It follows that F is projective and $F/H \cong F_0/(\sum Rx_i)$ is finitely presented.//

We will refer to a submodule N of an R-module M as a <u>tight</u> submodule (Fuchs [3]) if both

$$p.d.N \leq p.d.M \quad \text{and} \quad p.d.M/N \leq p.d.M.$$

<u>Theorem 3.</u> Let R be a Prüfer domain. In an R-module of projective dimension $\leq 1$, all finitely generated submodules are tight and finitely presented.

<u>Proof.</u> Let N be a finitely generated submodule of an R-module M with $p.d.M \leq 1$. We can write $M \cong F/H$ with F and H free. N is then of the form $N \cong G/H$ for some G between H and F. From the preceding lemma we conclude that N is finitely presented and G is projective. This completes the proof.//

The following consequence is most useful.

<u>Theorem 4.</u> Over a Prüfer domain R, a countably generated module has projective dimension $\leq 1$ if and only if it is the union of a countable ascending chain of finitely presented R-modules.

<u>Proof.</u> The necessity is a trivial corollary to Theorem 3. On the other hand, if M is the union of a chain

$$0 \leq M_1 \leq M_2 \leq \ldots \leq M_n \leq \ldots$$

where each $M_n$ is finitely presented, then all factors $M_{n+1}/M_n$ are likewise finitely presented, and thus of projective dimension $\leq 1$. An appeal to Auslander's Lemma concludes the proof.//

We now proceed to show that in an R-module of projective dimension 1, tight submodules are abundant. The next result, in a somewhat weaker form, was proved in Fuchs [3].

<u>Theorem 5.</u> Every R-module M of projective dimension 1 has a family $T = \{M_i \mid i \in I\}$ of submodules such that

(i) 0,M belong to T;

(ii) T is closed under unions of chains;

(iii) if $M_i < M_j$ in T, then p.d.$M_j/M_i \leq 1$;

(iv) given $M_i \in T$ and a countable subset $\Delta$ of M, there is an $M_j$ such that $\langle M_i, \Delta \rangle < M_j$ and $M_j/M_i$ is countably generated.

<u>Proof.</u> Start off with a presentation of M:

$$0 \longrightarrow H \longrightarrow F \xrightarrow{\phi} M \longrightarrow 0$$

where $F = \{Rx \mid x \in X\}$ is a free R-module on X and H is projective. By a well-known result of Kaplansky, $H = \{H_y \mid y \in Y\}$ where the $H_y$'s are countably generated projective R-modules.

Consider all pairs $(X_i, Y_i)$ of subsets $X_i \subseteq X$, $Y_i \subseteq Y$ such that $F_i = \oplus\{Rx \mid x \in X_i\}$ and $H_i = \oplus\{H_y \mid y \in Y_i\}$ satisfy

$$H_i = H \cap F_i.$$

Let i run over an index set I. Note that $H = H_i \oplus H_i^*$ where $H_i^* = \oplus\{H_y \mid y \in Y \setminus Y_i\}$. Consequently, the submodules $F_i + H = F_i \oplus H_i^*$ are projective.

Define

$$T = \{M_i \mid i \in I\} \text{ where } M_i = (F_i + H)/H.$$

Manifestly, $T$ satisfies (i) and (ii). As for $M_i < M_j$, the isomorphism $M_j/M_i \cong (F_j + H)/(F_i + H)$ holds, (iii) is likewise satisfied by $T$.

It remains to verify (iv) for $T$. The proof can be restricted to the case $M_i = 0$; in fact, the general result then follows by applying this special case to the R-module $M/M_i$ and to the family $\{M_j/M_i \mid M_i \leq M_j \in T\}$. Given a countable subset $\Delta$ of $M$, there is a countable subset $X^{(1)}$ of $X$ such that $\phi < X^{(1)} >$ contains $\Delta$. The submodule $< X^{(1)} > \cap H$ has at most countable rank, thus there is a countable subset $Y^{(1)}$ of $Y$ such that $< X^{(1)} > \cap H \leq < H_y \mid y \in Y^{(1)} >$. We can select a countable subset $X^{(2)}$ of $X$ that contains $X^{(1)}$ and satisfies $< H_y \mid y \in Y^{(1)} > \leq < X^{(2)} >$. Repeating this process, we obtain ascending chains of countable subsets

$$X^{(1)} \subseteq X^{(2)} \subseteq \ldots \subseteq X^{(n)} \subseteq \ldots \quad \text{and} \quad Y^{(1)} \subseteq Y^{(2)} \subseteq \ldots \subseteq Y^{(n)} \subseteq \ldots$$

of X and Y, respectively, such that

$$< X^{(n)} > \cap H \leq < H_y \mid y \in Y^{(n)} > \leq < X^{(n+1)} >$$

for each $n \geq 1$. Denoting by $X^*$ and $Y^*$ the unions of these chains, it is clear that they are countable and the pair $(X^*, Y^*)$ is one of the pairs $(X_i, Y_i)$ defined above. Thus $M^* = (F^* + H)/H$ belongs to $T$ and is countably generated; here $F^* = \oplus\{Rx \mid x \in X^*\}$.//

Starting from a family $T$ of R-submodules of M as in Theorem 5, we can construct a well-ordered continuous chain of tight submodules $M_\alpha \in T$,

(3) $0 = M_0 < M_1 < \ldots < M_\alpha < \ldots < M_\tau = M \quad (\alpha<\tau)$

such that, for each $\alpha<\tau$, $M_{\alpha+1}/M_\alpha$ is countably generated. Thus a comparison of Theorem 5 with Auslander's Lemma yields:

Corollary 6. An R-module M has projective dimension $\leq 1$ if and only if it has a well-ordered continuous ascending chain (3) of submodules such that, for each $\alpha<\tau$, $M_{\alpha+1}/M_\alpha$ is countably generated and has projective dimension $\leq 1$.//

Combining Theorem 4 and Corollary 6, we are led to

Corollary 7. A module M over a Prüfer domain has projective dimension $\leq 1$ exactly if it is the union of a well-ordered continuous chain (3) of submodules such that $M_{\alpha+1}/M_\alpha$ is finitely presented cyclic for each $\alpha<\tau$.//

The last result can be improved for valuation domains by demanding that the quotients $M_{\alpha+1}/M_\alpha$ be cyclically presented.

§3. The module $\partial$.

The next aim is to exhibit a divisible R-module which is a generator for the category of all divisible R-modules. R denotes an arbitrary domain.

Let the R-module $\partial$ be generated by all k-tuples $(r_1, \ldots, r_k)$ of non-zero and non-unit elements $r_i$ of R, for $k \geq 0$, subject to the defining relations

$$r_k\,(r_1,\dots,r_k) = (r_1,\dots,r_{k-1}) \qquad (k \geq 1).$$

The <u>length</u> of $(r_1,\dots,r_k)$ is defined to be k. The element $w = (\emptyset)$ of length 0 generates a submodule $\cong R$, and $\partial/Rw$ is a torsion module.

Let $\partial_k$ be the submodule generated by the generators of lengths $\leq k$. It is clear that $\partial$ is the union of the ascending chain $\partial_0 = Rw < \partial_1 < \dots < \partial_k < \dots$ such that $\partial_{k+1}/\partial_k$ is a direct sum of cyclically presented R-modules, viz. generated by $(r_1,\dots,r_{k+1}) + \partial_k$ with annihilators $Rr_{k+1}$. Hence for the Ulm submodule we have: $\partial^1_{k+1} = \partial_k$ $(k \geq 0)$ and $\text{p.d.}\,\partial_{k+1}/\partial_k = 1$. We conclude:

<u>Proposition 8</u>. $\partial$ is a divisible R-module of projective dimension 1.//

A noteworthy property of $\partial$ is as follows:

<u>Lemma 9</u>. Let M be an R-module and $a \in M$. There exists a homomorphism

$\eta : \partial \to M$ with $\eta w = a$ if and only if $a \in dM$.

<u>Proof.</u> As homomorphic images of divisible modules are divisible, this condition is necessary. To prove the converse, we construct successive homomorphisms $\eta_k : \partial_k \to M$. First, $\eta_0 : \partial_0 \to M$ is induced by $w \mapsto a$. If $\eta_{k-1}$ has been defined to map $\partial_{k-1}$ into dM, then let $\eta_k$ carry $(r_1, \dots, r_k)$ into an element $x \in dM$ such that $r_k x = \eta_{k-1}(r_1,\dots,r_{k-1}) \in dM$. This defines $\eta_k : \partial_k \to dM$ as desired.//

From Lemma C it follows that $\text{Tor}_n(R/Rr, C) = 0$ holds for $n \geq 1$ and for every torsion-free R-module C. Using this, one can easily derive the canonical isomorphism $\text{Tor}_n(\partial_k, C) \cong \text{Tor}_n(\partial_{k-1}, C)$ for every $k \geq 1$ and torsion-free C. As $\mathrm{Tor}_n(\partial_0, C) = 0$ holds trivially, we conclude (note that $\mathrm{Tor}_n$ commutes with direct limits):

<u>Proposition 10</u>. $\mathrm{Tor}_n^R(\partial, C) = 0$ holds for torsion-free R-modules C and for all $n \geq 1$.//

We proceed to prove:

<u>Proposition 11</u>. $\mathrm{Ext}_R^n(\partial, D) = 0$ holds for all divisible R-modules D and for all $n \geq 1$.

<u>Proof</u>. The modules $\partial_{k+1}/\partial_k$ are direct sums of cyclically presented R-modules, thus $\mathrm{Ext}^1(\partial_{k+1}/\partial_k, D) = 0$ by virtue of Lemma A. The claim now follows from Lemma D for $n = 1$, while for $n > 1$ it is a trivial consequence of Proposition 8.//

The following two results tell us about the extreme cases where $\partial$ is h-divisible, resp. h-reduced.

<u>Lemma 12.</u> The R-module $\partial$ is h-divisible if and only if $\mathrm{p.d.}_R Q = 1$.

<u>Proof.</u> From Lemma 9 it is clear that $\partial$ is h-divisible exactly if all divisible R-modules are h-divisible. As is shown by Lemma B, this happens if and only if $\mathrm{p.d.}Q = 1$.//

<u>Lemma 13.</u> Let R be a valuation domain and $\mathrm{p.d.}_R Q \geq 2$. Then $\partial$ is h-reduced, i.e. $\mathrm{Hom}(Q, \partial) = 0$.

<u>Proof</u>. Q is now uncountably generated. We may assume $Q = \bigcup Ra_\alpha$ with $\alpha$

running over the ordinals less then an initial ordinal $\Omega$ corresponding to the minimal cardinality of generating systems of Q. Suppose $\phi : Q \to \partial$ is a homomorphism $\neq 0$. As $\partial = \cup \partial_k$, it is clear that there is an index m such that $\partial_m$ contains $\phi a_\alpha$ for an index set of cardinality $\Omega$. By our choice, $\partial_m$ contains $\phi Q$ entirely. As $\partial_m$ is h-reduced, $\phi Q = 0$, as claimed.

Call an exact sequence $\partial$-exact if $\partial$ has the projective property relative to it.

Lemma 14. For every divisible R-module M, there is a $\partial$-exact sequence

(4) $$0 \longrightarrow N \longrightarrow D \xrightarrow{\eta} M \longrightarrow 0$$

of divisible R-modules such that D is a direct sum of copies of $\partial$.

Proof. Given a divisible R-module M for each pair (a,r) with $a \neq 0$ in M, $r \in R$ and $ra = 0$, select a copy $D_a$ of $\partial/Rrw$ along with a homomorphism $\eta_a : D_a \to M$ mapping the coset of w onto a. That this is possible should be clear from Lemma 9. The $\eta_a$'s induce a map $\eta : \oplus D_a = D \to M$ which is evidently surjective. It is readily seen that the arising exact sequence (4) is $\partial$-exact.

Next notice that each $\partial/Rrw$ $(r \in R)$ is isomorphic to a summand of $\partial$. In fact, it is readily checked that the correspondence

$$(r_1,\ldots,r_k) \longmapsto (r,r_1,\ldots,r_k) - s(rs, r_1,\ldots,r_k)$$

gives rise to a homomorphism of $\partial$ whose image is a summand of $\partial$ isomorphic to $\partial/Rrw$; here s is a fixed non-zero, non-unit element of R. We can now argue that D can be chosen so as to be a direct sum of copies of $\partial$.

It only remains to verify the divisibility of N. Evidently, $\partial$-exactness implies that cyclically presented R-modules have the projective property relative to the exact sequence. Hence (4) is RD-exact, and N as an RD-submodule of a divisible module is itself divisible.//

We obtain at once:

<u>Corollary 15</u>. $\partial$ is a generator of the category of divisible R-modules.//

The following interesting fact is worthwhile pointing out.

<u>Proposition 16.</u> Every R-module of projective dimension n can be embedded in a divisible R-module whose projective dimension is $\leq \max\{n, 1\}$.

<u>Proof.</u> Let F be a free R-module and $\phi : F \to M$ an epimorphism. The exact sequence $0 \to R \to \partial \to \partial/\partial_0 \to 0$ induces the exact sequence in the top row

$$\begin{array}{ccccccccc} 0 & \to & F & \to & F \otimes \partial & \to & F \otimes \partial/\partial_0 & \to & 0 \\ & & \downarrow \phi & & \downarrow \psi & & \| & & \\ 0 & \to & M & \to & \overline{M} & \to & F \otimes \partial/\partial_0 & \to & 0. \end{array}$$

By push-out construction we obtain this commutative diagram with exact bottom row. As $\psi$ is surjective, $\overline{M}$ is divisible. $F \otimes \partial/\partial_0$ has projective dimension 1, so it is clear that $\text{p.d.}\overline{M} \leq \max\{\text{p.d.}M, 1\}$, as claimed.//

## §4. Divisible modules over Prüfer domains

We now restrict our consideration to Prüfer domains, since we could prove the next result only for Prüfer domains. It will show that modules behave towards divisible modules as if their projective dimensions were one less.

Proposition 17. Let R be a Prüfer domain and M an R-module with $p.d._R M = m \geq 1$. Then

$$\mathrm{Ext}_R^m(M,D) = 0$$

for all divisible R-modules D.

Proof. We induct on m. If $m = 1$, then, by Corollary 7, M is the union of a well-ordered continuous ascending chain of submodules, $0 = M_0 < M_1 < \ldots < M_\alpha < \ldots$ $(\alpha < \tau)$ such that $M_{\alpha+1}/M_\alpha$ is finitely presented cyclic for each $\alpha < \tau$. Hence Lemma A implies $\mathrm{Ext}_R^1(M_{\alpha+1}/M_\alpha, D) = 0$ for $\alpha<\tau$ An appeal to Lemma D shows that our claim holds for $m = 1$.

If $m > 1$, let $0 \to N \to F \to M \to 0$ be an exact sequence with F projective. Evidently, $p.d.N = m - 1$. The exact sequence

$$\to \mathrm{Ext}^{m-1}(N, D) \to \mathrm{Ext}^m(M, D) \to \mathrm{Ext}^m(F, D) = 0$$

together with the induction hypothesis completes the proof.//

It is not difficult to identify the divisible modules of projective dimension one in the Prüfer case.

Theorem 18. A divisible module over a Prüfer domain has projective dimension 1 if and only if it is a summand of a direct sum of copies of $\partial$.

*Proof*. It suffices to show that if p.d.D = 1 holds for a divisible R-module D, then D is a summand of some $\oplus\partial$. By Lemma 14, there is an exact sequence $0 \to N \to \oplus\partial \to D \to 0$ where N is divisible. In view of p.d.D = 1, Proposition 17 implies that this sequence splits; thus the claim follows.//

Call a divisible R-module D *$\partial$-projective* if it has the projective property relative to all $\partial$-exact sequences. From Lemma 14, it is evident that the $\partial$-projective divisible modules are precisely the summands of direct sums of copies of $\partial$. In this terminology, Theorem 18 can be rephrased by stating that, over Prufer domains, the $\partial$-projective divisible modules are exactly the divisible modules of projective dimension 1.

We can now proceed and follow the pattern of [2] in introducing a dimension function for divisible modules. If D is divisible, then we can form a long $\partial$-exact sequence

$$(5) \qquad \cdots \longrightarrow P_n \xrightarrow{\delta_n} P_{n-1} \longrightarrow \cdots \longrightarrow P_0 \xrightarrow{\delta_0} D \longrightarrow 0$$

where the $P_i$ are $\partial$-projective. The *$\partial$-dimension* of D, denoted as $\partial$-dim D, is defined to be n if n is the smallest index such that Im $\delta_n$ is $\partial$-projective. An obvious version of Schanuel's lemma guarantees that this definition is independent of the particular choice of the $\partial$-exact sequence (5). It is easy to verify:

*Theorem 19*. Let R be a Prufer domain and D a divisible R-module. Then

$$\partial\text{-dim } D = \text{p.d.}D - 1.$$

Proof. We induct on p.d.D = n. As noted above, for n = 1 this is Theorem 18. If n > 1, then choose a $\partial$-exact sequence $0 \to N \to P \to D \to 0$ where P is $\partial$-projective (see Lemma 14). Here N is divisible of projective dimension n - 1. Hence $\partial$-dim N = n-2. If $n \geq 3$, $\partial$-dim D = n-1 follows at once. If n = 2, then only $\partial$-dim $D \leq n-1 = 1$ follows, but $\partial$-dim D = 0 can be ruled out in view of Theorem 18.

The case p.d.D = $\infty$ is easy, since then (5) implies that $\partial$-dim D cannot be finite.//

References

1. Auslander, L., On the dimension of modules and algebras, III, Nagoya Math. J. 9 (1955), 67-77.

2. de la Rosa, B. and Fuchs, L., On h-divisible torsion modules over domains, to appear.

3. Fuchs, L., On projective dimensions of modules over valuation domains, Abelian Group Theory, Lecture Notes in Math., Springer, 1006 (1983), 589-598.

4. Hamsher, R.M., On the structure of a one-dimensional quotient field, Journ. of Alg. 19 (1971), 416-425.

5. Kaplansky, I., The homological dimension of a quotient field, Nagoya Math. J. 27 (1966), 139-142.

6. Matlis, E., Divisible modules, Proc. Amer. Math. Soc. 11 (1960), 385-391.

7. ______, Cotorsion modules, Memoirs Amer. Math. Soc., No. 49 (1964).

# PROJECTIVE DIMENSIONS OF IDEALS OF PRUFER DOMAINS

Barbara L.Osofsky

Rutgers University

A Prüfer domain has all of its finitely generated ideals projective. Countably but not finitely generated ideals must be of projective dimension 1. A standard argument due to Auslander gives an upper bound for the projective dimension of an ideal I. If I is $\aleph_k$-generated, then I has projective dimension $\leq$ k+1. Here we look at how one might get the reverse inequality.

The major tools are two rather straight-forward propositions. As we shall see, it is applying them that is difficult.

Our first proposition requires some definitions to allow us to use induction to move back along a projective resolution.

<u>Definition</u>. A module is called $\aleph$-resolvable if it has a projective resolution consisting of $\aleph$-generated projectives.

<u>Definition</u>. A module M is called an $\aleph$-union if it is a direct union of $\aleph$-resolvable submodules $\{K_i\}$, where every subset of M of cardinality $\leq$ $\aleph$ is contained in some $K_i$.

Proposition. Let M be an $\aleph$-union of $\{K_i\}$. If M has projective dimension $\leq n < \infty$, and S is any subset of M with cardinality $S \leq \aleph$, then there exists a submodule K of M with K a well ordered union of some of the $K_i$ over a set of order type $\aleph$, $S \subseteq K$ and M/K has projective dimension $\leq n$.

Proof. We use induction on n.

If n = 0, then by a result of Kaplansky, M is a direct sum of countably generated projectives, say $M = \oplus C_j$. Since every $\aleph$-generated submodule of M is contained in some sum of $\aleph$ of the $C_j$ and every $\aleph$ of the $C_j$ is contained in some $K_i$, we can use a standard snaking argument to get an appropriate union of $K_i$'s which are also sums of $C_j$'s.

If n > 0, one maps a projective onto M by taking a coproduct of projectives $P_i$ mapping onto the $K_i$. Every subset S of M of cardinality $\leq \aleph$ is contained in a direct union of some $\aleph$ $K_i$, and the kernel of the map from the coproduct of the corresponding $P_i$ to M is $\aleph$-generated. Since the category of R-modules has exact direct limits, K will also be an $\aleph$-union of appropriate kernels, and induction applies.

Corollary. If M in the above proposition is not $\aleph$-generated, then there is a $K \subseteq M$ with K a well-ordered union of some of the $K_i$ over a set of order type $\aleph$ and M/K has projective dimension $\leq n$. If $\aleph$ is not singular, such a K cannot be generated by less than $\aleph$ elements.

Proposition (See [3]). Let N and L be submodules of a module M. If M, N + L and M/N all have projective dimension $\leq k < \infty$, but L has projective dimension $< k$, then $N \cap L$ also has projective dimension $< k$.

Proof. The exact sequence

$$0 \longrightarrow N + L \longrightarrow M \longrightarrow M/(N + L) \longrightarrow 0$$

shows $M/(N+L)$ has projective dimension $\leq k+1$. The exact sequence

$$0 \longrightarrow (N + L)/N \longrightarrow M/N \longrightarrow M/(N + L) \longrightarrow 0$$

shows $(N+L)/N$ has dimension $< k$. The exact sequence

$$0 \longrightarrow N \cap L \longrightarrow L \longrightarrow (N + L)/N \longrightarrow 0$$

then completes the proof.

We observe that these two propositions imply:

Theorem. Let R be a Prüfer domain with the property that if I is an ideal which cannot be generated by fewer than $\aleph_k$ elements, then for some x in R, $xR \cap I$ also cannot be generated by fewer than $\aleph_k$ elements. Then for any ideal I of R, I has projective dimension $\leq k + 1$ if and only if I is $\aleph_k$-generated.

Proof. The theorem is clearly true if k = 0, and the if is an immediate consequence of Auslander's Lemma [1]. For the only if portion, observe that an ideal not generated by $\aleph_k$ elements is an $\aleph_k$-union of ideals that are, and the second proposition enables us to reduce the projective dimension of some $\aleph_k$-generated ideal without reducing the number of generators and thus to apply induction.

The problem then reduces to that strange intersection property. There is one way that we can assure it. If for any $\aleph_k$-generated ideals I and I' with $I' \subseteq I$, $I/I'$ is an essential extension of some $\aleph_k$-generated submodule, then the intersection hypothesis of the theorem holds. This will occur, for example, if for every $\aleph_k$-generated ideal I', there are at most

$\aleph_k$ primes minimal over I′. For then those x in I which do not go to 0 in $(I/I')_P$ for some P minimal over I′ generate an essential submodule of I/I′. See [3] for details of the proof.

One might ask if the theorem is true in general, without any assumption of an intersection property. The following shows that this is not the case.

In the ring Z of integers, we have as first order properties:

i) Z is a domain whose finitely generated ideals are principal.

ii) There is a sequence $\{p_i\}$ of primes such that $p_i \neq p_j$ if $i \neq j$.

iii) For the above sequence and any natural number n, the set

$$\{ m_k = \prod_{i=1}^{k-1} p_i \prod_{i=k+1}^{n} p_i \mid 1 \leq k \leq n \}$$

satisfies $\{m_k Z\}$ are independent modulo $\prod_{i=1}^{n} p_i Z$. These properties will hold in any model R of the first order theory of Z. But property iii) implies that the ideal of R generated by $\{m_k \mid 1 \leq k \leq n^*\}$ has projective dimension 1 for $n^*$ any positive non-standard integer. Since there are non-standard models of Z with arbitrarily large cardinality, the theorem is not true in general.

References

[1] Auslander, M., On the dimension of modules and algebras. III: Global dimension, *Nagoya Math J.*,9,1955,67.

[2] Kaplansky, I., Projective modules, *Ann. of Math.*,68,1958,372.

[3] Osofsky, B., Projective dimension of "nice" directed unions, *J. Pure and Applied Algebra*,13,1978,179.

# ON MODULES OF FINITE PROJECTIVE DIMENSION OVER VALUATION DOMAINS

S. BAZZONI
Università di Padova

L. FUCHS
Tulane University
New Orleans

## INTRODUCTION

In this note we consider only modules over valuation domains R.

In [3] Osofsky proved that the projective dimension of an ideal I of R is $\leq n+1$ if and only if I can be generated by $\aleph_n$ elements ( where $n \geq -1$, and $\aleph_{-1}$ means finite ).

In [2] this was generalized to torsion free modules whose generator systems are of larger cardinalities than their ranks. In [2] modules of projective dimension one are also studied. These are shown to be coherent ( finitely generated submodules are finitely presented ) and to contain a large supply of the so called tight submodules.

In this note we extend the last mentioned results to modules of any finite projective dimension.
The major tool is the notion of tight system for an R-module ( See section 2 ) which gives a converse of Auslander's Lemma [1] : every module of projective dimension $\leq$ n is the union of a well ordered continuous ascending chain of submodules $M_\alpha$ such that the factors $M_{\alpha+1}/M_\alpha$ are of projective dimension $\leq$ n and $\aleph_{n-1}$-generated.

In Section 3 it is shown that in a module of projective dimension n+1 all $\aleph_{n-1}$-generated submodules are $\aleph_{n-1}$-presented and that every module of projective dimension n+1 is the union of a well ordered continuous ascending chain of submodules $M_\alpha$ such that the factors $M_{\alpha+1}/M_\alpha$ are cyclic $\aleph_{n-1}$-presented.

Some of our results can also be derived from Osofsky [4] where the setting is much more general than valuation domains.

## 1. PRELIMINARIES

R will always denote a valuation domain and Q its quotient field.

The rank of a torsion free R-module M is the Q-dimension of the Q-vector space $M \otimes_R Q$.

A submodule N of a torsion free module M is pure if M/N is torsion free.

For any subset X of M, $\langle X \rangle$ denotes the submodule of M generated by X.

An R-module M is said to be $\kappa$-generated ($\kappa$ a cardinal) in case M can be generated by a set of cardinality $\leq \kappa$. An R-module M is called $\kappa$-presented if it is of the form F/H where F is a free R-module and both F and H are $\kappa$-generated.

We use the notion of projective dimension of an R-module. We will make a frequent use of the following well known Lemma (valid over any ring)

KAPLANSKY's LEMMA Let $0 \to A \to B \to C \to 0$ be an exact sequence of R-modules. If two of proj. dim. A, proj. dim. B, proj. dim. C are finite, then so is the third and only the following cases can occur:

(i) proj. dim. A < proj. dim. B = proj. dim. C;

(ii) proj. dim. B < proj. dim. A = proj. dim. C - 1;

(iii) proj. dim. A = proj. dim. B $\geqslant$ proj. dim. C -1.

We recall also the following.

AUSLANDER's LEMMA Let $0=M_0 < M_1 < \ldots < M_\alpha < \ldots \quad (\alpha < \tau)$ be a well ordered ascending chain of submodules of a module M such that:

(i) $M = \bigcup_{\alpha<\tau} M_\alpha$

(ii) the chain is continuous (i.e. $\forall\, \alpha < \tau$, $\alpha$ limit ordinal, $M_\alpha = \bigcup_{\beta<\alpha} M_\beta$)

(iii) for some integer n, proj. dim. $M_{\alpha+1}/M_\alpha \leq n \quad (\forall\, 1 \leq \alpha+1 < \tau)$.

Then proj. dim. M n.

1.2.

We list here some lemmas which will be used in the following and whose proofs can be found in [2].

LEMMA A. If M is a torsion free $\aleph_n$-generated R-module, then the same holds for all its pure submodules.

LEMMA B. An $\aleph_n$-generated torsion free module M satisfies proj. dim.M $\leqslant$n+1.

LEMMA C. If M is a torsion free R-module of rank $\aleph_m$ which can be generated by $\aleph_n$, but not by fewer elements where $m<n$, then proj. dim.M = n+1.

LEMMA D. Let M be a torsion free module of rank $\aleph_n$. Then proj. dim. M $\leqslant$n if and only if all pure submodules of rank $<\aleph_n$ in M have proj. dim. $\leqslant$n.

## 2. TIGHT SYSTEMS.

In view of the results in [2] on projective dimension, one can expect that modules with small projective dimension, but with a large number of generators can be built up of modules of the same projective dimension, but with smaller numbers of generators.

This is in fact the case and this is what we intend to prove in this section.

First we need the notion of tight submodule of a module.

A tight submodule of an R-module M is a submodule N such that both proj. dim. N and proj. dim. M/N are $\leqslant$ proj. dim. M.

We now introduce the following terminology.

A tight system for a module M is a family $\mathcal{T}$ of submodules $M_i$ $(i \varepsilon I)$ of M such that:

(i) $0, M \varepsilon \mathcal{T}$;

(ii) $\mathcal{T}$ is closed under unions of chains;

(iii) if $M_i, M_j \varepsilon \mathcal{T}$ and $M_i \leq M_j$, then proj. dim. $M_j/M_i \leq$ proj. dim. $M = n$;

(iv) if $M_i \varepsilon \mathcal{T}$ and $\Delta$ is a subset of M of cardinality $\leq \aleph_{n-1}$, then there is $M_j \varepsilon \mathcal{T}$ satisfying:

(a) $< M_i, \Delta > \leq M_j$;

(b) $M_j/M_i$ is $\aleph_{n-1}$-generated.

From (i) and (iii) it is clear that all submodules in $\mathcal{T}$ are tight in M. It is readily cecked, by using (ii), that (iv) continues to hold if $\aleph_{n-1}$ is replaced by any cardinal number larger than $\aleph_{n-1}$.

The main result of this section is as follows.

THEOREM 2.1. The following assertions are true for every $n \geq 0$:

$(A_n)$ Every R-module M of projective dimension n has a tight system.

$(B_n)$ Every torsion free R-module of projective dimension n has a tight system consisting of pure submodules.

$(C_n)$ In every torsion free R-module of projective dimension n, a pure submodule of rank $\leq \aleph_{n-1}$ is $\aleph_{n-1}$-generated.

Proof. We induct on n in the following way.

First we prove $A_0$, and then verify the implications $(A_n) \Longrightarrow (C_n)$; $(A_n) + (C_n) \Longrightarrow (B_n)$ and $(B_n) \Longrightarrow (A_{n+1})$.

It is easy to see that $(A_0)$ holds. In fact, let M be free ; we can write $M = \oplus\{Rx \mid x \in X\}$. Let I denote the power set of X and let $M_i = \oplus\{Rx \mid x \in i\}$ for $i \in I$. Then $\mathcal{T} = \{M_i \mid i \in I\}$ is evidently a tight system for M.

$(A_n) \Longrightarrow (C_n)$ Let M be torsion free of projective dimension n, and N a pure submodule of rank $\leq \aleph_{n-1}$. By way of contradiction, suppose that N requires $\aleph_k$ $(k \geq n)$ generators. By Lemma C, proj. dim. N = k+1. From $(A_n)$ it follows that N can be embedded in a submodule $\bar{N}$ of M such that proj. dim. $\bar{N} \leq n$ and $\bar{N}$ is $\aleph_k$-generated. Lemma A shows that $\bar{N}$ cannot be generated by fewer than $\aleph_k$-elements. If the rank of $\bar{N}$ were $< \aleph_k$, then Lemma C would imply proj. dim. $\bar{N}$ = k+1 > n, impossible. Hence $\bar{N}$ has rank $\aleph_k$. Now, Lemma D applied to $\bar{N}$ leads us to the conclusion proj. dim. $N \leq k$ which is a contradiction.

$(A_n) + (C_n) \Longrightarrow (B_n)$ Let again M be torsion free of projective dimension n, and $\mathcal{T}$ a tight system for M as stipulated by $(A_n)$. Let $\mathcal{T}'$ be the subsystem of $\mathcal{T}$ consisting of those $M_i \in \mathcal{T}$ which are pure in M. We claim that $\mathcal{T}'$ is likewise a tight system for M. It is clearly enough to show that (iv) holds for $\mathcal{T}'$, and it suffices to do this for $M_i = 0$. In fact the general result then follows by applying this special case to $M/M_i$ and to the family $\{M_j/M_i \mid M_i \leq M_j \in \mathcal{T}'\}$.

Any subset $\Delta$ of cardinality $\leq \aleph_{n-1}$ is contained in an $\aleph_{n-1}$-generated member $M^{(1)}$ of $\mathcal{T}$. In view of $(C_n)$, the purification $M_*^{(1)}$ of $M^{(1)}$ is again $\aleph_{n-1}$-generated.

There is an $\aleph_{n-1}$-generated $M^{(2)} \in \mathcal{T}$ that contains $M_*^{(1)}$ whose purification $M_*^{(2)}$ is again $\aleph_{n-1}$-generated. Thus proceeding we obtain an ascending chain $M^{(1)} \leq M_*^{(1)} \leq M^{(2)} \leq M_*^{(2)} \leq \dots$ whose union is in $\mathcal{T}'$, as it belongs to $\mathcal{T}$ and is pure in M. Manifestly it is $\aleph_{n-1}$-generated.

$(B_n) \Longrightarrow (A_{n+1})$. Let M be an R-module of projective dimension n+1, and choose a presentation of M, $0 \to H \to F \to M \to 0$ where $F = \oplus \{Rx \mid x \in X\}$ is a free R-module and proj. dim. $H = n$. By $(B_n)$, H has a tight system $\{H_t \mid t \in T\}$ consisting of pure submodules. Consider pairs $(X_i', T_i)$ of subsets $X_i \subseteq X$, $T_i \subseteq T$ satisfying $H_i = H \cap F_i$, where $H_i = \langle H_t \mid t \in T_i \rangle$ and $F_i = \langle X_i \rangle$ ; here i runs over a suitable index set I.
We claim that $\mathcal{T} = \{M_i \mid i \in I\}$, where $M_i = (F_i + H)/H$, is a tight system for M. Properties (i) and (ii) are clear for $\mathcal{T}$.
Since $(F_i + H_j)/F_i \cong H_j / H_i$ has projective dimension $\leq n$ ($H_i$ being tight in $H_j$), we have proj. dim. $(F_i + H_j) \leq n$. Now, if $M_i \leq M_j$ in $\mathcal{T}$ then

$$M_j / M_i \cong (F_j + H)/(F_i + H) \cong F_j / \{F_j \cap (F_i + H)\} \cong F_j / (F_i + H_j)$$

Therefore proj. dim. $M_j / M_i \leq n+1$, establishing (iii).
It remains to verify (iv) for $\mathcal{T}$. The proof can be restricted to the case $M_i = 0$ ( as before ). Given a subset $\Delta$ of cardinality $\leq \aleph_{n-1}$ there is a subset $X^{(1)}$ of X of cardinality $\leq \aleph_{n-1}$ such that $X^{(1)}$ contains $\Delta$. The submodule $\langle X^{(1)} \rangle \cap H$ in H, has rank $\leq \aleph_{n-1}$. Thus, by $(C_n)$ and the property (ii) of a tight system, there is a $t_1 \in T$ such that $\langle X^{(1)} \rangle \cap H \leq H_{t_1}$ and $H_{t_1}$ $\aleph_{n-1}$-generated. We can select a subset $X^{(2)}$ in X of cardinality $\leq \aleph_{n-1}$ that contains $X^{(1)}$ and satisfies $H_{t_1} \leq \langle X^{(2)} \rangle$. Repeating this process, we obtain ascending chains of $\aleph_{n-1}$-generated submodules:

$$\langle X^{(1)} \rangle \leq \langle X^{(2)} \rangle \leq \dots \quad \text{and} \quad H_{t_1} \leq H_{t_2} \leq \dots \text{ of F and H respectively}$$

such that :

$$\langle X^{(n)} \rangle \cap H \leq H_{t_n} \leq \langle X^{(n+1)} \rangle, \text{ for each } n \geq 1.$$

Denote by $F^{*}$ and $H^{*}$ the unions of the chains. It is clear that they are $\aleph_{n-1}$-generated and putting $X^{*} = \bigcup_{n} X^{(n)}$ , $T^{(*)} = \{t_n\}$, the pair $(X^{*}, T^{*})$ is one of the pairs $(X_i, T_i)$ defined above. Thus $M^{*} = (F^{*} + H)/ H$ belongs to $\mathcal{T}$ and is $\aleph_{n-1}$-generated.□

Starting from a tight system $\mathcal{T}$ for an R-module M of projective dimension n we can construct a well ordered continuous ascending chain of tight submodules $M_\alpha$ :

$$0 = M_0 < M_1 < \ldots < M_\alpha < \ldots < M_\tau = M \qquad (\alpha < \tau)$$

such that, for each $\alpha < \tau$, $M_{\alpha+1} / M_\alpha$ is $\aleph_{n-1}$-generated. Thus a comparison of Theorem 2.1. with Auslander's Lemma yields:

COROLLARY 2.2. An R-module M has projective dimension $\leq n$ if and only if it has a well ordered continuous ascending chain of submodules $\{M_\alpha \mid \alpha < \tau\}$ such that, for each $\alpha < \tau$, $M_{\alpha+1} / M_\alpha$ is $\aleph_{n-1}$-generated and has projective dimension $\leq n$.□

## 3. TIGHT SUBMODULES AND $\aleph_n$-PRESENTED MODULES.

In this section we extend the results proved in [ 2 ] about modules of projective dimension one, to modules of any projective dimension n.

PROPOSITION 3.1. Let M be a torsion free R-module of projective dimension n. Every pure submodule of rank $\leq \aleph_{n-1}$ in M, is tight in M.

Proof. Let A be a pure submodule of M of rank $\aleph_k$, $k \leq n-1$. If A is $\aleph_m$-generated, then $k \leq m$.

1st case: $m < n-1$. By Lemma B, proj. dim. $A \leq m+1 < n$. Kaplansky's Lemma applied to $0 \to A \to M \to M/A \to 0$, implies proj. dim. $M/A \leq n$ and thus A is tight in M.

Note that, if $m \geq n-1$ there exists an $\aleph_m$-generated tight submodule B of M containing A (by Theorem 2.1. ). Thus:

$2^{nd}$ case: $m = n-1$. Here proj. dim. $A \leq m+1 = n$ and Kaplansky's Lemma applied to the sequence $0 \rightarrow B/A \rightarrow M/A \rightarrow M/B \rightarrow 0$ yields proj. dim. $M/A \leq n$, since B is tight in M and B/A is $\aleph_m$-generated torsion free and thus of projective dimension $\leq m+1 = n$(Lemma B ). Hence A is again tight in M.

The third alternative $m > n-1$ cannot occur. In fact, then $k \leq n-1 < m$ implies proj. dim. $A = m+1$ ( by Lemma C ). Now B is $\aleph_m$-generated of projective dimension $\leq n \leq m$. Since A is pure in B and of rank $\aleph_k < \aleph_m$, Lemma D can be applied to obtain the wanted contradiction proj. dim. $A \leq m$.□

COROLLARY 3.2. If M is a torsion free R-module of projective dimension n, then every submodule of M of rank $\aleph_k$, $k \geq n$ is contained in a tight submodule of the same rank.

Proof. An $\aleph_m$-generated submodule A of rank $\aleph_k$ ( $k \geq n$ ) is contained in an $\aleph_m$-generated tight submodule B in M. A can be assumed to be pure in M, since the pure submodule generated by A in M is of the same rank as A. Now, if $k < m$, proj. dim. $A = m+1$ ( Lemma C ). But Lemma D applied to B leads to the contradiction proj. dim. $A \leq n < m$. So $k = m$. □

We'll see now that the results proved in Lemma 4.1., Theorem 4.2. and Theorem 4.3. of [2] can be extended to R-modules of any projective dimension n.

PROPOSITION 3.3. Let G be a torsion free R-module and H a submodule such that proj.dim. $H \leq n$ and G/H is $\aleph_{n-1}$-generated.
Then proj.dim. $G \leq n$.

Proof. Let X be a set of elements of G such that $|X| \leq \aleph_{n-1}$ and

$G = \langle X \rangle + H$. Denote by $Y$ the pure submodule in $G$ generated by $X$, and by $A$ the pure submodule in $H$ generated by $\langle X \rangle \cap H$. It is easily checked that $Y = \langle X \rangle + A$ and $H \cap Y = A$. Now $A$ is of rank at most $\aleph_{n-1}$, hence by Proposition 3.1., $A$ is tight in $H$. This means $A$ is $\aleph_{n-1}$-generated ( by Lemma C ), and thus the same holds for $Y = \langle X \rangle + A$.
Now proj. dim. $Y \leq n$ ( by Lemma B ) and proj. dim. $(H+Y)/Y$ = proj. dim. $H/A \leq n$ . Kaplansky's Lemma applied to the sequence $0 \to Y \to G \to (H+Y)/Y \to 0$ yields proj. dim. $G \leq n$.□

PROPOSITION 3.4. Let $M$ be a module of projective dimension $n+1$. Then every $\aleph_{n-1}$-generated submodule is tight in $M$.

Proof. Let $A$ be an $\aleph_{n-1}$-generated submodule of $M$ and $0 \to H \to F \to M \to 0$ an exact sequence with $F$ free ( and proj. dim. $H = n$ ). Take $G \leq F$ such that $G/H \cong A$. By Proposition 3.3., proj. dim. $G/H \leq n+1$ and proj. dim. $M/A$ = proj. dim. $F/G \leq n+1$.□

COROLLARY 3.5. In a module $M$ of projective dimension $n+1$, the annihilators of elements are of projective dimension at most $n$, and thus $\aleph_{n-1}$-generated.

Proof. Let $\langle x \rangle \cong R/I$ be a cyclic submodule of $M$. Then proj. dim. $R/I \leq n+1$ (by Proposition 3.4. ) which means ( by Kaplansky's Lemma ) proj. dim. $I \leq n$ and $I$ $\aleph_{n-1}$-generated ( by Lemma C ).□

LEMMA 3.6. Let $M$ be an $\aleph_{n-1}$-generated module. proj. dim. $M \leq n+1$ if and only if $M$ is $\aleph_{n-1}$-presented.

Proof. Consider an exact sequence $0 \to H \to F \to M \to 0$ with $F$ free of rank $\leq \aleph_{n-1}$. Then proj. dim. $M \leq n+1$ if and only if proj. dim. $H \leq n$ ( by Kaplansky's Lemma ).

Now, if M is $\aleph_{n-1}$-presented, H can be chosen to be $\aleph_{n-1}$-generated. Hence proj. dim. H $\leq$ n by Lemma A.
Conversely, suppose proj. dim. H $\leq$ n, then since H is of rank $\leq \aleph_{n-1}$, it is also $\aleph_{n-1}$-generated ( by Lemma B ).□

THEOREM 3.7. Let M be an $\aleph_n$-generated R-module. Then proj. dim. M $\leq$ n+1 if and only if M is the union of a well ordered continuous ascending chain ( of cardinality at most $\aleph_n$ ) of $\aleph_{n-1}$-presented submodules.

Proof. Necessity: M is the union of a well ordered continuous ascending chain of $\aleph_{n-1}$-generated submodules. By Proposition 3.4. and Lemma 3.6. we are done.
For the converse : let $0 = M_0 < M_1 < \ldots < M_\tau = M$ $(\tau \leq \aleph_n)$ be a well ordered continuous ascending chain of $\aleph_{n-1}$-presented submodules of M. Then all the factors $M_{\alpha+1}/M_\alpha$ $(0<\alpha+1 \leq \tau)$ are $\aleph_{n-1}$-presented. An application of Auslander's Lemma yields the conclusion.□

Combining Corollary 2.2. with Theorem 3.7. and Corollary 3.5. we get:

COROLLARY 3.8. An R-module M is of projective dimension $\leq$ n+1 if and only if it is the union of a well ordered continuous ascending chain $M_\alpha$ $(\alpha < \tau)$ of submodules such that all the factors $M_{\alpha+1}/M_\alpha$ are cyclic $\aleph_{n-1}$-presented.□

REFERENCES

[1] Auslander, L., On the dimension of modules and algebras. III, Nagoya Math. J. 9 (1955), 67-77.

[2] Fuchs, L., On projective dimensions of modules over valuation domains, Abelian Group Theory, Lecture Notes in Math. 1006 (1983); 589-598.

[3] Osofsky, B., Global dimension of valuation rings. Trans. Amer. Math. Soc. 127 (1967), 136-149.

[4] ______, Projective dimension of "nice" directed unions. J. Pure and and Applied Alg. 13 (1978), 179-219.

# ON PURE SUBMODULES OF FREE MODULES
# AND $\kappa$-FREE MODULES

Radoslav Dimitrić

Beograd, Yugoslavia

In this note we give a few results concerning pure submodules of free modules over commutative valuation domains, formulated in a slightly more general way than in [1]. Also, we define $\kappa$-free modules and prove several results on smooth ascending chains of such modules. It is evident that there are some important properties common both to pure submodules of free modules and $\kappa$-free modules, one of them being the fact that if such a module is of rank $\aleph_n$, then its projective dimension is $\leq n$ .

Conventions. R denotes throughout a commutative valuation domain and Q its field of quotients. $N \leq_* M$ means that N is a pure submodule of M. By rk M and gen M we shall denote the cardinalities of a maximal independent set and a generating set of minimal cardinality respectively (or one of either of those sets, when necessary). We call a submodule H of a free R-module $F = \bigoplus_{i \in I} Rx_i$ a slice of F, if $H = \bigoplus_{i \in J} Rx_i$, $J \subseteq I$.

A torsion free R-module M is separable if its every finite subset is contained in a summand of M that is the direct sum of rank one submodules.

## Pure submodules of free modules

Pure submodules of free R-modules are far from being free. The following theorem shows how far, and describes (for n = 2) the category of R-modules in which pure submodules of free modules are likewise free.

Theorem 1. For a valuation domain R and $n \geq 2$, the following are equivalent:

(1) pure submodules of free modules have projective dimension $\leq n - 2$;

(2) gld $R \leq n$ and pd $Q \leq n - 1$;

(3) rank one torsion free R-modules have projective dimension $\leq n - 1$;

(4) torsion free R-modules have projective dimension $\leq n - 1$.

<u>Proof.</u> (1) $\Rightarrow$ (2): For every ideal $I \lhd R$ there is a canonical free resolution $0 \to H_I \to F_I \to I \to 0$, where $H_I$ is a pure submodule of the free module $F_I$, such that $\mathrm{pd}\, H_I \leq n-2$. As $\mathrm{pd}\, I \leq n-1$ we get (by Auslander's formula $\mathrm{gld}\, R = 1 + \sup_{I \lhd R} \mathrm{pd}\, I$) $\mathrm{gld}\, R \leq n$. Similarly, from the free resolution $0 \to H \to F \to Q \to 0$ we get $\mathrm{pd}\, Q \leq n-1$ (which is equivalent to $\mathrm{gen}\, Q \leq \aleph_{n-2}$, by results of Kaplansky and Small);

(2) $\Rightarrow$ (1): Let $H \leq_* F$. If $\mathrm{pd}\, H > n-2$, then from the exact sequence $0 \to H \to F \to F/H \to 0$, we get $\mathrm{pd}\, F/H > n-1$, so $\mathrm{gld}\, R \leq n-1$ is impossible. If $\mathrm{gld}\, R = n$, then $\mathrm{pd}\, F/H = n$ i.e. the $\mathrm{gld}\, R$ is attained by this torsion free module, which, by Corollary 2.3 in [2] is possible only if $\mathrm{gld}\, R = \mathrm{pd}\, Q$ and this gives a contradiction too;

(2) $\Leftrightarrow$ (3) follows from the fact that a rank one torsion free module over a valuation domain is isomorphic either to an ideal of $R$ or to the $Q$;

(2) $\Rightarrow$ (4): If for a torsion free $R$-module $M$ we had $\mathrm{pd}\, M = n$, then $\mathrm{gld}\, R = \mathrm{pd}\, M = n$ and, as above, $\mathrm{pd}\, Q = n$ - a contradiction. $\triangle$

Notice that if $H \leq_* F$ and $\mathrm{rk}\, H = |\{a_i\}_{i \in I}|$, then all the $a_i$'s are contained in a $|J|$-generated slice $F'$ of $F$ and thus $H \leq_* F'$. By Lemma 1.1 in [2] we conclude that $\mathrm{gen}\, H = |J|$. Combining this with Propositions 7 and 8 in [1] we have

Proposition 2. Let H be a pure submodule of a free R-module. Then:

(a) H is separable ,

(b) countable rank pure submodules of H are free ,

(c) rk H = gen H and if rk H = $\aleph_n$ , then pd H $\leqslant$ n. $\triangle$

The next lemma may be useful in applications:

Lemma 3. Let $0 \to A \to B \to C \to 0$ be an exact sequence where A and C are (purely) embeddable in free modules $F_1$ and $F_2$ respectively. Then, there is a (pure) embedding k : B$\to F_1 \oplus F_2$ making the following diagram commutative

$$
\begin{array}{ccccccccc}
0 & \to & A & \xrightarrow{f} & B & \xrightarrow{g} & C & \to & 0 \\
 & & i\downarrow & \swarrow^{p} & \downarrow k & & \downarrow j & & \\
0 & \to & F_1 & \to & F_1 \oplus F_2 & \to & F_2 & \to & 0
\end{array}
$$

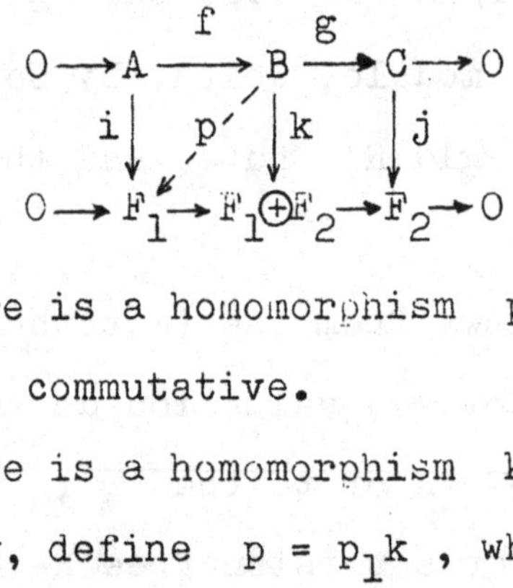

if and only if there is a homomorphism p : B$\to F_1$ making the triangle $ABF_1$ commutative.

Proof. If there is a homomorphism k : B$\to F_1 \oplus F_2$ with the stated property, define p = $p_1 k$ , where $p_1 : F_1 \oplus F_2 \to F_1$ is the canonical projection. Assume now that given p:B$\to F_1$ makes the triangle commutative. Define k = p $\oplus$ jg ; it is not difficult to see that this k makes the diagram commutative and therefore is an embedding.

Let us show that if i and j are pure embeddings, then so is k : If for some $f_1 \in F_1$ , $f_2 \in F_2$ , b $\in$ B $r(f_1+f_2)$ = kb = pb+jgb , we get (1) $rf_1$ = pb, (2) $rf_2$ = jgb.

From (2), there is a $b' \in B$ such that (3) $f_2 = jgb'$ and thus there is an $a \in A$ such that $rb' - b = fa$ i.e. by (1) $r(pb' - f_1) = ia$. Therefore, there is an $a' \in A$ such that $pb' - f_1 = ia'$ i.e. $f_1 = pb' - ia' = pb' - pfa' = p(b'-fa')$. By (3) also $f_2 = jg(b'-fa')$. So indeed $f_1 + f_2 \in kB$. △

## $\kappa$-free modules

Theorem 1 shows that the familiar definition of $\kappa$-freeness for abelian groups cannot be carried over verbatim to modules, since free modules would not be $\kappa$-free. So we have

Definition 4. For an infinite cardinal $\kappa$, call an R-module M $\kappa$-free if every (pure) submodule K of M of rank $< \kappa$ can be embedded in a free pure submodule F of M.

From Proposition 2 we derive

Corollary 5. a) If M is a $\kappa$-free module, then for a pure submodule K of M with rk $K < \kappa$ we have
rk K = gen K ,
b) Every pure submodule of a free module is $\aleph_1$-free . △

There is an example in [1] showing that the converse of b) does not hold even if the $\aleph_1$-free module is $\aleph_1$-generated and separable .

Proposition 6. Let M be a $\kappa$-free module. Then rk M = gen M and if rk M = $\aleph_n$ , then pd M $\leq$ n .

<u>Proof.</u> Let $\{a_i\}_{i \in I}$ be a maximal independent system of $M$. Since every $x \in M$ depends on a finite number of $a_i$'s, $M$ can be represented as the union of finite rank pure submodules of $M$ (whose number equals $|I| = \text{rk } M$). Thus, by $\aleph$-freeness, $M$ is the union of as many as $|I|$ finite rank pure free submodules. This proves gen $M$ = rk $M$.

Let us now assume that rk $M = \aleph_n$ = gen $M$. By Lemma 2.2 in [2] pd $M \leq n + 1$. If pd $M = n + 1$, then by Corollary 2.7 (ibidem) there is a finite rank $K \leq_* M$ with pd $K = n + 1$. By the $\aleph$-freeness of $M$ and Proposition 2.b), $K$ has to be free. This contradiction shows that pd $M \leq n$. △

We shall need the following lemma in the sequel:

<u>Lemma 7.</u> Let $0 \to A \to B \to C \to 0$ be an exact sequence such that $A$, $C$ are $\kappa$-free modules. Then $B$ is $\kappa$-free.

<u>Proof.</u> Let $K \leq_* B$ with rk $K < \kappa$. Since $(K+A)/A \leq B/A \cong C$, rk $(K+A)/A \leq$ rk $K$, by the $\kappa$-freeness of $C$, there is a free module $F$ of rank $< \kappa$ such that $(K+A)/A \leq (F \oplus A)/A \leq_* B/A$; this implies $K \leq F \oplus A \leq_* B$. rk $(A \cap (K+F)) \leq$ rk $K$ so by $\kappa$-freeness of $A$, there is a free submodule $F' \leq_* A$ such that $A \cap (K+F) \leq F'$. It becomes clear now that $K \leq F \oplus F' \leq_* B$ so $B$ is indeed $\kappa$-free. △

## Pure ascending chains of modules

We shall use in this section the following two results from [1] :

Lemma A. The union of a countable ascending chain $0 = F_0 \leq F_1 \leq \dots \leq F_n \leq \dots$ of the free R-modules $F_n$ is again free, provided each $F_n$ is pure in $F_{n+1}$.

Lemma B. If $0 = F_0 \leq F_1 \leq \dots \leq F_\alpha \leq \dots$ $(\alpha < \omega_1)$ is a continuous chain of $\aleph_1$-generated free R-modules $F_\alpha$ such that for each $\alpha$, $F_\alpha$ is pure in $F_{\alpha+1}$ and in $F_{\alpha+1}/F_\alpha$ the finite rank submodules are finitely generated, then the union $\bigcup F_\alpha$ is again a free R-module.

Proposition 8. Given a pure countable ascending chain $0 = M_0 \leq_* M_1 \leq_* \dots \leq_* M_n \leq_* \dots \leq_* M_{\omega_0} = M$ of R-modules where every $M_n$ is $\kappa$-free, then $M = \bigcup M_n$ is also $\kappa$-free.

Proof. Choose any $K \leq_* M$ with $\operatorname{rk} K < \kappa$ ; then $K = \bigcup_{n<\omega_0} (K \cap M_n)$. By the $\kappa$-freeness of every $M_n$, we can inductively choose pure free modules $F_n \leq_* M_n$ of rank $< \kappa$ such that $\langle F_{n-1}, K \cap M_n \rangle \leq F_n$. Hence $K \leq \bigcup_{n<\omega_0} F_n \leq_* M$ and by Lemma A, $\bigcup_{n<\omega_0} F_n = F$ is a free module, which proves $\kappa$-freeness of $M$. $\triangle$

As an immediate application we get

Corollary 9. Let $M$ be the union of a pure ascending chain (of any cardinality $\mu$) of $\aleph_1$-free modules $M_\alpha$. Then $M$ is likewise $\aleph_1$-free.

<u>Proof.</u> If $K \leq_* M$ is of countable rank, then $K$ is contained in a countable subchain $\{M_{\alpha(n)}\}_{n<\omega_0}$ of the chain $\{M_\alpha\}_{\alpha<\mu}$. We finish the proof by noticing that Proposition 8 implies that $\bigcup_{n<\omega_0} M_{\alpha(n)}$ is $\aleph_1$-free. $\triangle$

We point out that by Pontryagin's criterion (see Corollary 2.6 in [2]) $\aleph_0$- and $\aleph_1$-freeness coincide.

In order to get stronger results, we use more restrictive hypotheses:

<u>Theorem 10.</u> Let $0 = F_0 \leq_* F_1 \leq_* \ldots \leq_* F_\alpha \leq_* \ldots \leq_* F$ ($\alpha < \omega_1$) be a continuous chain of pure free submodules $F_\alpha$ such that every $F_{\alpha+1}/F_\alpha$ is $\aleph_0$-free. Then $F$ is $\aleph_2$-free.

<u>Proof.</u> Let $K \leq_* F$ be such that $\mathrm{rk}\, K \leq \aleph_1$. Every $K \cap F_\alpha$ is a pure submodule of an $\aleph_1$-generated slice $F'_\alpha$ of $F_\alpha$, thus $K \leq_* \sum_{\alpha<\omega_1} F'_\alpha$ i.e., by Lemma 1.1 in [2], $K$ must also be $\aleph_1$-generated. By Lemma 2 in [1], we construct a submodule $L$ of $F$ such that:

(a) $K$ is contained in $L$, (b) $\mathrm{rk}\, L \leq \aleph_1$,

(c) for every $\alpha < \omega_1$ $L \cap F_\alpha$ is a slice of $F_\alpha$,

(d) for every $\alpha < \omega_1$ $L + F_\alpha$ is pure in $F$.

Notice that from (c) and (d) we easily get that $L$ is pure in $F$. Since $L = \bigcup_{\alpha<\omega_1} (L \cap F_\alpha) = \bigcup_{\alpha<\omega_1} H_\alpha$, where by (c) and (b) every $H_\alpha$ is a rank $\aleph_1$ free module and $H_{\alpha+1}/H_\alpha = ((F_\alpha + L) \cap F_{\alpha+1})/F_\alpha \leq_* F_{\alpha+1}/F_\alpha$, we can apply Lemma B to conclude that $L$ is free. This proves $\aleph_2$-freeness of $F$. $\triangle$

We can improve the last theorem by assuming no restriction on the lenght of the chain involved:

Theorem 11. Let $0 = F_0 \leq_* F_1 \leq_* \dots \leq_* F_\alpha \leq_* \dots \leq_* F$ ($\alpha<\mu$ , $\mu$ - any ordinal) be a continuous chain of pure free submodules $F_\alpha$ such that every $F_{\alpha+1}/F_\alpha$ is $\aleph_0$-free . Then $F$ is $\aleph_2$-free.

Proof. If $K$ is a pure submodule of $F$ of the rank $\aleph_1$ , then it is contained in a smooth subchain of the lenght not exceeding $\omega_1$. To finish the proof, by applying Theorem 10, we only need to show that every $F_\beta/F_\alpha$ $(\alpha<\beta<\mu)$ is $\aleph_0$-free. This is done by transfinite induction in $\beta>\alpha$ , by the use of Lemma 7 in case of an isolated $\beta$ and Corollary 9 for a limit $\beta$ . $\triangle$

## References

[1] R. Dimitrić and L. Fuchs, "On torsion-free modules over valuation domains" , to appear.

[2] L. Fuchs, "On projective dimensions of modules over valuation domains" , Abelian Group Theory, Lecture Notes in Math. 1006 (1983), 589-598 .

# ON PURE-INJECTIVE MODULES

E. MONARI MARTINEZ

UNIVERSITA' DI PADOVA

## SUMMARY

Some results on pure-injective modules over a commutative ring with 1, proved by Ziegler using model theory, are proved here through algebraic methods. As application of these results we obtain again the structure of indecomposable pure-injective modules over a valuation domain, showing that their elements have constant indicator.

## INTRODUCTION

In the first section all modules are over a commutative ring with 1; partial homomorphisms, partial isomorphisms and smallness are defined and it is proved that a module M is pure-injective if and only if every partial homomorphism in M is extendible. If A is a submodule of a pure-injective module M , the hull H(A) of A in M is a minimal pure-injective summand of M containing A. The existence and the uniqueness (up to isomorphisms) of H(A) is proved and, using these results, a characterization of the pure-injective envelope of a module is given.

In the second section, indecomposable pure-injective modules over a valuation domain are studied: we prove that a pure-injective module has no elements of limit height (this result was independently obtained by Fuchs and Salce), that in an indecomposable pure-injective module the indicator of every element is constant and that an indecomposable pure-injective module is isomorphic to the pure-injective envelope of J/I, where $I \leq J$ are submodules of Q, the field of quotients of R.

All results in this paper, except the one regarding the indicators of the elements in an indecomposable pure-injective module over a valuation domain, have been obtained by Ziegler [3] using model theory; here model theory is completely avoided and the new notions and results are given in a purely algebraic setting.

In the following the terminology and notations will be like in [2] .

## 1. PURE-INJECTIVE MODULES.

Let R be a commutative ring with 1 and let all R-modules be unital.

Let M and N be R-modules and M' a submodule of M.

A homomorphism f: M' → N is called a <u>partial homomorphism</u> (<u>isomorphism</u>) <u>from</u> M <u>to</u> N if, for every finite set of linear equations over R with constants in M'

$$\sum_{1}^{n} r_{ij} x_j = a_i \qquad (1 \leq i \leq m)$$

which is soluble in M, the set of equations

$$\sum_{1}^{n} r_{ij} x_j = f(a_i) \qquad (1 \leq i \leq m)$$

is soluble in N (and vice versa). M' is denoted by dom f .

If f is a partial isomorphism from M to N, f is monic and there is a partial isomorphism $f^{-1}$ from N to M such that $f^{-1} \circ f = 1_{\text{dom } f}$ and dom $f^{-1} =$ = Im f ; $f^{-1}$ is the inverse partial isomorphism of f .

Recall that M' is a pure submodule of M if every finite set of linear equations over R with constants in M', which is soluble in M, is soluble in M'.

It is trivial that:

If dom f is a pure submodule of M , f is a partial homomorphism (isomorphism) from M to N if and only if f is a homomorphism (isomorphism with Im f pure in N) from dom f to N .

Let M be a R-module and $A \leq B \leq M$ .

B is small over A in M if every partial homomorphism f from M to a R-module N with dom f = B , whose restriction to A is a partial isomorphism from M to N , is a partial isomorphism from M to N .

The transitiveness of the smallness is obvious i.e., given $A \leq B \leq C \leq$ $\leq M$ , if B is small over A in M and C is small over B in M , then C is small over A in M .

If B is pure in M , then B is small over A in M if and only if B is small over A in B .

From the definition of pure-essentiality (see [2]) it follows that: A is a pure-essential submodule of the R-module B if and only if every $f \in \text{Hom}(B,N)$, where N is a R-module, such that $f_{|A}$ is monic and f(A) is pure in N , is monic. Hence, if A is pure in B and B is small over A in B, then A is pure-essential in B .

Warfield proved in [2] that a R-module M is pure-injective if and only if it is algebraically compact , i.e. every finitely soluble set of linear equations over R with constants in M has a simultaneous solution. Using this result, we are able to prove :

THEOREM 1.1 (Ziegler [3])

A R-module M is pure-injective if and only if every partial homomorphism from a R-module N to M can be extended to a homomorphism from N to M.

Proof : Let A be pure in B and $f \in \mathrm{Hom}(A,M)$; f is a partial homomorphism from B to M and hence there is a $f' \in \mathrm{Hom}(B,M)$ whose restriction to A is f.

Conversely, let f be a partial homomorphism from a R-module N to M with dom f = N'. If $x \in M^N$, we denote the $a^{th}$-coordinate of x ($a \in N$) by x(a). Consider the following system of equations, with x(a) ($a \in N$) as variables : for all a , $b \in N$ , for all $r \in R$ , for all $c \in N'$

$$rx(a) = x(ra) \ , \ x(a)+x(b)-x(a+b) = 0 \ , \ x(c) = f(c) \ .$$

Every finite subset of equations has solution in M : in fact if ra = c , with $a \in N$ and $c \in N'$, then there is $y \in M$ such that ry = f(c), since f is a partial homomorphism ; a solution of the equations

$rx(a) = x(ra)$ , $x(c) = f(c)$, where ra = c ,is x(a) = y and x(ra) = = x(c) = f(c) . M is algebraically compact, therefore there is a common solution $(\bar{x}(a))_{a \in N}$ of the system of equations. The map $\bar{f} : N \rightarrow M$ defined by $\bar{f}(a) = \bar{x}(a)$, for all $a \in N$ , is a homomorphism extending f .

Let M be a pure-injective R-module and A a submodule of M . $H_M(A)$ is called a hull of A in M if :

i) $H_M(A)$ is a pure-injective submodule of M , pure in M and containing A ,

ii) if B is a pure-injective submodule of M , pure in M and $A \leq B \leq H_M(A)$, then $B = H_M(A)$ .

From the definition it follows that $H_M(A)$ is a minimal summand of M , which contains A .

THEOREM 1.2 (Ziegler [3] and Fischer)

Let A be a submodule of a pure-injective module M . Then there is a hull $H_M(A)$ of A in M . Moreover if $H_N(B)$ is a hull of B in a pure-injective R-module N , then any partial isomorphism f from M to N , with dom f = A and f(A) = B , can be extended to an isomorphism from $H_M(A)$ to $H_N(B)$. Hence $H_M(A)$ is unique up to isomorphism.

Proof of existence: Let $\mathcal{B} = \{B_i\}_{i\in I}$ be the set of all submodules of M which contain A and are small over A in M (I is a set of indices). $\mathcal{B}$ is partially ordered by inclusion and is an inductive set . By Zorn's lemma, there exists a maximal element $B \in \mathcal{B}$ , called a maximal small extension of A in M. We prove at ones that B is algebraically compact and pure in M , by showing that every system of linear equations over R with constants in B , which is finitely soluble in M , is soluble in B .

Let $\mathcal{S}$ be the set defined by

$\mathcal{S} = \{S \mid S$ is a system of linear equations over R with constants in B, finitely soluble in M$\}$ .

Let $S_o \in \mathcal{S}$ be a fixed system. $\mathcal{S}$ is partially ordered set by inclusion and is inductive. By Zorn's lemma, there exists a maximal element $\bar{S} \in \mathcal{S}$ such that $S_o \subseteq \bar{S}$ . Since M is algebraically compact, $\bar{S}$ is soluble in M and let $(y_i)_{i\in K}$ be a solution of $\bar{S}$ in M , where K is a set of indices. Now we consider the submodule $B' = B + \sum_{i\in K} Ry_i$ and we prove that B' is small over B in M. Let N be a R-module and let $f \in Hom(B',N)$ be a partial homomorphism from M to N, such that $f_{|B}$ is a partial isomorphism from M to N. Let $g \in Hom(f(B),M)$ be the inverse partial isomorphism of $f_{|B}$ . Since M is pure-injective, by theor. 1.1 there exists $g' \in Hom(N,M)$ such that $g'_{|f(B)} = g$ . Let g'f = k . k is a partial homomorphism from M to M with $k_{|B} = 1_B$ and dom k = B' . Let $k(y_i) = y'_i$ for every $i \in K$ . It is easy to check that $(y'_i)_{i\in K}$ is an other solution of $\bar{S}$, since $k_{|B} = 1_B$. Now we prove that k is a partial isomorphism from M to M. From this result it will follow that f is a partial isomorphism from M to N and B' is small over B in M .

If k is not a partial isomorphism, there is a set of equations

$$\sum_{1}^{n} {}_{v}\, r_{uv} x_{v} = \sum_{1}^{m} {}_{t}\, s_{ut} y'_{i_t} + b_{u} \qquad ( 1 \leqq u \leqq p ),$$

(where n , m , p are positive integers , $r_{uv}$ and $s_{ut} \in R$ , $b_u \in B$ and $x_v$ are unknowns) which is soluble in M , but such that the set of equations

$$\sum_{1}^{n} {}_{v}\, r_{uv} x_{v} = \sum_{1}^{m} {}_{t}\, s_{ut} y_{i_t} + b_{u} \qquad ( 1 \leqq u \leqq p )$$

is not soluble in M . This means that the finite set E of equations

$$\sum_{1}^{n} {}_{v}\, r_{uv} x_{v} - \sum_{1}^{m} {}_{t}\, s_{ut} x_{i_t} = b_{u} \qquad ( 1 \leqq u \leqq p )$$

is not a subset of $\bar{S}$ . Hence the set of equations $\bar{S} \cup E$ has solution in M and $\bar{S}$ is not maximal . This fact is absurd .

By transitiveness of the smallness, B' is small over A in M and then B' = B , since B is maximal; hence $y_i \in B$ for all $i \in K$ . This means that every set $S_0$ of linear equations over R with constants in B, finitely soluble in M , is soluble in B and therefore B is pure in M and algebraically compact .

From this point on , the proof is due to Ziegler and it is included here for sake of completeness.

To prove $B = H_M(A)$ we must show that B satisfies the condition ii) in the definition of $H_M(A)$. Let C be a summand of M such that $A \leqq C \leqq B \leqq M$ and let p be the projection of M over C : $p_{|A}$ is a partial isomorphism from M to C and , since B is small over A , $p_{|B}$ is a partial isomorphism from M to C . It follows that B = C .

<u>Proof of the uniqueness</u> : Let N be a pure-injective R-module, let B be a submodule of N and let $f \in Hom(A,B)$ be a partial isomorphism from M to N. Since $H_N(B)$ is pure in N , f is a partial isomorphism from M to $H_N(B)$ and ,

since $H_N(B)$ is pure-injective, there is a homomorphism $f' \in Hom(M, H_N(B))$ such that $f'_{|A} = f$ . $f'_{|H_M(A)}$ is a partial isomorphism from M to $H_N(B)$, because $H_M(A)$ is small over A in M and then $f'(H_M(A))$ is a pure submodule of $H_N(B)$ and pure-injective, i.e. it is a summand of $H_N(B)$. We have $B \leq f'(H_M(A))$ which is a summand in $H_N(B)$ and then $f'(H_M(A)) = H_N(B)$.

From the foregoing proof, it follows that :

COROLLARY 1.3 (Ziegler [3])

Let A be a submodule of a pure-injective R-module M and let $A \leqq B \leqq M$ . The following are equivalent :

i) $B = H_M(A)$ ;

ii) B is a maximal small extension of A in M ;

iii) B is small over A in M , is pure-injective and pure in M .

Recall that <u>the pure-injective envelope</u> , PE(M), of M is a pure-injective pure-essential extension of M ; pure-injective envelopes exist and are unique up to isomorphism (see [2]) .

The relation between smallness and pure-essentiality for pure-injective modules is given in the following

PROPOSITION 1.4 (Ziegler [3])

B = PE(M) if and only if M is pure in B , B is pure-injective and B is small over M in B .

Proof : Assume B = PE(M) ; $H_B(M)$ is a summand of B and then $H_B(M) = B$. Hence B is small over M in B. Conversely, as we noted before, pure small extensions are pure-essential.

## 2. INDECOMPOSABLE PURE-INJECTIVE MODULES OVER VALUATION DOMAINS.

Let R be a valuation domain i.e. a commutative domain with 1 in which the ideals are totally ordered by inclusion. All R-modules will be unital. The maximal ideal of R is denoted by P . Let Q be the field of quotients of R and let $\Gamma$ be the value group of Q .

For every element x belonging to a R-module M , the height ideal $H_M(x)$, the height $h_M(x)$ and the indicator $i_M(x)$ of x in M are defined in [1]. We recall here the definitions :

the height ideal $H_M(x)$ is the union of all R-submodules J of Q such that $R \leq J$ and there exists $f \in \mathrm{Hom}(J,M)$ with $f(1) = x$ ;

let $H_M(x)/R = U$ ; the height $h_M(x)$ is defined as $h_M(x) = U^-$ or $h_M(x)=U$ according as there is not or there is $f \in \mathrm{Hom}(H_M(x),M)$ such that $f(1) = x$ and we call them respectively a limit or a non-limit height ; we set $h_M(0)= =\infty$ and denote the ordered set of all heights by $\Sigma$ . If $h_M(x) = J/R$ , then, for all $0 \neq r \in R$ , $rh_M(x) = (rJ + R)/R$ . If $h_M(x) = (J/R)^-$, $rh_M(x) = = ((rJ + R)/R)^-$ ;

the indicator of x in M , $i_M(x)$, is a function from the positivity domain $\Gamma^+$ of $\Gamma$ into $\Sigma$ defined by $i_M(x)(\gamma) = r_\gamma h_M(r_\gamma x)$ where $r_\gamma \in R$ is a fixed element of value $\gamma \in \Gamma^+$ ; $i_M(x)$ is constant if , for all $r \in R \sim \mathrm{Ann}(x)$, $rh_M(rx) = h_M(x)$ . It is clear that the indicator is a non- decreasing function .

Let M be a R-module and $M' \leqslant M$ . In this case it is well known (see [3]) that M' is pure in M if and only if, for all $r \in R$, $rM \cap M' = rM'$ .

Let M and N be R-modules . It is clear that f is a partial homomorphism ( isomorphism) from M to N with dom $f = M' \leqslant M$ if and only if, $f \in \mathrm{Hom}(M',N)$ and, for all $x \in M'$, $H_M(x) \leq H_N(f(x))$ ($H_M(x) = H_N(f(x))$ and f is monic) .

The following result was proved independently by Fuchs and Salce .

THEOREM 2.1

If M is a pure-injective R-module, then, for every $a \in M$ , $h_M(a)$ is a non-limit height .

Proof : Let $a \in M$ and $H_M(a) = \sum_{i \in I} Rr_i^{-1}$ where I is an ordinal, $r_i \in R$ , $i < j$ if and only if $r_j = r_i s_{ij}$ with $s_{ij} \in P$ $(i,j \in I)$. Let $i_0$ be the minimum of I. Let S be the set of linear equations over R defined by :

$$r_{i_0} x_{i_0} = a \ , \quad r_i r_j^{-1} x_j = x_i \quad \text{for every } i \ , \ j \in I \text{ such that } i < j \ .$$

If F is any finite subset of S and $x_{i_1}, x_{i_2}, \dots, x_{i_n}$ are all its variables with $i_1 < i_2 < \dots < i_n$ , then there is a homomorphism $f\colon Rr_{i_n}^{-1} \to M$ with $f(1) = a$ and $x_{i_1} = f(r_{i_1}^{-1}), x_{i_2} = f(r_{i_2}^{-1}), \dots, x_{i_n} = f(r_{i_n}^{-1})$ is a solution of F. M is algebraically compact, therefore S is soluble in M and let $(\bar{x}_i)_{i \in I}$ be a solution . We can define the homomorphism $g\colon H_M(a) \to M$ by setting $g(r_i^{-1}) = \bar{x}_i$ for all $i \in I$ . We have $g(1) = g(r_{i_0} r_{i_0}^{-1}) = r_{i_0} g(r_{i_0}^{-1}) = r_{i_0} \bar{x}_{i_0} = a$ and then $h_M(a)$ is not a limit height .

B. Zimmermann-Huisgen and W. Zimmermann proved in [4] that a non-zero pure-injective R-module U is indecomposable if and only if its endomorphism ring is local. Using this result, we can prove :

THEOREM 2.2

Let V be an indecomposable pure-injective R-module. For every $a \in V$ , $i_V(a)$ is constant .

Proof : Assume, by way of contradiction, that there is $a \in V$ such that, for some $r \in R \setminus \mathrm{Ann}(a)$, $rH_V(ra) > H_V(a)$ . By theor. 2.1 , there is $b \in V$ such that $rb = ra$ and $H_V(b) = rH_V(ra)$ . Let $f \in \mathrm{Hom}(Ra,V)$ defined by $f(a) = b$ .
f is a partial homomorphism from V to V , because, for all $s \in R$ , $H_V(sa) \leqslant H_V(sb)$ and then, by theor. 1.1 , there is an endomorphism f' of V extending f . f' is not an automorphism, because $H_V(b) > H_V(a)$, and then, being End V a local ring , $g = 1 - f'$ must be an automorphism. But this is false: in fact $g(a) = a - f'(a) = a - f(a) = a - b$ , $g(ra) = rg(a) = r(a-b) = 0$ and

$0 \neq ra \in \ker g$ .

The preceding result is used in proving that the only indecomposable pure-injective R-modules are the pure-injective envelopes of uniserial modules . This result was obtained through model- theoretical methods by Ziegler in [3] .

THEOREM 2.3 (Ziegler [3])
A pure-injective R-module V is indecomposable if and only if $V \cong PE(J/I)$ for some $0 \leqslant I \leqslant R \leqslant J \leqslant Q$ .

Proof: Assume V indecomposable pure-injective. Let $0 \neq a \in V$ , $H_V(a) = J$ and $Ann(a) = I$ . By theor. 2.1 , $h_V(a)$ is a non-limit height and there is a homomorphism $f : J/I \to V$ such that $f(1+I) = a$ and $\ker f = 0$ . Let $f(J/I) = W$ . W is pure in V , because $i_V(a)$ is constant by theor. 2.2 , therefore $V = PE(W)$ because V is indecomposable . Then $V = PE(W) \cong PE(J/I)$.

Conversely, let $V = PE(J/I)$. If $V = A \oplus B$ , $1 + I = a+b$ with $a \in A$ and $b \in B$ . Since J/I is pure in V , $i_V(a+b)$ is constant and $h_V(a+b) = J/I$ with $Ann_V(a+b) = I$ . It is easy to check that either $H_A(a) = J$ and $Ann_A(a) = I$ with $i_A(a)$ constant or $H_B(B) = J$ and $Ann_B(b) = I$ with $i_B(b)$ constant . We suppose that $H_A(a) = J$ and $Ann_A(a) = I$ with $i_A(a)$ constant and consider the projection p of $A \oplus B$ over A : $p(a+b) = a$ and $p_{|R(a+b)}$ is a partial isomorphism from $A \oplus B$ to A . It is obvious that J/I is small over $R(1+I)$ and, by prop. 1.4 , V is small over J/I , hence, by transitiveness of the smallness , V is small over $R(a+b)$ and then p is an isomorphism . Hence $B =$ $= \ker p = 0$ .

## REFERENCES


[1] L. Fuchs and L. Salce "Prebasic submodules over valuation rings" Annali Mat. Pura Appl. 32 (1982) 257-274 .

[2] R.B. Warfield, jr. "Purity and algebraic compactness for modules" Pacific J. Math. 28 , No.3 , 1969 , 699-719 .

[3] M. Ziegler "Model theory of modules" Annals Pure Appl. Logic, 26 (1984), 149-213.

[4] B. Zimmermann-Huisgen and W. Zimmermann "Algebraically compact ring and modules" Math. Zeitschrift 161 (1978) 81-93 .

# MORITA DUALITY - A SURVEY

Bruno J. Müller
McMaster University
Hamilton, Ontario, L8S 4K1, Canada

We leave aside the prolific and interesting investigations concerned with dualities between categories of topological modules, be they extensions or generalizations of Morita dualities, between more general categories, or between module categories with less perfect closure properties. Instead we confine ourselves totally to the original setting, as introduced by Morita himself, and defined in Section 1 . We feel that the core problems, regarding existence and selfduality in natural situations, present themselves most clearly here. Some progress has been made in recent years, mainly through the contributions of Schofield (division rings), Vamos (commutative and PI-rings), Haack and Dischinger - Müller (artinian serial rings), Jategaonkar, Musson and Donkin (noetherian rings), and MacDonald and Anh (generalized Morita dualities).

We consider only rings with identity element. $J(R)$ denotes the Jacobson radical of the ring $R$. A module $M$ is a right-module unless specified differently, and $E(M)$ is its injective hull.

## 1. MORITA DUALITIES

A (Morita) duality is a contravariant equivalence between full subcategories of S-left- and R-right-modules, which are closed under submodules, factormodules and finite direct sums, and contain the modules $S$ and $R$ respectively.

The most immediate instance of such a duality is the one between the finite-dimensional left- and right-vectorspaces over a division ring. Other special cases were considered, before Morita, by Kaplansky [35], Dieudonné [17], Tachikawa [66], Azumaya [6] and Matlis [41].

Morita [45] showed that a duality is representable by a bi-module ${}_SU_R$ which is a faithfully balanced injective cogenerator on either side. The rings $S$ and $R$ are semiperfect (Osofsky [56]), in fact left- respectively right-linearly compact (Müller [47]). They determine each other uniquely up to Morita equivalence, and hence up to isomorphism if we assume (as we may, and usually tacitly do) that they are basic. $U$ is then the minimal (injective) cogenerator, on either side.

$S$ and $R$ have a number of isomorphic ingrediences: their centres, their radical factorrings, and their ideal lattices (under annihilators, also compatible with products). (Many of these fundamental facts may be found in Anderson-Fuller [4].)

## 2 . LINEAR COMPACTNESS

The U-reflexive modules are precisely the linearly compact ones (Müller [47]), or equivalently the U-complete ones (Upham [69]). The natural domain of a duality is, therefore, also closed under extensions, and finite limits and colimits.

Every artinian module is linearly compact. All the linearly compact modules are explicitly known in the following cases:

Over a two-sided perfect ring, they are the modules of finite length (Sandomierski [63]).

Over a commutative noetherian ring, they are the extensions of finitely generated linearly compact modules by artinian ones (Zöschinger [76] . Note that, if the ring is in addition local and complete, then every finitely generated module is linearly compact.)

## 3 . EXISTENCE CRITERIA

We say that a ring R <u>has duality</u> if there is a Morita duality for R-right-modules and S-left-modules for some ring S .

A more explicit criterion for the existence of such a duality than Morita's, ie. the existence of a faithfully balanced left- and right-cogenerator bi-module U , was given by Müller [47] : it suffices that $U_R$ is linearly compact and R is U-complete. A variant by Lemonnier [37] requires that $U_R$ satisfies AB5* and R is right-linearly compact.

More readily verifiable criteria are known for special classes of rings, and will be discussed below. The problem of classifying all rings with duality is still very much open. Equally open is the much vaguer question of whether one can attach rings with duality to (more or less) arbitrary rings, for instance by the localization-completion procedure which works in the commutative noetherian case (cf. the comments in Sections 6 and 10 ).

## 4 . SELFDUALITIES

Perhaps the most intriguing puzzle about Morita dualities is to determine when they are selfdualities, ie. when the rings S and R are isomorphic. This has been established in a number of instances, and is expected to be the "normal" behavior.

The only known examples of dualities which are not selfdualities, are based on division ring extensions $T \subset D$ with different dimensions on the two sides. Cohn [12],[13] constructed such extensions which are $\infty$-dimensional on the left and 2-dimensional on the right. Recently Schofield [64] appears to have solved the old Artin problem completely, by manufactoring extensions of arbitrary finite dimensions $m, n \geqq 2$ on the left and right.

The examples of rings with dualities which are not selfdualities are variants of $R = \begin{pmatrix} T & D \\ 0 & D \end{pmatrix}$ . This ring is right- but not left-artinian for Cohn's example, and (left-and right-)artinian for Schofield's. Moreover, for Cohn's example, the ring $R = \begin{pmatrix} D & D \\ 0 & T \end{pmatrix}$ is artinian without duality.

A duality is called (<u>weakly</u>) <u>symmetric</u> if the isomorphism between S and R is compatible with the (restriction to the maximal ideals of the) ideal lattice isomorphism described in Section 1 .

## 5 . MATRIX REPRESENTATIONS

The next considerations, which follow the procedure in [48] , reduce the study of dualities, to a certain extend, from semiperfect to local rings.

A semiperfect ring R has a <u>matrix representation</u> $(R_{ij})_{i,j=1}^{n}$ , where the $R_{ii}$ are local rings, the $R_{ij}$ are $R_{ii}$-$R_{jj}$-bi-modules, and the multiplication is given by bi-module homomorphisms

$$R_{ij} \otimes_{R_{jj}} R_{jk} \rightarrow R_{ik} \qquad (i \neq j \neq k)$$

satisfying appropriate associativity conditions. Such a representation is unique, up to an obvious concept of isomorphism. Every representation (involving local rings) leads to a semiperfect ring.

The ring R has duality iff the $R_{ii}$ have dualities (with rings $S_i$ induced by bi-modules $U_i$ ) and the $R_{ij}$ and $R_{ij}^* = \hom_{R_{jj}}(R_{ij}, U_j)$ are reflexive (ie. linearly compact) as right-modules. In particular, the existence of a duality does not depend on the multiplication maps. The correspnding second ring is $S = (R_{ij}^{**})$ .

R has a weakly symmetric selfduality iff the $R_{ii}$ have selfdualities, and there are bi-module isomorphisms between $R_{ij}$ and $R_{ij}^{**}$ which are compatible with the multiplication maps. (These bi-module isomorphisms are closely related to the "duality conditions" employed in [54] and [5].)

These observations illustrate that the matrix examples in the preceeding section are, in a sense, general.

## 6 . COMMUTATIVE RINGS

In this section only, all rings R are commutative. If R has duality, then it is linearly compact, hence a finite product of local rings (Zelinsky [75]). Therefore we assume wlog that R is local.

Every commutative ring with duality, has a symmetric selfduality [47].

It is conjectured that every linearly compact commutative ring has duality. This has been verified for artinian rings (which are always linearly compact; Azumaya [6]), for noetherian rings (which are linearly compact iff they are complete; Matlis [41]), for valuation domains (which are linearly compact iff they are maximal; [75], [41]), more generally for chain rings (Vamos [72]) and for rings with linearly compact quotient field (which are always linearly compact, have as integral closure a maximal valuation ring and hence linearly ordered spectrum; Goblot [23], Vamos [72]).

Further examples can be constructed using ([72], 2.14): For a ring extension $R \subset T$ with $T_R$ linearly compact, $R$ has duality iff $T$ does.

Another construction involves waist prime ideals, ie. non-zero prime ideals comparable with all ideals. $P$ is such a waist of $R$ iff $R$ is the pullback of $R/P$ and $R_P$. Then, $R$ is linearly compact iff $R/P$ has linearly compact quotient field and $R_P$ is linearly compact; and $R$ has duality iff $R_P$ does. This reduces the existence problem for dualities to the class of linearly compact rings without waist (since, in case of a minimal waist, the pullback results can be applied, and in case of waists but no minimal one duality always exists [72]).

Wiseman [74] has constructed an example of a linearly compact ring with waists but no minimal one whose quotient field is not linearly compact. (Note that this type of ring has again linearly ordered spectrum.)

In general, a linearly compact commutative ring $R$ has duality iff the endomorphism ring of the minimal cogenerator is also commutative. Every such ring $R$ has an ideal $I$ such that $R/I'$ has duality iff $I' \supset I$.

One might wonder whether a (properly defined) linear compactification exists for every commutative local ring, as suggested by the classes of noetherian rings (completions) and valuation domains (maximal valuation domains). In conjunction with the above conjecture concerning the existence of dualities, this would attach a ring with duality to an arbitrary commutative local ring.

## 7.. ASCENT AND DESCENT

Let $A \subset R$ be a ring extension. If $A$ has duality induced by $V$, and if $R_A$ and $R^*_A$ are linearly compact, then $R$ has duality induced by $U = R^*$. In particular, if $R$ is an algebra over a commutative ring $A$ with duality, and if $R$ is linearly compact as $A$-module, then $R$ has symmetric selfduality.

This applies to finite-dimensional algebras over a field, more generally to Artin algebras, and to group and semigroup rings [22] over commutative rings with duality.

On the other hand, if $R$ has duality, only few results are known regarding duality for $A$; cf. Vamos [72], Kitamura [36] and Fuller-Haack [22]. The following is an unpublished observation of the author: If $R$ is one-sided noetherian with duality, and module-finite over its centre $A$, then $A$ has duality (and is noetherian, and $R$ has symmetric selfdualtiy and is noetherian).

## 8 . (RIGHT-)ARTINIAN RINGS

A right-artinian ring has duality iff the minimal cogenerator $U_R$ has finite length, iff $\hom(J/J^2_R, R/J_R)$ has finite length as R-right-module (Rosenberg-Zelinsky [59]). The last condition depends only on $R/J^2$. The corresponding ring $S$ is left-artinian. A one-sided perfect ring with duality is right-artinian [56].

If the minimal cogenerator $U_R$ has finite length but $R$ is arbitrary, then $R$ need not have duality. Counterexamples were provided by Cozzens [14] and Anh [3]; Cozzens' examples are non-artinian simple noetherian domains with simple $U_R$. However if $R$ is assumed semilocal, (then it is automatically semiprimary and) the question is open and is equivalent to whether finite-dimensionality of $X_T^*$ implies that of $X_D$, for any bi-vectorspace $_TX_D$ over division rings.

If one compares rings which are only right-artinian with artinian ones, one observes a stricking difference: All known examples (cf. Section 4) of only right-artinian rings with duality, as well as of artinian rings without duality, or with duality but without selfduality, are based on Cohn's and Schofield's division ring extensions. Exciting as these examples are, we cannot help feeling that the constructions are highly artificial, the division rings are unreasonably large, and that "reasonable smallness conditions" should prevent these phenomena. We are thus led to expect that, in "natural circumstances", artinian rings should have selfduality, and right- but not left-artinian rings should not have duality.

On the positive side, an artinian ring $R$ has duality iff all the (right- and left-finite-dimensional) bi-vectorspaces $_TX_D$ which arise from $J/J^2$, have finite-dimensional duals $X_T^*$, or equivalently iff the simple artinian rings $\mathrm{endo}(X_D)$ which are left-finite-dimensional over $T$, are also right-finite-dimensional over $T$. In particular, an artinian ring has duality if it satisfies a polynomial identity [59], or is of finite module type.

After several special cases had been handled by Haack [25], [26], Dischinger-Müller [19] have now proved that every artinian serial ring has weakly symmetric selfduality.

Further instances of artinian rings with selfduality appear in Roux 62], Auslander-Platzek-Reiten [5], Haack [24], [27] and Fuller-Haack [21], [22].Obvious examples are commutative artinian rings and quasi-Frobenius rings. Azumaya [7] conjectures selfduality for his "exact" rings.

## 9 . (GENERALIZED) QUASI-FROBENIUS RINGS

Clearly a quasi-Frobenius ring R has selfduality induced by U = R . It is weakly symmetric iff soc(eR) is isomorphic to top(eR) for every primitive idempotent, ie. if the Nakayama permutation is trivial. It is symmetric iff the left- and right-annihilators of each ideal coincide. (Even if these properties fail, R can have a (weakly) symmetric selfduality induced by a modified U ; for instance if R is a finite-dimensional algebra over a field, then U = R* always induces a symmetric selfduality, even if R is not quasi-Frobenius; cf. Section 7 . )

Similar statements hold for generalized quasi-Frobenius rings, ie. rings which are injective cogenerators on both sides.

If R is only an injective cogenerator on the right, then it is semiperfect and right-essential over a finite socle, hence similar to the minimal cogenerator (Utumi [71], Osofsky[56]). (Self)duality exists iff it is left-injective, iff it is right-linearly compact. Dischinger-Müller [18] have now obtained an example of a right-injective cogenerator ring without duality.

The structure of QF-3 rings (on which there exists an extensive literature, cf. Tachikawa [67]) crucially involves the relationship with a subring with duality.

## 10 . NOETHERIAN RINGS

The ring $R = \begin{pmatrix} D & Q \\ 0 & Q \end{pmatrix}$ where $D$ is a complete discrete valuation ring with quotient field $Q$ , is right- but not left-noetherian, and has self-duality. It is right- but not left-strictly linearly compact, and satisfies $\cap J(R)^n \neq 0$ .

An arbitrary ring $R$ with $\cap J(R)^n = 0$ has duality iff it is semilocal right-noetherian complete (in the $J(R)$-adic topology) and has artinian minimal cogenerator $U_R$ (Müller [46]).

A semilocal right-noetherian ring $R$ is complete iff it is right-(strictly-)linearly compact and $J(R)$ has the right-AR property (Hinohara [28], Deshpande [16]); the right-AR property is equivalent to $U = \bigcup_n \mathrm{ann}_U(J^n)$ (Jategaonkar [29]). The completion $\hat{R}$ of a semilocal right-noetherian ring $R$ is $J(\hat{R})$-adically complete, satisfies $\cap J(\hat{R})^n = 0$ , and equals the second commutator $\mathrm{biendo}(U_R)$ (Hinohara [28], Jategaonkar [29]). $U$ is also the minimal $\hat{R}$-cogenerator, provided $\hat{R}$ is right-noetherian [29].

It follows that, for a semilocal right-noetherian ring $R$ , the completion $\hat{R}$ has duality iff it is right-noetherian and $U_R$ is artinian. The validity of these two conditions are famous open questions, to which many partial answers are known, some of which are discussed below.

If $R$ is right-noetherian and $S$ is a right-classical semiprime ideal (ie. $\mathcal{C}(S)$ is a right-Ore set and $E(R/S) = \bigcup_n \mathrm{ann}(S^n)$ ), then the localization $R_S$ is semilocal right-noetherian, and therefore the completion $\hat{R}_S$ (which equals the second commutator biendo( $E(R/S)_R$ ) ) has duality iff it is right-noetherian and $E(R/S)$ has descending chain condition on $E(R/S)$-closed submodules. In this manner one can often attach a Morita duality to a (right-)classical semiprime ideal of a (right-)noetherian ring, quite as in the commutative situation [41].

In the sequel we shall discuss the four best understood classes of noetherian rings, viz. rings of Krull dimension one, PI-rings, envelopinga algebras of finite-dimensional solvable Lie algebras over an algebraically closed field of characteristic zero, and group rings of polycyclic-by-finite groups over a field or the ring of integers, from this point of view.

## 11 . KRULL DIMENSION ONE

For a semilocal HNP ring $R$ , the minimal cogenerator $U_R$ is artinian (as it is a finite direct sum of serial modules) and the completion $\hat{R}$ is noetherian (Upham [68], Deshpande [16], Marubayashi [40]); hence $\hat{R}$ has duality. Since, explicitly, $U = Q/R$ , where $Q$ is the quotient ring of $R$ , and $\text{endo}(Q/R_R) = \hat{R}$ , this is a selfduality.

Interestingly enough, this natural selfduality is not weakly symmetric, as each maximal ideal corresponds under annihilators to its cycle-neighbor (Eisenbud-Robson [15]) rather than to itself. In the same vein, the induced duality for any proper factorring $R/X$ is with $R/Y$ , for the cycle-neighbor $Y$ of the ideal $X$ . Since these factors are serial artinian, the Dischinger-Müller result [19] implies that $R/X \cong R/Y$ holds along the cycles. Whether an automorphism of $R$ exists which leads to a (weakly) symmetric selfdualtiy, is open at this time.

A generalization (Jansen, cf. Müller [52]) concerns semilocal prime noetherian 1-Gorenstein rings R . These rings are still of Krull dimension one, have Q/R as minimal cogenerator, and satisfy $\hat{R}$ = endo(Q/R). They have (self)duality iff Q/R is (right-)artinian, iff $\hat{R}$ is (right-) noetherian. These conditions are, for instance, fulfilled if R satisfies a polynomial identity.

If one starts with a (not necessarily semilocal) bounded HNP ring or a prime noetherian 1-Gorenstein ring with polynomial identity, then it has enough finite clans (Müller [49], [50]), and the preceeding considerations apply to these localizations.

12 . RINGS WITH POLYNOMIAL IDENTITY

For a semilocal noetherian PI-ring R , the minimal cogenerator is artinian and the completion $\acute{R}$ is noetherian (Vamos [73], Jategaonkar [31]); and hence $\acute{R}$ has duality (on both sides).

If one starts with a (not necessarily semilocal) noetherian PI-ring R and a finite clan, then the localization is as in the last paragraph, and therefore its completion has duality. If we also assume that R is an affine algebra over a field k , and restrict attention to clans of maximal ideals, then the factors R/P are algebraic over k (Procesi [58]) hence finite-dimensional (Kaplansky [34]), and Jategaonkar's patching argument [32] shows that we obtain a selfduality. In general, this question is open.

## 13 . ENVELOPING ALGEBRAS

The enveloping algebra R of a finite-dimensional solvable Lie algebra over an algebraically closed field is a noetherian ring in which every ideal has a normalizing sequence of generators (McConnell [42]). A prime ideal is either localizable, or belongs to an infinite clan (Brown [10]). All prime ideals are localizable, iff all maximal ideals of finite codimension are localizable, iff the Lie algebra is nilpotent (Brown [9]). It follows then from Jategaonkar [32] that the completion of the localization at each localizable prime ideal has duality. It is known to be a selfduality for the maximal ideals of finite codimension.

## 14 . NOETHERIAN GROUP RINGS

The relevant information concerning group rings of polycyclic-by-finite groups over a field k or the ring of integers $\mathbb{Z}$ is found in Jategaonkar [30], [32], Musson [55], Donkin [20], Brown-Lenagan-Stafford [8] and Passman [57]. (Some of the arguments here are quite sophisticated.)

For the integral group ring $\mathbb{Z}G$, every maximal ideal is of finite index and belongs to a finite clan. The completion of the localization is noetherian. The patching argument produces a selfduality. As a Corollary, the injective hulls of simple modules are artinian.

For the group ring $kG$ over a field k, we restrict attention to maximal ideals of finite codimension. (This covers all primitive ideals, iff G is abelian-by-finite or k is algebraic of positive characteristic.)

In positive characteristic p, $kG$ has enough finite clans. The completion of the localization at any clan of maximal ideals of finite codimension is noetherian and has selfduality. As before it follows that the injective hulls of the finite-dimensional simple modules are artinian. (All clans are singletons iff G is p-nilpotent.)

In characteristic zero, Hopf algebra methods [20] show that the injective hull of every finite dimensional module is artinian, and the completion at every ideal of finite codimesion is noetherian. One deduces that the finite clans of maximal ideals of finite codimension lead to selfdualities. (If all maximal ideals belong to finite clans, then G is nilpotent-by-finite. If G is finite-by-nilpotent, then all prime ideals are localizable [57].)

Little to nothing is known regarding dualities attached to other semiprime ideals.

## 15 . GENERALIZED MORITA DUALITIES

We conclude by mentioning that the recent successes at localizing noetherian rings at infinite clans (Müller [51], Jategaonkar [33], Brown [10], [11], Stafford [65]), which produce localizations with countably many maximal ideals but still with many local-like properties, and the desire to attach dualities to the completions of these localizations, suggest a detailed analysis of a generalization of Morita duality which was initiated by Goblot [23], MacDonald [38], [39], Anh [2]; cf. also Sandomierski [63], Upham [70], Jategaonkar [33].

These are dualities between full subcategories of S-left- and R-right-modules, which are closed under submodules, factormodules and finite direct sums, but need not contain the modules S or R . One associates topologies on the rings S and R (by taking a right-ideal

I of R open if R/I lies in the subcategory), and considers the categories S dis and dis R of those discrete (!) modules which allow the topologized rings.

Such a duality is representable iff the subcategories consist of linearly compact modules. Then they are induced by a unique bi-module ${}_S U_R$ which is similar to the minimal (injective) cogenerator in S dis and dis R (and is consequently a selfinjective selfcogenerator in S mod and mod R ). The endomorphism rings of U are the topological completions $\hat{S}$ and $\hat{R}$ ; they are linearly topologized linearly compact and F-semiperfect. The duality extends naturally to one between all the linearly compact (discrete !) modules over these complete topological rings. All linearly topologized linearly compact rings appear in these roles.

Thus this theory constitutes a very natural generalization of the ordinary Morita duality, and avoids the limitations imposed by semilocality. The analogy is most striking if (as is often the case) the injective hulls of the simple modules are still linearly compact. In particular the development of Section 5 carries then over, and leads to infinite matrix representations of these rings (over local rings with Morita duality).

In certain cases, for instance for infinite clans of maximal ideals in affine noetherian PI-rings, or of maximal ideals of finite codimension in enveloping algebras of solvable Lie algebras over an uncountable algebraically closed field of characteristic zero, one obtains "generalized selfdualities".

REFERENCES

1. Anh,Pham Ngoc, Duality over noetherian rings with a Morita duality, J. Algebra 75 (1982), 275-285.

2. Anh, Pham Ngoc, Duality of modules over topological rings, J. Algebra 75 (1982), 395-425.

3. Anh, Pham Ngoc, On a problem of B. J. Müller, Archiv Math. 39 (1982), 303-305.

4. Anderson, F. W. and K. R. Fuller, Rings and Categories of Modules, Springer Verlag 1974.

5. Auslander, M., M. I. Platzeck and I. Reiten, Coxeter functors without diagrams, Trans. Amer. Math. Soc. 250 (1979), 1-46.

6. Azumaya, G., A duality theory for injective modules, Amer. J. Math. 81 (1959), 249-278.

7. Azumaya, G., Exact and serial rings, J. Algebra 85 (1983), 477-489.

8. Brown, K. A., T. H. Lenagan and J. T. Stafford, K-theory and stable structure of some noetherian group rings, Proc. London Math. Soc. 42 (1981), 193-230.

9. Brown, K. A., Localisation, bimodules and injective modules for enveloping algebras of solvable Lie algebras, Bull. Sci. Math. 107 (1983), 225-251.

10. Brown, K. A., Ore sets in enveloping algebras, Comp. Math. (to appear).

11. Brown, K. A., Ore sets in noetherian rings (preprint).

12. Cohn, P. M., Quadratic extensions of skew fields, Proc. London Math. Soc. 11 (1961), 531-556.

13. Cohn, P. M., On a class of binomial extensions, Ill. J. Math. 10 (1966), 418-424.

14. Cozzens, J. H., Homological properties of the ring of differential polynomials, Bull. Amer. Math. Soc. 76 (1970), 75-79.

15. Eisenbud, D. and J. C. Robson, Hereditay noetherian prime rings, J. Algebra 16 (1970), 86-104.

16. Deshpande, V. K., Completions of noetherian hereditary prime rings, Pacific J. Math. 90 (1980), 285-297.

17. Dieudonné, J., Remarks on quasi-Frobenius rings, Ill. J. Math. 2 (1958), 346-354.

18. Dischinger, F. and W. Müller, Die triviale Ringerweiterung $R \rtimes {}_R E_R$ ist links PF aber nicht rechts PF (preprint).

19. Dischinger, F. and W. Müller, Einreihig zerlegbare artinsche Ringe sind selbstdual (preprint).

20. Donkin, S., Locally finite representations of polycyclic-by-finite groups, Proc. London Math. Soc. 44 (1982), 333-348.

21. Fuller, K. R. and J. Haack, Rings with quivers that are trees, Pacific J. Math. 76 (1978), 371-379.

22. Fuller, K. R. and J. K. Haack, Duality for semigroup rings, J. Pure Applied Algebra 22 (1981), 113-119.

23. Goblot, R., Sur les anneaux linéairement compacts, C. R. Acad. Sci. Paris 270 (1970), A 1212-1215.

24. Haack, J. K., Incidence rings with self-duality, Proc. Amer. Math. Soc. 78 (1980), 165-169.

25. Haack, J. K., Self-duality and serial rings, J. Algebra 59 (1979), 345-363.

26. Haack, J. K., Serial rings and sudirect products, J. Pure Applied Algebra (to appear).

27. Haack, J. K., V. P. Camillo and K. R. Fuller, Azumaya's exact rings and a problem in linear algebra (preprint).

28. Hinohara, Y., Note on non-commutative semi-local rings, Nagoya J. Math. 17 (1960), 161-166.

29. Jategaonkar, A. V., Injective modules and localization in noncommutative noetherian rings, Trans. Amer. Math. Soc. 188 (1974), 109-123.

30. Jategaonkar, A. V., Integral group rings of polycyclic-by-finite groups, J. Pure Applied Algebra 4 (1974), 337-343.

31. Jategaonkar, A. V., Certain injectives are artinian, Lecture Notes Math. 545 (1976), Springer Verlag, 128-139.

32. Jategaonkar, A. V., Morita duality and noetherian rings, J. Algebra 69 (1981), 358-371.

33. Jategaonkar, A. V., Localization in noetherian rings (preprint).

34. Kaplansky, I., On a problem of Kurosch and Jacobson, Bull. Amer. Math. Soc. 52 (1946), 496-500.

35. Kaplansky, I., Dual modules over valuation rings, Proc. Amer. Math. Soc. 4 (1953), 213-219.

36. Kitamura, Y., Quasi-Frobenius extensions with Morita duality, J. Algebra 73 (1981), 275-286.

37. Lemonnier, B., AB5* et la dualité de Morita, C. R. Acad. Sci. Paris 289 (1979), A 47-50.

38. Macdonald, R. N. S., Dualities between finitely closed subcategories of modules, Ph.D. thesis, MacMaster University (1977).

39. Macdonald, R. N. S., Representable dualities between finitely closed subcategories of modules, Canad. J. Math. 31 (1979), 465-475.

40. Marubayashi, H., Completions of hereditary noetherian prime rings, Osaka J. Math. 17 (1980), 391-406.

41. Matlis, E., Injective modules over noetherian rings, Pacific J. Math. 8 (1958), 511-528.

42. McConnell, J. C., Localisation in enveloping rings, J. London Math. Soc. 43 (1968), 421-428.

43. McConnell, J. C., The noetherian property in complete rings and modules, J. Algebra 12 (1969), 143-153.

44. McConnell, J. C., On completions of non-commutative noetherian rings, Comm. Algebra 6 (1978), 1485-1488.

45. Morita, K., Duality for modules and its applications to the theory of rings with minimum condition, Sci. Report Tokyo Kyoiku Daigaku 6 (1958), 83-142.

46. Müller, B. J., On Morita duality, Canad. J. Math 21 (1969), 1338-1347.

47. Müller, B. J., Linear compactness and Morita duality, J. Algebra 16 (1970), 60-66.

48. Müller, B. J., The structure of quasi-Frobenius rings, Canad. J. Math. 26 (1974), 1141-1151.

49. Müller, B. J., Localization in non-commutative noetherian rings, Canad. J. Math. 28 (1976), 600-610.

50. Müller, B. J., Localization in fully bounded noetherian rings, Pacific J. Math 67 (1976), 233-245.

51. Müller, B. J., Two-sided localization in noetherian PI-rings, J. Algebra 63 (1980), 359-373.

52. Müller, B.J., Links between maximal ideals in bounded noetherian prime rings of Krull dimension one, Proc. Conf. Methods in Ring Theory, Antwerp (1983) (to appear).

53. Müller, B. J., Affine noetherian PI-rings have enough clans, J. Algebra (to appear).

54. Müller, W., Unzerlegbare Moduln über artinschen Ringen, Math. Z. 137 (1974), 197-226.

55. Musson, I. M., Injective modules for group rings in polycyclic groups I, Quart. J. Math. Oxford 31 (1980), 429-448.

56. Osofsky, B. L., A generalization of quasi-Frobenius rings, J. Algebra 4 (1966), 373-387.

57. Passman, D. S., Universal fields of fractions for polycyclic group algebras, Glasgow J. Math. 23 (1982), 103-113.

58. Procesi, C., Rings with Polynomial Identities, Marcel Dekker (1973).

59. Rosenberg, A. and D. Zelinsky, Finiteness of the injective hull, Math. Z. 70 (1959), 372-380.

60. Roux, B., Sur la dualité de Morita, Tohoku J. Math 23 (1971), 457-472.

61. Roux, B., Anneaux artiniens et extensions d'anneaux semi-simples, J. Algebra 25 (1973), 295-306.

62. Roux, B., Modules injectifs indécomposables sur les anneaux artiniens et dualité de Morita, Séminaire P. Dubreil 1972/73, Secrétariat Math. Paris (1973), pp. 19 .

63. Sandomierski, F. L., Linearly compact modules and local Morita duality, Proc. Conf. Ring Theory Park City 1971, Academic Press (1972), 333-346.

64. Schofield, A. H., (I've heard rumors).

65. Stafford, J. T., The Goldie rank of a module (preprint).

66. Tachikawa, H., Duality theorem of character modules for rings with minimum condition, Math. Z. 68 (1958), 479-487.

67. Tachikawa, H., Quasi-Frobenius Rings and Generalizations, Lecture Notes in Math. 351, Springer Verlag (1973).

68. Upham, M. H., Localization and completion of FBN hereditary rings, Comm. Algebra 7 (1979), 1269-1307.

69. Upham, M. H., Two remarks on duality, Houston J. Math. 5 (1979), 437-443.

70. Upham, M. H., The first and second endomorphism rings of an injective module over an FBN ring (prerpint).

71. Utumi, Y., Self-injective rings, J. Algebra 6 (1967), 56-64.

72. Vamos, P., Rings with duality, Proc. London Math. Soc. 35 (1977), 275-289.

73. Vamos, P., Semi-local noetherian PI-rings, Bull. London Math. Soc. 9 (1977), 251-256.

74. Wiseman, A. N., Integral extensions of linearly compact domains, Comm. Algebra 11 (1983), 1099-1121.

75. Zelinsky, D., Linearly compact modules and rings, Amer. J. Math. 75 (1953), 79-90.

76. Zöschinger,H., Linear-kompakte Moduln über noetherschen Ringen (preprint).

# ON THE STRUCTURE OF LINEARLY COMPACT RINGS AND THEIR DUALITIES *) **)

DIKRAN DIKRANJAN

Bulgarian Academy of Sciences
Sofia 1090,Bulgaria

ADALBERTO ORSATTI

Istituto di Algebra e Geometria
Via Belzoni,7,35100 Padova,Italia

## 0. Introduction

0.1. In this work we study (left) linearly compact (l.c.) rings giving contributions in the following three directions.

- A theorem of representation of any l.c. ring as the endomorphism ring of a module canonically associated to the ring. Then the structure of the module gives useful informations on the structure of the ring.
- A duality theory which characterizes l.c. rings.
- The existence of a pair of l.c. rings (the cobasic ring and the basic ring) canonically associated to a given l.c. ring.

Some applications of these results is given.

*) This work was partially supported by Ministero della Pubblica Istruzione

**) While working in this paper,the first author had a grant from italian C.N.R.

0.2. A wider and almost selfcontained version of this work will appear in "Rendiconti Accademia Nazionale delle Scienze detta dei XL. Memorie di Matematica".

In the sequel that paper will be referred to as [DO].

0.3. Some of the above results and others about primary l.c. rings have been announced in the Conference on Topology held at L'Aquila, March 1983. See [2].

0.4. Acknowledgements. It is a pleasure to thank P.N. Ánh, A. Facchini, E. Gregorio, V. Roselli and expecially C. Menini for helpful suggestions and improvements.

## 1. Preliminaries

1.1. All rings considered in this paper have a non zero identity and all modules are unital. Let R be a ring. We denote by R-Mod (Mod-R) the category of left (right) R-modules. Morphisms between modules will be written on the opposite side of scalars. If $M \in$ R-Mod, E(M) is the injective hull of M in R-Mod.

1.2. All ring and module topologies are linear and Hausdorff. Let $(R, \tau)$ be a left linearly topologized (l.t.) ring R with topology $\tau$, and let $R_\tau$-LT be the category of l.t. left modules over the topological ring $(R, \tau)$. Writing $(M, \mathcal{E}) \in R_\tau$-LT we mean that the module M endowed with the topology $\mathcal{E}$ is an object of $R_\tau$-LT. If $L, M \in R_\tau$-LT, $\mathrm{Chom}_R(L,M)$ is the group of continuous morphisms of L in M.

Recall that a module $M \in R_\tau$-LT is linearly compact (l.c.) if every family of closed cosets, with finite intersection property, has non empty intersection.
M is strictly linearly compact (s.l.c.) if M is complete in its canonical uniformity and for every open submodule H of M, M/H is artinian. Then M is l.c. since artinian modules are

l.c.d. ( = linearly compact in the discrete topology).

Recall that a module $H \in R\text{-Mod}$ is finitely cogenerated (f.c.) if the socle Soc(H) of H is finitely generated (f.g.) and essential in H. A submodule H of a module M is cofinite if M/L is f.c.

Let $(M,\mathcal{E}) \in R_{\tau}\text{-LT}$. The Leptin topology $\mathcal{E}_*$ of $(M,\mathcal{E})$ is the topology having as a basis of neighbourhoods of 0 in M the cofinite $\mathcal{E}$-open submodules of $(M,\mathcal{E})$.

Clearly $(M,\mathcal{E}_*) \in R_{\tau}\text{-LT}$ and $\mathcal{E}_* \le \mathcal{E}$. $\mathcal{E}$ and $\mathcal{E}_*$ are equivalent topologies in the sense that they have the same closed submodules. If $(M,\mathcal{E})$ is l.c., $\mathcal{E}_*$ is a minimal topology.

If $(M,\mathcal{E})$ is s.l.c., then $\mathcal{E} = \mathcal{E}_*$. $\tau_*$ is a ring topology.

1.3. Let R be a ring, $L, K \in R\text{-Mod}$. For every subset F of L set:

$$O(F) = \{ \varphi \in \mathrm{Hom}_R(L,K) \ : \ F\varphi = 0 \}$$

O(F) is a subgroup of $\mathrm{Hom}_R(L,K)$. Note that $O(F) = O(RF)$ where RF is the submodule of L generated by F.

The subgroups O(F) with F finite can be assumed as a basis of neighbourhoods of 0 for a group topology on $\mathrm{Hom}_R(L,K)$ called the finite topology. In this topology $\mathrm{Hom}_R(L,K)$ is l.t. complete and Hausdorff. If L = K , the O(F)'s are left ideals of $\mathrm{End}({}_RK)$ and $\mathrm{End}({}_RK)$ is a topological ring.

Let ${}_RK \in R\text{-Mod}$. The K-topology of R is obtained by taking as a basis of neighbourhoods of 0 in R the left ideals $\mathrm{Ann}_R(F)$ where F is a finite subset of K. This topology is Hausdorff iff ${}_RK$ is faithful. If $B = \mathrm{End}({}_RK)$ then, considering K as a right B-module, the finite topology of B coincides with its K-topology.

1.4. We conclude this preliminaries with a definition.

A bimodule ${}_RK_B$ is faithfully balanced if the canonical morphisms $R \longrightarrow \mathrm{End}(K_B)$ and $B \longrightarrow \mathrm{End}({}_RK)$ are ring isomorphisms.

## 2. Linear compactness conditions and a representation theorem

2.1. Throughout this Section $(R,\tau)$ will be a fixed, but arbitrary, left l.t. ring. Denote by $\mathfrak{F}_\tau$ the filter of left open ideals of $(R,\tau)$ and by $\mathcal{T}_\tau$ the class of $\tau$-torsion modules

$$\mathcal{T}_\tau = \{M \in R\text{-Mod} : \forall x \in M, \mathrm{Ann}_R(x) \in \mathfrak{F}_\tau\}$$

Then a left ideal $I$ of $R$ is open iff $R/I \in \mathcal{T}_\tau$.

$\mathcal{T}_\tau$ is a Grothendieck category so that it has enough injectives. If $M \in \mathcal{T}_\tau$, $E_\tau(M)$ is the injective hull of $M$ in $\mathcal{T}_\tau$. $E_\tau(M)$ is the $\tau$-torsion submodule of $E(M)$.

a) A module $M \in R$-Mod, endowed with the discrete topology, is an object of $R_\tau$-LT iff $M \in \mathcal{T}_\tau$.

2.2. Lemma. Let $R$ be a ring and let ${}_RK$ be a module with essential socle such that $R$ is l.c. in the K-topology $w$. Then $w = w_*$.

Proof. $w \geq w_*$. Conversely, it is sufficient to prove that for every $x \in K, x \neq 0$, $Rx \cong R/\mathrm{Ann}_R(x)$ is f.c. Indeed since $\mathrm{Ann}_R(x) \in \mathfrak{F}_w$, $Rx$ is l.c.d., hence $\mathrm{Soc}(Rx)$ is f.g. and, since $\mathrm{Soc}(Rx)$ is essential in $Rx$, $Rx$ is f.c.

2.3. Lemma ([8]. Lemma 6). Let ${}_RK$ be a cogenerator of $\mathcal{T}_\tau$. Then the K-topology of $R$ is equivalent to $\tau$ and coarser than $\tau$.

2.4. Let ${}_RK$ be a faithful module and endow $R$ with the K-topology $w$. Then ${}_RK \in \mathcal{T}_w$. We say that ${}_RK$ is quasi-injective (q.i.) if ${}_RK$ is an injective object in $\mathcal{T}_w$ and that ${}_RK$ is strongly quasi-injective (s.q.i.) if ${}_RK$ is an injective cogenerator in $\mathcal{T}_w$.

By 2.3 we obtain the following statement :

a) Suppose that ${}_RK \in \mathcal{T}_\tau$. If ${}_RK$ is an injective (cogenerator) in $\mathcal{T}_\tau$, then ${}_RK$ is q.i. (s.q.i.).

Let ${}_RK \in \mathcal{T}_\tau$. Then ${}_RK$, with the discrete topology, is an object of $R_\tau$-LT. ${}_RK$ is a cogenerator of $R_\tau$-LT if for every $M \in R_\tau$-LT and $x \in M, x \neq 0$, there exists a continuous morphism $f : M \longrightarrow {}_R K$ such that $f(x) \neq 0$. ${}_R K$ is an injective object in $R_\tau$-LT if, for every topological submodule H of a module $M \in R_\tau$-LT, any continuous morphism of H in ${}_R K$ extends to a continuous morphism of M in ${}_R K$.

It is easy to prove that

b) ${}_R K$ is an (injective) cogenerator in $R_\tau$-LT iff ${}_R K$ is an (injective) cogenerator in $\mathcal{C}_\tau$.

2.5. Let ${}_R K$ be an injective cogenerator of $\mathcal{C}_\tau$, $B = \mathrm{End}({}_R K)$, V a simple submodule of ${}_R K$. It is known that $V^* = \mathrm{Hom}_R(V, {}_R K)$ is a simple submodule of $K_B$ and every simple submodule of $K_B$ has this form ( see [10] 6.9). Denote by $\Sigma(V)$ the isotypic component of $\mathrm{Soc}({}_R K)$ relative to V and define in a similar manner $\Sigma(V^*)$.

2.6. Theorem. In the situation 2.5 we have

a) $\mathrm{Soc}({}_R K) = \mathrm{Soc}(K_B)$.

b) $\mathrm{Soc}(K_B)$ is essential in $K_B$.

v) For every simple submodule of ${}_R K$, $\Sigma(V) = \Sigma(V^*)$.

The proof of this theorem is analogous to the proof of Proposition 6.10 of [10].

The following proposition is an easy consequence of the structure theorem of the endomorphism ring of a q.i. module (see e.g. [4]).

2.7. Proposition. Let ${}_R K$ be an injective cogenerator with essential socle of $\mathcal{C}_\tau$, $B = \mathrm{End}({}_R K)$, J(B) the Jacobson radical of B. Then

$$\mathrm{Ann}_B(\mathrm{Soc}({}_R K)) = J(B)$$

and

$$\mathrm{End}_R(\mathrm{Soc}({}_R K)) \cong B/J(B) \quad \text{canonically}.$$

We now give some conditions in order that the left l.t. ring $(R, \tau)$ be l.c.

2.8. *Proposition*. *Let* ${}_RK$ *be a cogenerator of* $\mathcal{T}_\tau$, $B=\mathrm{End}({}_RK)$. *Then the following conditions are equivalent:*

(a) $(R,\tau)$ *is l.c.*

(b) *The bimodule* ${}_RK_B$ *is faithfully balanced and* $K_B$ *is quasi-injective.*

*Proof*. See Main Theorem of [8].

2.9. *Proposition*. *Let* ${}_RM_B$ *be a faithfully balanced bimodule and suppose that* ${}_RM$ *is s.q.i. Then the following conditions are equivalent:*

(a) $M_B$ *is s.q.i.*

(b) *R is l.c. in its M-topology and* $\mathrm{Soc}({}_RM)$ *is essential in* ${}_RM$.

*Proof*. See Theorem 10 of [8].

The following theorem is a slight generalization of a result due to C. Menini ([8], Theorem 10).

2.10. *Theorem*. *Let* ${}_RK$ *be an injective cogenerator of* $\mathcal{T}_\tau$, $B=\mathrm{End}({}_RK)$, $\beta$ *the finite topology of* $B$. *Then the following conditions are equivalent:*

(a) $(R,\tau)$ *is l.c. and* $\mathrm{Soc}({}_RK)$ *is essential in* ${}_RK$.

(b) $(R,\tau)$ *and* $(B,\beta)$ *are both l.c. and* $\mathrm{Soc}({}_RK)$ *is essential in* ${}_RK$.

(c) *The bimodule* ${}_RK_B$ *is faithfully balanced and* $K_B$ *is an injective cogenerator of* $\mathcal{T}_\beta$.

*If these conditions hold then* $\beta=\beta_*$, $\mathrm{Soc}({}_RK)=\mathrm{Soc}(K_B)$ *and* $\mathrm{Soc}(K_B)$ *is essential in* $K_B$.

*Proof*. (a) $\Longrightarrow$ (c) By 2.8 ${}_RK_B$ is faithfully balanced. Since ${}_RK$ is an injective cogenerator of $\mathcal{T}_\tau$, ${}_RK$ is s.q.i. By 2.3, R is l.c. in its K-topology. Then, by 2.9, $K_B$ is s.q.i. On the other hand $\beta$ is the K-topology of B, hence $K_B$ is an injective cogenerator of $\mathcal{T}_\beta$.

(c) $\Longrightarrow$ (b) $K_B$ is s.q.i. By 2.8 and 2.6, $(R,\tau)$ is l.c. and $\mathrm{Soc}({}_RK)$ is essential in ${}_RK$. On the other hand, ${}_RK$ is a cogenerator of $\mathcal{T}_\tau$, ${}_RK_B$ is faithfully balanced and ${}_RK$ is quasi-injective. Then, by 2.8, $(B,\beta)$ is l.c.

(b) $\Longrightarrow$ (a) is trivial.

Suppose that the above conditions hold. Then by 2.6 $Soc({}_R K) = Soc(K_B)$ and $Soc(K_B)$ is essential in $K_B$. It follows from 2.2 that $\beta = \beta_*$.

2.11. Remark. In the notations of 2.10, suppose that ${}_R K$ is an injective cogenerator with essential socle of $\mathcal{C}_\tau$ and that ${}_R K_B$ is faithfully balanced. Then if $(R,\tau)$ is l.c., $(B,\beta)$ is l.c. too. The converse is in general not true, as the following example shows.

Let p be a prime number, $\mathbb{Z}(p^\infty)$ the Prüfer p-group, $\mathbb{Z}_p$ the ring of rationals whose denominator is prime to p, $J_p$ the ring of p-adic integers. Consider the trivial extensions $R = \mathbb{Z}(p^\infty) \oplus 1\mathbb{Z}_p$, $\bar{R} = \mathbb{Z}(p^\infty) \oplus 1 \cdot J_p$. Then $R \hookrightarrow \bar{R}$ in an obvious way. R is a valuation ring with a simple essential socle S and the unique linear Hausdorff topology on R is the discrete one. It is easy to see that $E(S) = E(R) = {}_R\bar{R}$. Set $A = End({}_R\bar{R})$. The bimodule ${}_R\bar{R}_A$ is faithfully balanced. Clearly R is not l.c. but A with the finite topology is l.c. Indeed, since $\bar{R}$ is commutative, $\bar{R}$ is a subring of A. Let F be a finite subset of $\bar{R}$. Then $A/O(F)$ is l.c.d. as an $\bar{R}$-module, hence it is l.c.d. as an A-module. Thus A is l.c. since $A = \varprojlim A/O(F)$.

2.12. Definition. The left l.t. ring $(R,\tau)$ is said to be topologically artinian (noetherian) if for every $I \in \mathcal{F}_\tau$, the module R/I is artinian (noetherian).

2.13. Lemma. Let ${}_R K_B$ be a faithfully balanced bimodule. Suppose that $\tau$ is the K-topology of R and endow B with the K-topology $\beta$. Suppose that $K_B$ is a cogenerator of $\mathcal{C}_\beta$. If $(R,\tau)$ is topologically artinian (noetherian) then $(B,\beta)$ is topologically noetherian (artinian).

The easy proof of this lemma is similar to the proof of Proposition 2.13 of [12].

2.14. Let $(V_\gamma)_{\gamma\in\Gamma}$ be a fixed system of representatives of the isomorphism classes of simple modules belonging to $\mathcal{T}_\tau$. Set $D_\gamma = \mathrm{End}_R(V_\gamma)$. Then $V_\gamma$ is a right vector space over the division ring $D_\gamma$. Denote by $\nu_\gamma$ the dimension of $V_\gamma$ over $D_\gamma$. Put

$$_RW = E_\tau(\bigoplus_{\gamma\in\Gamma} V_\gamma) \cong E_\tau(\bigoplus_{\gamma\in\Gamma} E_\tau(V_\gamma))$$

$_RW$ is the minimal injective cogenerator of $\mathcal{T}_\tau$. $_RW$, with the discrete topology is an object of $R_\tau$-LT and $_RW$ is a faithful R-module since $(R,\tau)$ is Hausdorff. Let $A=\mathrm{End}(_RW)$ and denote by $\sigma$ the finite topology of A ; $(A,\sigma)$ is right l.c. and $\sigma = \sigma_*$ by 2.10. The meaning of the symbols $\mathcal{F}_\sigma$, $\mathcal{T}_\sigma$, $E_\sigma$ is clear.

The notations just established will be of current use and, in general, their meaning will not be recalled.

2.15. Representation theorem for l.c. rings. Suppose that the ring $(R,\tau)$ is left l.c. Then:

a) The bimodule $_RW_A$ is faithfully balanced, $\mathrm{Soc}(_RW) = \mathrm{Soc}(W_A)$, $W_A$ is an injective cogenerator of $\mathcal{T}_\sigma$ with essential socle, $(A,\sigma)$ is right l.c. and $\sigma = \sigma_*$.

b) $(R,\tau_*)$ is topologically isomorphic, in a natural way, to the ring $\mathrm{End}(W_A)$ endowed with the finite topology.

c) $W_A \cong E_\sigma(\bigoplus_\gamma D_\gamma^{(\nu_\gamma)})$

d) $A/J(A) \cong \prod_\gamma D_\gamma$ and the right A-modules $D_\gamma$ are a system of representatives of simple non isomorphic modules in $\mathcal{T}_\sigma$.

e) If $(R,\tau)$ is s.l.c. then :

$$W_A \cong \bigoplus_{\gamma\in\Gamma} E_\sigma(D_\gamma^{(\nu_\gamma)}) \cong \bigoplus_{\gamma\in\Gamma} (E_\sigma(D_\gamma))^{(\nu_\gamma)}$$

Proof. a) Follows from 2.10.

b) Follows from a) and 2.2.

c) $V_\gamma^* = \mathrm{Hom}_R(V_\gamma, {_RW}) \cong D_\gamma$. Then by 2.6 $\mathrm{Soc}(W_A) = \bigoplus_\gamma D_\gamma^{(\nu_\gamma)}$ and the inclusions

$$\mathrm{Soc}(W_A) \leq W_A \leq E_\sigma(\mathrm{Soc}(W_A))$$

are essential. Since $W_A$ is an injective object in $\mathcal{T}_\sigma$, $W_A = E_\sigma(\mathrm{Soc}(W_A))$.

d) By 2.7 , $A/J(A) \cong \prod_{\gamma} D_{\gamma}$ . Since $W_A$ is a cogenerator of $\mathcal{C}_{\sigma}$, the conclusion follows from the structure of $W_A$.

e) If $(R,\tau)$ is s.l.c. then, by 2.13, $(A,\sigma)$ is topologically noetherian. In this case, applying classical methods (see e.g. [1]), it can be proved that in $\mathcal{C}_{\sigma}$ direct sums of injectives are injective.

## 3. Some applications

In this section we apply the representation Theorem 2.15 to obtain very short and natural proofs of two classical theorems of Leptin and Zelinsky.

**3.1. Theorem of Leptin on semiprimitive l.c. rings** [6]

Let $(R,\tau)$ be a left l.c. ring, $J(R)$ its Jacobson radical, ${}_RW$ the minimal injective cogenerator of $\mathcal{C}_{\tau}$. By 2.6 $Soc(W_A) = Soc({}_RW)$. Since both $Soc(W_A)$ and $Soc({}_RW)$ are A-submodules of $W_A$, we have by 2.7 :

$$J(R) = Ann_R(Soc(W_A)) = Ann_R(Soc({}_RW))$$

Assume now that R is semiprimitive, i.e. $J(R)=0$. Then

$$W_A = Soc(W_A) \quad \text{and} \quad {}_RW = Soc({}_RW)$$

since $W_A$ is s.q.i. It follows from 2.7 :

$$J(A) = 0 \quad \text{so that} \quad A \cong \prod_{\gamma\in\Gamma} D_{\gamma} .$$

By 2.15 we get the topological isomorphisms :

$$(R,\tau_*) \cong End_A(Soc(W_A)) \cong End_A(\bigoplus_{\gamma\in\Gamma} V_{\gamma}) \cong \prod_{\gamma\in\Gamma} End_{D_{\gamma}}(V_{\gamma})$$

which prove the classical Leptin's theorem on the structure of semiprimitive l.c. rings.

**Remark.** It may be shown that if $(R,\tau)$ is semiprimitive and left l.c. then $\tau = \tau_*$ . For more informations about semiprimitive l.c. rings see [16].

**3.2. Zelinsky's theorem on the structure of commutative l.c. rings**

Let $(R,\tau)$ be a commutative l.c. ring. Since $R \cong End(W_A)$ ,

$W_A \cong E_\sigma(\oplus D_\gamma^{(\nu_\gamma)})$ is q.i. and R is commutative, it is obvious that $\nu_\gamma = 1$ for every $\gamma \in \Gamma$. Thus :

$$W_A = E_\sigma(\bigoplus_{\gamma\in\Gamma} D_\gamma) \cong E_\sigma(\bigoplus_{\gamma\in\Gamma} E_\sigma(D_\gamma))$$

Let $\gamma, \gamma' \in \Gamma$, $\gamma \neq \gamma'$. The ring $\mathrm{End}_A(E_\sigma(D_\gamma) \oplus E_\sigma(D_{\gamma'}))$ is commutative since every such endomorphism extends to an endomorphism of $W_A$. Then every endomorphic image of $E_\sigma(D_\gamma) \oplus E_\sigma(D_{\gamma'})$ is fully invariant, so that we have $\mathrm{Hom}_A(E_\sigma(D_\gamma), E_\sigma(D_{\gamma'})) = 0$. Denote by $E(D_\gamma)$ the usual injective hull of $D_\gamma$ in Mod-A. By the observation above and since $\bigoplus_{\gamma\in\Gamma} E_\sigma(D_\gamma)$ is $\sigma$-torsion, we have :

$$\mathrm{Hom}_A(\bigoplus_\gamma E_\sigma(D_\gamma), \prod_\gamma E(D_\gamma)) \cong \mathrm{Hom}_A(\bigoplus_\gamma E_\sigma(D_\gamma), \prod_\gamma E_\sigma(D_\gamma)) \cong$$

$$\cong \mathrm{Hom}_A(\bigoplus_\gamma E_\sigma(D_\gamma), \bigoplus_\gamma E_\sigma(D_\gamma)) .$$

It follows that $\bigoplus_{\gamma\in\Gamma} E_\sigma(D_\gamma)$ is q.i. Now $\bigoplus_{\gamma\in\Gamma} E_\sigma(D_\gamma)$ is a cogenerator of $\mathcal{T}_\sigma$ and $\sigma = \sigma_*$ is minimal, $(A, \sigma)$ being l.c. Then, by 2.2, $\sigma$ coincides with the $\bigoplus_\gamma E_\sigma(D_\gamma)$-topology of $W_A$ so that $\bigoplus_{\gamma\in\Gamma} E_\sigma(D_\gamma)$ is an injective object in $\mathcal{T}_\sigma$, hence $W_A = \bigoplus_\gamma E_\sigma(D_\gamma)$. We have the topological isomorphisms

$$(R, \tau_*) \cong \mathrm{End}(W_A) \cong \mathrm{End}_A(\bigoplus_\gamma E_\sigma(D_\gamma)) \cong \prod_{\gamma\in\Gamma} \mathrm{End}_A(E_\sigma(D_\gamma)) ,$$

and $\mathrm{End}_A(E_\sigma(D_\gamma))$ is local since $E_\sigma(D_\gamma)$ is the injective hull in $\mathcal{T}_\sigma$ of a simple module. We obtain in this way the Zelinsky's theorem : any commutative l.c. ring is a topological product of local l.c. rings ( cf. [19]). A more precise result can be obtained using [3]. See [DO].

<u>Remark</u>. Let $(R, \tau)$ be a commutative l.c. ring. We do not know if $A = \mathrm{End}(_R W)$ is commutative. An affirmative answer to this question would imply that every commutative local l.c.d. ring has a Morita duality : this is the well known Zelinsky-Müller conjecture [13].

## 4. <u>Duality</u>

<u>4.1</u>. In a recent paper , [1] , P.N. Ánh introduced the notion of <u>Topological Morita Duality</u> (briefly TMD) and proved that a l.t. ring is l.c. iff it admits a TMD.

More precisely, let $(R,\tau)$ be a left l.t. ring and $(B,\beta)$ a right l.t. ring. By definition $(R,\tau)$ has a TMD with $(B,\beta)$ if there exists a faithfully balanced bimodule ${}_RK_B$ such that ${}_RK$ and $K_B$ are injective cogenerators of $\mathcal{C}_\tau$ and $\mathcal{C}_\beta$ respectively.

Observe that in this case, according to 2.6, $\mathrm{Soc}({}_RK)=\mathrm{Soc}(K_B)$ and this bimodule is essential in ${}_RK$ and in $K_B$. Suppose that $(R,\tau)$ has a TMD with $(B,\beta)$ induced by ${}_RK_B$. Then, by 2.8, $(R,\tau)$ is l.c. Conversely, if $(R,\tau)$ is l.c. then, using the bimodule ${}_RW_A$, it follows from 2.15 that $(R,\tau)$ has a TMD with $(A,\sigma)$.

In this section we study a duality for l.t. rings using methods from [10] and [11], obtaining in this way a characterization of l.c. rings which sharpens that of Ánh.

<u>4.2</u>. Throughout this section $(R,\tau)$ is a fixed, but arbitrary, left l.t. ring and ${}_RK$ a given faithful module in $\mathcal{C}_\tau$. Put $B = \mathrm{End}({}_RK)$ and endow B with the finite topology $\beta$. The modules ${}_RK$ and $K_B$ are supposed endowed with the discrete topology so that ${}_RK \in R_\tau\text{-LT}$ and $K_B \in \text{LT-}B_\beta$.

<u>4.3</u>. Let $M \in R_\tau\text{-LT}$ ($M \in \text{LT-}B_\beta$). A <u>character</u> of M is a continuous morphism of M in ${}_RK$ (in $K_B$).
A module $M \in R_\tau\text{-LT}$ is said to be ${}_RK$-<u>completely regular</u> if M is a topological submodule of a topological power of ${}_RK$. Clearly M is ${}_RK$-completely regular iff the topology of M coincides with the weak topology of its characters. Denote by $\mathcal{B}({}_RK)$ the full subcategory of $R_\tau$-LT consisting of all ${}_RK$-completely regular modules and define in a similar manner the category $\mathcal{B}(K_B)$.

Let $M \in R_\tau\text{-LT}$. The <u>character module</u> $M^*$ of M is the module $\mathrm{Chom}_R(M,{}_RK)$ endowed with the finite topology. Clearly $M^* \in \mathcal{B}(K_B)$. If $f : L \longrightarrow M$ is a continuous morphism in $R_\tau$-LT, its transpose $f^* : M^* \longrightarrow L^*$ is continuous.
If $M \in \text{LT-}B_\beta$, $M^*$ is defined in a similar manner.

Let $M \in R_\tau$-LT ($M \in$ LT-$B_\beta$) and let $x \in M$. Denote by $\tilde{x}$ the evaluation in x of the characters of M. Then $\tilde{x} \in M^{**} = (M^*)^*$ . The morphism $\omega_M : M \longrightarrow M^{**}$ given by $\omega_M(x) = \tilde{x}$ is called the canonical morphism.

4.4. Denote by $D_1 : R_\tau\text{-LT} \longrightarrow \text{LT-}B_\beta$ the contravariant functor which associates to every $M \in R_\tau$-LT the module $M^*$ and to every continuous morphism f in $R_\tau$-LT its transpose $f^*$. The functor $D_2 : \text{LT-}B_\beta \longrightarrow R_\tau\text{-LT}$ is defined in a similar manner. Put $D_K = (D_1, D_2)$. Let ${}_R\mathcal{M}$ and $\mathcal{M}_B$ be full subcategories of $R_\tau$-LT and LT-$B_\beta$ respectively. We say that $D_K$ induces a duality between ${}_R\mathcal{M}$ and $\mathcal{M}_B$ if $D_1({}_R\mathcal{M}) = \mathcal{M}_B$, $D_2(\mathcal{M}_B) = {}_R\mathcal{M}$ and for every $L \in {}_R\mathcal{M}$ and $M \in \mathcal{M}_B$ the canonical morphisms $\omega_L$ and $\omega_M$ are topological isomorphisms.
The following proposition is the main result on $D_K$. Namely, by Theorem 5.3 in [11] and by 6.1, 6.2 in the same paper we get :

4.5. Proposition. *In the situation 4.2 the following conditions are equivalent:*

(a) $D_K$ *induces a duality between* $\mathcal{B}({}_RK)$ *and* $\mathcal{B}(K_B)$ .

(b) *The bimodule* ${}_RK_B$ *is faithfully balanced and the modules* ${}_RK$ *and* $K_B$ *are both quasi-injective.*

4.6. Denote by $\mathcal{C}({}_RK)$ the full subcategory of $\mathcal{B}({}_RK)$ consisting of these ${}_RK$-completely regular modules which are complete in their canonical uniformity. A module $M \in \mathcal{C}({}_RK)$ iff M is a closed submodule of a topological power of ${}_RK$. The modules in $\mathcal{C}({}_RK)$ are called ${}_RK$-compact.

Let $M \in$ R-Mod such that $\text{Hom}_R(M, {}_RK)$ separates points of M. Denote by $\chi_M$ the weak topology of $\text{Hom}_R(M, {}_RK)$. Clearly $(M, \chi_M) \in \mathcal{B}({}_RK)$. A module $(M, \mathcal{E}) \in \mathcal{B}({}_RK)$ is called ${}_RK$-discrete if $\mathcal{E} = \chi_M$. Denote by $\mathcal{D}({}_RK)$ the full subcategory of $\mathcal{B}({}_RK)$ consisting of all ${}_RK$-discrete modules. A module $M \in \mathcal{B}({}_RK)$ is ${}_RK$-discrete iff $\text{Chom}_R(M, {}_RK) = \text{Hom}_R(M, {}_RK)$. Define in a similar manner the categories $\mathcal{C}(K_B)$ and $\mathcal{D}(K_B)$.

It is clear that:

$$M \in \mathcal{D}({}_R K) \Longrightarrow M^* \in \mathcal{C}(K_B) \quad ; \quad M \in \mathcal{D}(K_B) \Longrightarrow M^* \in \mathcal{C}({}_R K) .$$

The following theorem is an improvement of Theorem 6.6 of [11] and characterizes l.c. rings.

4.7. Theorem. In the situation 4.2, suppose that ${}_R K$ is an injective cogenerator with essential socle of $\mathcal{T}_\tau$. Then the following conditions are equivalent:

(a) $(R, \tau)$ is l.c.

(b) $D_K$ induces a duality between $\mathcal{B}({}_R K)$ and $\mathcal{B}(K_B)$.

(c) $K_B$ is q.i. and $D_K$ induces a duality between $\mathcal{D}({}_R K)$ and $\mathcal{C}(K_B)$.

(d) For every $L \in R_\tau$-LT and every $M \in$ LT-$B_\beta$ the canonical morphisms $\omega_L$ and $\omega_M$ are continuous isomorphisms.

(e) ${}_R K_B$ induces a TMD between $(R, \tau)$ and $(B, \beta)$.

If these conditions hold then

1) For every $M \in R_\tau$-LT ( $M \in$ LT-$B_\beta$ ) the topology of $M$ is equivalent to the weak topology of $\omega_M$.

2) $D_K$ induces a duality between $\mathcal{C}({}_R K)$ and $\mathcal{D}(K_B)$.

Proof. (a)$\Longleftrightarrow$(e) by 2.10.

(a)$\Longleftrightarrow$(b) by 2.8 and 4.5.

(b) $\Longrightarrow$ (c). Since (b)$\Longleftrightarrow$(a), $K_B$ is s.q.i. by 2.10 and 2.4 a). Hence the conclusion follows from 4.12 of [11].

(c) $\Longrightarrow$(b). Since $D_K$ induces a duality between $\mathcal{D}({}_R K)$ and $\mathcal{C}(K_B)$, the bimodule ${}_R K_B$ is faithfully balanced. Now the conclusion follows from 4.5.

(b)$\Longrightarrow$(d) and (b)$\Longrightarrow$ 1). Let $(L, \mathcal{E}) \in R_\tau$-LT. Since ${}_R K$ is a cogenerator of $R_\tau$-LT (see 2.4 b) ), the characters of $(L, \mathcal{E})$ separate points of $L$. Denote by $\mathcal{E}_1$ the weak topology of characters of $(L, \mathcal{E})$. Then $(L, \mathcal{E}_1) \in \mathcal{B}({}_R K)$ and $(L, \mathcal{E})^* = (L, \mathcal{E}_1)^*$ so that $(L, \mathcal{E})^{**} = (L, \mathcal{E}_1)^{**}$. The canonical morphism $\omega_1 : (L, \mathcal{E}_1) \longrightarrow (L, \mathcal{E})^{**}$ is a topological isomorphism so that the canonical morphism $\omega_L : (L, \mathcal{E}) \longrightarrow (L, \mathcal{E})^{**}$ is a continuous isomorphism. Let $M \in$ LT-$B_\beta$. Since (b) implies (e), $K_B$ is a cogenerator of LT-$B_\beta$ and the above argument works.

Let us prove 1). Let $(L,\mathcal{E}) \in R_\tau$-LT. Since $\omega_1$ above is a topological isomorphism and $(L,\mathcal{E}_1) \in \mathcal{B}({}_RK)$ the weak topology of $\omega_L$ coincides with $\mathcal{E}_1$. Now $\mathcal{E}_1 \leq \mathcal{E}$ and every $\mathcal{E}_1$-closed submodule of L is $\mathcal{E}$-closed. On the other hand, since ${}_RK$ is a cogenerator of $R_\tau$-LT, every $\mathcal{E}$-open submodule of L is intersection of $\mathcal{E}_1$-open ones. Therefore every $\mathcal{E}$-closed submodule of L is intersection of $\mathcal{E}_1$-open submodules so that it is $\mathcal{E}_1$-closed. For $M \in$ LT-$B_\beta$ we use a similar argument.

(d) $\Longrightarrow$ (b) is clear since for every $M \in \mathcal{B}({}_RK)$ ( $M \in \mathcal{B}(K_B)$) $\omega_M$ is a topological embedding.

<u>4.8</u>. In the situation 4.2, assume that ${}_RK$ is a cogenerator of $\mathcal{T}_\tau$ and $K_B$ is a cogenerator of $\mathcal{T}_\beta$.

Denote by $\overline{\mathcal{T}}_\tau$ the subclass of $\mathcal{B}({}_RK)$ consisting of the modules $(M,\mathcal{E})$ such that $M \in \mathcal{T}_\tau$ and $\mathcal{E} = \chi_M$.

Clearly $\overline{\mathcal{T}}_\tau \leq \mathcal{D}({}_RK)$. A module $(M,\mathcal{E}) \in \mathcal{D}({}_RK)$ belongs to $\overline{\mathcal{T}}_\tau$ iff $M \in \mathcal{T}_\tau$. ${}_RK$ with the discrete topology is in $\overline{\mathcal{T}}_\tau$.

Similar statements hold for the class $\overline{\mathcal{T}}_\beta$.

Let $M \in \mathcal{T}_\tau (M \in \mathcal{T}_\beta)$. Since every submodule of M is closed in $(M,\chi_M)$ , $\chi_M$ is equivalent to the discrete topology.

## 5. <u>Applications of the duality</u>

<u>5.1</u>. Throughout this section $(R,\tau)$ is a left l.c. ring, ${}_RK$ a fixed injective cogenerator with essential socle of $\mathcal{T}_\tau$, $B = \mathrm{End}({}_RK)$, $\beta$ the finite topology of B. Then Theorem 2.10 applies to $(R,\tau)$. The couple of functors $D_K = (D_1, D_2)$ has the same meaning as in Section 4.

<u>5.2</u>. Let $R_\tau$-$\mathcal{A}$ be a full subcategory of $R_\tau$-LT. Put :

$$R_\tau\text{-}\mathcal{A}_* = \{(M,\mathcal{E}_*) \ : \ (M,\mathcal{E}) \in R_\tau\text{-}\mathcal{A}\}$$

If $\mathcal{B}$-$B_\beta$ is a subcategory of LT-$B_\beta$ , $\mathcal{B}_*$-$B_\beta$ is defined in a similar way.

Let $R_\tau$-LC be the category of left l.c. modules over $(R,\tau)$

and $R_\tau$-SLC the subcategory of s.l.c. modules.
The meaning of LC-$B_\beta$ and SLC-$B_\beta$ is obvious.

5.3. Proposition.

a) Let $(M,\mathcal{E}) \in R_\tau$-LT. Then the weak topology of characters of $(M,\mathcal{E})$ coincides with $\mathcal{E}_*$ so that $(M,\mathcal{E})^* = (M,\mathcal{E}_*)^*$. Moreover $(M,\mathcal{E}_*) \in \mathcal{C}({}_R K)$.

b) $R_\tau\text{-SLC} \subseteq R_\tau\text{-LC}_* \subseteq \mathcal{C}({}_R K)$.

c) If $(R,\tau)$ is s.l.c. then $R_\tau$-LC = $R_\tau$-SLC.

Similar results hold for $(B,\beta)$.

Proof. a) and b) are obvious.

c) Let $M \in R_\tau$-LT , L an open submodule of M. Then $M/L \in \mathcal{G}_\tau$ and M/L is l.c.d. Since $(R,\tau)$ is s.l.c. every f.g. submodule of M/L is artinian. Then M/L is artinian by Lemma 1.4 of [13].

Our main purpose is to determine the image of $R_\tau$-LC (LC-$B_\beta$) under $D_K$. By 5.3 :

$$D_1(R_\tau\text{-LC}) = D_1(R_\tau\text{-LC}_*) \quad \text{and} \quad D_2(\text{LC-}B_\beta) = D_2(\text{LC}_*\text{-}B_\beta)$$

5.4. Let $(M,\mathcal{E}) \in R_\tau$-LC. It is known that, among all the topologies on M equivalent to $\mathcal{E}$ , there exists a finest one which will be denoted by $\mathcal{E}^*$. The existence of $\mathcal{E}^*$ was established by several authors ( [18], [1], [9]).

Following [18] the topology $\mathcal{E}^*$ has as a basis of neighbourhoods of O in M the closed submodules H of M such that M/H is l.c.d. Obviously $\mathcal{E}_* \leq \mathcal{E} \leq \mathcal{E}^*$. The topology $\tau^*$ is a ring topology. By Theorem 1.6 of [9], a basis of neighbourhoods of O in $(R,\tau^*)$ is given by $\{\text{Ann}_R(L)\}$ where L is a l.c.d. submodule of $K_B$. Moreover it is easy to show that every $I \in \mathcal{F}_{\tau^*}$ is of the form $\text{Ann}_R(L)$.

Let $\mathcal{G}(\tau^*)$ the subcategory of $\mathcal{B}({}_R K)$ defined as follows:

$$\mathcal{G}(\tau^*) = \{ M \in \mathcal{B}({}_R K) \ : \ \forall x \in M \quad \text{Ann}_R(x) \in \mathcal{F}_{\tau^*} \}$$

and define in a similar manner the subcategory $\mathcal{G}(\beta^*)$ of $\mathcal{B}(K_B)$.

5.5. Theorem. Under assumptions 5.1 we have :

a) $D_K$ induces a duality between $R_\tau$-$LC_*$ and $\mathcal{C}(\beta^*) \cap \mathcal{D}(K_B)$.

b) $D_K$ induces a duality between $\mathcal{C}(\tau^*) \cap \mathcal{D}({}_RK)$ and $LC_*$-$B_\beta$ .

Proof. a) Let $M \in R_\tau$-$LC_*$ , $\varphi \in M^*$ . Then $M\varphi$ is a l.c.d. submodule of ${}_RK$, hence $I = \mathrm{Ann}_B(M\varphi) \in \mathcal{F}_{\beta^*}$ and $\varphi I = 0$. It follows $M^* \in \mathcal{C}(\beta^*)$. On the other hand $M \in \mathcal{C}({}_RK)$ and thus, by 4.7, $M^* \in \mathcal{D}(K_B)$. Conversely, let F be a f.g. module in $\mathcal{C}(\beta^*) \cap \mathcal{D}(K_B)$. There exists an exact sequence

$$\bigoplus_{j=1}^{m} B/I_j \longrightarrow F \longrightarrow 0$$

with $I_j \in \mathcal{F}_{\beta^*}$. Applying $\mathrm{Hom}(\,-\,,K_B)$ we obtain an injection $\mathrm{Hom}_B(F,K_B) \hookrightarrow \bigoplus_{j=1}^{m} \mathrm{Hom}_B(B/I_j,K_B)$. Since F is f.g., $F^*$ is discrete so that it can be identified with a submodule of $\bigoplus_{j=1}^{m} \mathrm{Hom}_B(B/I_j,K_B)$. Since $B/I_j$ is cyclic, $\mathrm{Hom}_B(B/I_j,K_B) \cong \cong \mathrm{Ann}_K(I_j)$. For every j, there exists a l.c.d. submodule $L_j$ of ${}_RK$ such that $I_j = \mathrm{Ann}_B(L_j)$. Consequently $\mathrm{Hom}_B(B/I_j,K_B) \cong = L_j$ hence $F^* \leq \bigoplus_{j=1}^{m} L_j$, so that $F^*$ is l.c.d. Let $M \in \mathcal{C}(\beta^*) \cap \mathcal{D}(K_B)$ and let F be a f.g. submodule of M. Now $M^*/O(F)$ is topologically isomorphic to $F^*$ which is l.c.d. and, by 4.7, $M^* \in \mathcal{C}({}_RK)$. Thus $M^* \in R_\tau$-$LC_*$.

The proof of statement b) is analogous.

5.6. Corollary. In the situation 5.1 the following conditions are equivalent:

(a) $R_\tau$-$LC_* = \mathcal{C}({}_RK)$

(b) ${}_RK$ is l.c.d.

(c) $K_B$ is an injective cogenerator of Mod-B.

(d) $B_B$ is l.c.d.

Proof. (a)$\Longleftrightarrow$(b) follows from 5.3 since ${}_RK \in \mathcal{C}({}_RK)$.

(b)$\Longleftrightarrow$(c) and (b) $\Longrightarrow$(d) follow from Proposition 4.3 of [17] .

(d) $\Longrightarrow$(b). Let d be the discrete topology of $B_B$. Clearly $d = \beta^*$ so that there exists a l.c.d. submodule L of ${}_RK$ such that $\mathrm{Ann}_B(L) = 0$. Then $L = {}_RK$.

5.7. Denote by $\mathcal{N}_\tau$ the subcategory of $\mathcal{C}_\tau$ consisting of all submodules of f.g. modules belonging to $\mathcal{C}_\tau$ . $\mathcal{N}_\tau$ is a finitely closed subcategory of $\mathcal{C}_\tau$. Since $(R,\tau)$ is l.c. every object of $\mathcal{N}\mathcal{C}_\tau$ is l.c.d. Let $R_\tau$-FLC be the subcategory of $R_\tau$-LC consisting of all modules $M \in R_\tau$-LC such that for every open submodule H of M , $M/H \in \mathcal{N}\mathcal{C}_\tau$. The meaning of $\mathcal{N}\mathcal{C}_\beta$ and FLC-$B_\beta$ is clear. These categories have an interesting behaviour under $D_K$.

5.8. Proposition. In the situation 5.1 we have :

a) $D_K$ induces a duality between $R_\tau$-$FLC_*$ and $\bar{\mathcal{C}}_\beta$.

b) If $\tau = \tau_*$, $D_K$ induces a duality between $\bar{\mathcal{C}}_\tau$ and $FLC_*$-$B_\beta$.

Proof. a) Let $(M,\mathcal{E}) \in R_\tau$-$FLC_*$, $\varphi \in M^*$. Then $M\varphi$ is a submodule of ${}_RK$ and $M\varphi \in \mathcal{N}\mathcal{C}_\tau$. Since ${}_RK$ is injective in $\mathcal{C}_\tau$, $M\varphi$ is a submodule of a f.g. submodule of ${}_RK$. Hence $M\varphi \leq \sum_{i=1}^{n} Rx_i$, $x_i \in K$. Let $I = \bigcap_{i=1}^{n} \mathrm{Ann}_B(x_i)$. Then $I \in \mathfrak{F}_\beta$ and $\varphi I = 0$ therefore $\mathrm{Chom}_R(M, {}_RK) \in \mathcal{C}_\beta$. By 5.3 $(M,\mathcal{E})^* = (M,\mathcal{E}_*)^* \in \mathcal{C}({}_RK)$ hence, by 4.7, $(M,\mathcal{E})^* \in \mathcal{D}(K_B)$. Therefore $M^* \in \bar{\mathcal{C}}_\beta$.

Let $M \in \bar{\mathcal{C}}_\beta$. Then $M^* = \varprojlim M^*/O(F)$ where F runs over all f.g. submodules of M, the isomorphism being topological. On the other hand $M^*/O(F) \cong F^*$ as discrete modules. We now prove that $F^* \in \mathcal{N}\mathcal{C}_\tau$ from which the conclusion will follow. Suppose, without loss of generality, that F is cyclic : $F = xB$ , $x \in F$. Since $\beta$ is the K-topology of B there exists a f.g. submodule H of ${}_RK$ such that $\mathrm{Ann}_B(x) \supseteq \mathrm{Ann}_B(H)$. Then we have a natural surjection $B/\mathrm{Ann}_B(H) \longrightarrow B/\mathrm{Ann}_B(x)$ and hence an injection $(xB)^* \hookrightarrow (B/\mathrm{Ann}_B(H))^* = \mathrm{Ann}_K \mathrm{Ann}_B(H) = H$. Since H is f.g. and $H \in \mathcal{C}_\tau$, $F^* = (xB)^* \in \mathcal{N}\mathcal{C}_\tau$.

b) If $\tau = \tau_*$ the previous argument works.

The couple $D_K$ enables us to characterize s.l.c. rings $(R,\tau)$ by an explicit description of the duals of the objects of $\bar{\mathcal{C}}_\tau$. Let $R_\tau$-NLC the subcategory of all topologically noetherian modules of $R_\tau$-LC. The meaning on NLC-$B_\beta$ is clear.

5.9. Theorem. In the situation 5.1 the following conditions are equivalent:

(a) $(R,\tau)$ is s.l.c.

(b) $\tau = \tau_*$ and $(B, \beta)$ is topologically noetherian (and right l.c.).

(c) For every f.g. module $F \in \mathcal{C}_\tau$ there exists $n \in \mathbb{N}$ such that $F \hookrightarrow {}_RK^n$.

(d) For every $I \in \mathfrak{F}_\tau$ there exists a finite subset F of K such that $I = \mathrm{Ann}_R(F)$.

(e) For every $M \in \overline{\mathcal{C}}_\tau$, $M^*$ is l.c. and its discrete quotients are f.g.

(f) $D_K$ induces a duality between $\overline{\mathcal{C}}_\tau$ and $NLC_*$-$B_\beta$.

Proof. (a)$\Longleftrightarrow$(b) by 2.2 and 2.13.

(a)$\Longrightarrow$(c). F is artinian, hence $F \hookrightarrow \bigoplus_{i=1}^{n} E_\tau(V_{\gamma_i})$. Thus $F \hookrightarrow {}_RK^n$.

(c)$\Longrightarrow$(d). $R/I \hookrightarrow {}_RK^n$, hence $I = \bigcap_{i=1}^{n} \mathrm{Ann}_R(x_i)$ where $(x_1,\dots,x_n)$ is the image of $1+I$ in $K^n$.

(d)$\Longrightarrow$(a). It is well known that a module is artinian iff each one of its homomorphic images is f.c. Thus it is enough to show that for every $I \in \mathfrak{F}_\tau$, R/I is f.c. It is $I = \bigcap_{i=1}^{n} \mathrm{Ann}_R(x_i)$, $x_i \in K$. Then $R/I \hookrightarrow \bigoplus_{i=1}^{n} Rx_i \leq K^n$. Hence Soc(R/I) is essential in R/I and, since R/I is l.c.d., Soc(R/I) is f.g.

(a)$\Longrightarrow$(f). By 5.8 $D_K$ induces a duality between $\overline{\mathcal{C}}_\tau$ and $FLC_*$-$B_\beta$. Since (a)$\Longleftrightarrow$(b), $FLC_*$-$B_\beta$ = $NLC_*$-$B_\beta$.

(f)$\Longrightarrow$(e) is obvious.

(e)$\Longrightarrow$(d). Let $I \in \mathfrak{F}_\tau$. Then $(R/I)^* = \mathrm{Hom}_R(R/I, K_B) \cong \mathrm{Ann}_K(I)$ which is f.g. by assumption. Now $I = \mathrm{Ann}_R\mathrm{Ann}_K(I)$ and $\mathrm{Ann}_R(I) = \sum_{i=1}^{n} x_iB$, $x_i \in K$. Thus $I = \mathrm{Ann}_R(\{x_1,\dots,x_n\})$.

5.10. A module $M \in R_\tau$-LC is absolutely l.c. (a.l.c.) if for every open submodule H of M, M/H has finite length. Note that if R is commutative every s.l.c. module is a.l.c.
Denote by $R_\tau$-ALC the category of a.l.c. modules.

Suppose that $(R, \tau)$ is a.l.c. Then :
$R_\tau$-LC = $R_\tau$-$LC_*$ = $R_\tau$-SLC since a.l.c. implies s.l.c. and
$R_\tau$-ALC = $R_\tau$-FLC = $R_\tau$-NLC.

The category $R_\tau$-ALC, where $(R, \tau)$ is a.l.c., was studied by Gabriel (cf. [5] page 395). Gabriel's results on $R_\tau$-ALC can be easily derived from ours. See [DO].

5.11. Remark. It can be shown that the Gabriel-Oberst duality, applied to the Grothendieck category $\mathcal{C}_\tau$ of a left s.l.c. ring $(R,\tau)$, maps $\mathcal{C}_\tau$ onto SNLC-$B_\beta$. $M \in$ SNLC-$B_\beta$ if $M \in$ NLC-$B_\beta$ and M is strict i.e. : if H is a closed submodule of M and $M/H \in \mathcal{K}_\beta$, then H is open in M. See [14].

## 6. An example

The existence of s.l.c. rings which are not a.l.c. was proved by Leptin (see [7] page 298). Nevertheless we think that the following example has some interest. In particular it solves Problem 2 of [12].

6.1. Let R be a commutative, local noetherian ring with maximal ideal $\mathfrak{m}$. Suppose that R is not artinian, equicharacteristic and complete in its $\mathfrak{m}$-adic topology. Put $k = R/\mathfrak{m}$. Then $R = k \oplus \mathfrak{m}$, so that every $r \in R$ may be written, in a unique way, in the form

$$r = r_1 + r_2 \qquad (r_1 \in k \ , \ r_2 \in \mathfrak{m}) \tag{1}$$

Let H be the injective hull in R-Mod of the simple module $R/\mathfrak{m}$ and consider the bimodule ${}_RH_R$ where R acts on the left as the endomorphism ring of H, on the right by means of $xr = xr_1$ $(x \in H, r \in R)$.

Let $\overline{R} = H \oplus 1R$ be the trivial extension of H by R. $\overline{R}$ is the ring consisting of the couples $(x,r)$, $x \in H, r \in R$, where the addition is defined pointwise and the multiplication is given by $(x,r)(y,s) = (ry+xs, rs)$; $\overline{R}$ is non commutative with identity $(0,1)$. Set :

$$\overline{H} = \{(x,0) : x \in H\}, \qquad \overline{\mathfrak{m}} = \{(0,r) : r \in \mathfrak{m}\}.$$

$\overline{H}$ is a two-sided ideal and $\overline{\mathfrak{m}}$ is a left ideal of $\overline{R}$. Note that $\overline{R}/\overline{H} \cong R$, hence R is an $\overline{R}$-module. Clearly $J(\overline{R}) \cong \overline{H} \oplus \overline{\mathfrak{m}}$ and $\overline{R}$ is a local ring.

a) For every $n \in \mathbb{N}$ the cyclic $\bar{R}$-module $\bar{R}/\bar{\mathfrak{m}}^n$ is artinian not noetherian.

Indeed $J(\bar{R})/\bar{\mathfrak{m}}^n \cong (\bar{H} \oplus \bar{\mathfrak{m}})/\bar{\mathfrak{m}}^n \cong \bar{H} \oplus \bar{\mathfrak{m}}/\bar{\mathfrak{m}}^n$, and ${}_{\bar{R}}\bar{H}$ is artinian not noetherian, while $\bar{\mathfrak{m}}/\bar{\mathfrak{m}}^n$ has finite length. Endow $\bar{R}$ with its $\bar{\mathfrak{m}}$-adic topology $\tau$. It is easily checked that $\tau$ is a Hausdorff ring topology. $(\bar{R},\tau)$ is complete since $\tau$ coincides with the product topology of the discrete topology on H by the $\mathfrak{m}$-adic topology on R. By a) $(R,\tau)$ is s.l.c. not a.l.c.

6.2. The module ${}_{\bar{R}}W$. Put ${}_{\bar{R}}W = E_{\tau}(V)$, where V is the unique simple $\bar{R}$-module. $V \cong R/\mathfrak{m}$ and H is the injective hull of V in R-Mod. Consider the module ${}_{\bar{R}}E = \mathrm{Hom}_R(\bar{R},H)$. Since ${}_R H$ is injective, ${}_{\bar{R}}E$ is injective in $\bar{R}$-Mod. Using the following isomorphisms of R-modules

$$\mathrm{Hom}_R(\bar{R},H) \cong \mathrm{Hom}_R(H \oplus R,H) \cong R \oplus H$$

it can be shown that the elements of ${}_{\bar{R}}E$ are the couples $[r,h]$ $r \in R, h \in H$ with the scalar multiplication

$$(y,b)[r,h] = [b_1 r,\ ry+bh]$$

where $b=b_1+b_2$ according to (1) of 6.1. It follows that ${}_{\bar{R}}E$ is $\tau$-torsion. Since ${}_{\bar{R}}E$ contains a copy of V as a $\bar{R}$-submodule, we see that ${}_{\bar{R}}W$ is the minimal (injective) cogenerator in $\bar{R}$-Mod.

Thus $\bar{R}$ is a left pseudo-artinian ring which is not strongly pseudo-artinian. This gives a negative answer to Problem 2 of [12].

Remark. For details and more informations about $\bar{R}$, see [DO].

## 7. The cobasic ring, the grade and the basic ring of a l.c. ring

7.1. Let $(R,\tau)$ be a left l.c. ring, ${}_RW = E_{\tau}(\bigoplus_{\gamma\in\Gamma} V_\gamma)$, $D_\gamma = \mathrm{End}_R(V_\gamma)$, $A = \mathrm{End}({}_RW)$, $\sigma$ the finite topology of A. Then $(A,\sigma)$ is right l.c., $\sigma = \sigma_*$, $W_A = E_\sigma(\bigoplus_{\gamma\in\Gamma} D_\gamma^{(\nu_\gamma)})$ and

$A/J(A) \cong \prod_{\gamma\in\Gamma} D_\gamma$. Set $\nu_\Gamma = (\nu_\gamma)_{\gamma\in\Gamma}$.

Definition. The right l.c. ring $(A,\sigma)$ will be called the cobasic ring of $(R,\tau)$ and $\nu_\Gamma$ the grade of $(R,\tau)$.

7.2. Lemma. In the situation 7.1, let $\tau_*$ be the Leptin topology of $(R,\tau)$ and ${}_RK$ the minimal injective cogenerator of $\mathcal{C}_{\tau_*}$. Then, in R-Mod, ${}_RK \cong {}_RW$.

Proof. Since $\tau$ and $\tau_*$ are equivalent, $\mathcal{F}_\tau$ and $\mathcal{F}_{\tau_*}$ have the same closed left maximal ideals. Since such ideals are open, $(V_\gamma)_{\gamma\in\Gamma}$ is a system of representatives of the non isomorphic simple modules in $\mathcal{C}_{\tau_*}$. Put $E = E(\bigoplus_{\gamma\in\Gamma} V_\gamma)$ in R-Mod and recall that ${}_RW$ coincides with the $\sigma$-torsion part $t_\tau(E)$ of E. Then:

$${}_RK = t_{\tau_*}(E) \leq t_\tau(E) = {}_RW$$

On the other hand, by Lemma 2.2, the ${}_RW$-topology of R coincides with $\tau_*$, so that ${}_RW$ is $\tau_*$-torsion. Thus ${}_RK = {}_RW$.

An important consequence of this lemma is that $(R,\tau)$ and $(R,\tau_*)$ have the same cobasic ring and the same grade.

Our purpose is to show that $(R,\tau_*)$ is uniquely determined, up to topological isomorphisms, by its cobasic ring and its grade.

The following two lemmata are easily established. For a proof see [DO].

7.3. Lemma. Let $(A,\sigma)$ be a right l.c. ring, with $\sigma = \sigma_*$, such that $A/J(A) \cong \prod_{\gamma\in\Gamma} D_\gamma$ where the $D_\gamma$'s are division rings. Then the $D_\gamma$'s, as simple right A-modules, are a system of representatives of the non isomorphic simple modules in $\mathcal{C}_\sigma$.

7.4. Lemma. Let $M \in$ R-Mod be a semisimple module whose endomorphism ring is a cartesian product of division rings. Then M is a direct sum of non isomorphic simple modules.

7.5. Theorem: Let $(A,\sigma)$ be a right l.c. ring such that $\sigma = \sigma_*$ and $A/J(A) = \prod_{\gamma\in\Gamma} D_\gamma$ where the $D_\gamma$'s are division rings. Let $\nu_\Gamma = (\nu_\gamma)_{\gamma\in\Gamma}$ be a $\Gamma$-tuple of cardinal numbers $\geq 1$. Then there exists a left l.c. ring $(R,\tau)$, with $\tau = \tau_*$, having grade

$\nu_\Gamma$ <u>and cobasic ring $(A, \sigma)$. Such $(R, \tau)$ is unique up to topological isomorphisms.</u>

<u>Proof</u>. Consider the right A-module $W_A = E_\sigma(\bigoplus_{\gamma\in\Gamma} D_\gamma^{(\nu_\gamma)})$. By 7.3 $W_A$ is an injective cogenerator of $\mathcal{C}_\sigma$. Clearly $Soc(W_A)$ is essential in $W_A$. By 2.2, $\sigma = \sigma_*$ coincides with the W-topology of A. Put $R = End(W_A)$ and let $\tau$ be the W-topology of R.

By 2.10 and 2.2 we have :

1) $(R, \tau)$ is left l.c. ; 2) ${}_RW_A$ is faithfully balanced and ${}_RW$ is an injective cogenerator of $\mathcal{C}_\tau$ ; 3) $Soc({}_RW) = Soc(W_A)$ and $Soc({}_RW)$ is essential in ${}_RW$ ; 4) $\tau = \tau_*$ .

By 2) ${}_RW$ has a submodule ${}_RK$ which is isomorphic to the minimal injective cogenerator of $\mathcal{C}_\tau$. Let us prove that ${}_RK = {}_RW$.

By 2.7 , $End_R(Soc({}_RW)) \cong A/J(A) = \prod_{\gamma\in\Gamma} D_\gamma$. Thus, by 7.4, $Soc({}_RW)$ is a direct sum of non isomorphic simple modules.

Since $Soc({}_RK) = {}_RK \cap Soc({}_RW)$ it follows $Soc({}_RK) = Soc({}_RW)$.

By 3) we have the essential inclusions $Soc({}_RW) \leq {}_RK \leq {}_RW$ . Then, since ${}_RK$ is a direct summand of ${}_RW$, we have ${}_RK = {}_RW$.

Now ${}_RW = E_\tau(\bigoplus_{\lambda\in\Lambda} V_\lambda)$ where $(V_\lambda)_{\lambda\in\Lambda}$ is a system of representatives of the non isomorphic simple modules in $\mathcal{C}_\tau$ . Putting $S_\lambda = End_R(V_\lambda)$, we have $V_\lambda = S_\lambda^{(\mu_\lambda)}$ where the $\mu_\lambda$'s are suitable cardinal numbers and the $S_\lambda$'s are non isomorphic simple right A-modules. Therefore $Soc({}_RW) = Soc(W_A) = \bigoplus_{\lambda\in\Lambda} S_\lambda^{(\mu_\lambda)} = \bigoplus_{\gamma\in\Gamma} D_\gamma^{(\nu_\gamma)}$, where $S_\lambda^{(\mu_\lambda)}$ and $D_\gamma^{(\nu_\gamma)}$ are isotypical components. Then, up to a bijection, $S_\gamma^{(\mu_\gamma)} = D_\gamma^{(\nu_\gamma)} = V_\gamma$. It follows that $S_\gamma = D_\gamma$ and $\mu_\gamma = \nu_\gamma$ for every $\gamma \in \Gamma$ . It is now proved that $(R, \tau)$ is a left l.c. ring, with $\tau = \tau_*$ , having grade $\nu_\Gamma = (\nu_\gamma)_{\gamma\in\Gamma}$ and cobasic ring $(A, \sigma)$.

Finally, let $(R, \tau)$ and $(R', \tau')$ be left l.c. rings such that $\tau = \tau_*$ and $\tau' = \tau'_*$ . Suppose that they have the same grade and that the respective cobasic rings are topologically isomorphic. Let ${}_RW, {}_{R'}W$ be the minimal injective cogenerators of $\mathcal{C}_\tau$ and $\mathcal{C}_{\tau'}$ respectively. Then :

$W_A \cong E_\sigma(\bigoplus_\gamma D_\gamma^{(\nu_\gamma)})$ and $A/J(A) \cong \prod D_\gamma$ ; $W_{A'} \cong E_{\sigma'}(\bigoplus_\gamma D'^{(\nu_\gamma)}_\gamma)$ and $A'/J(A') \cong \prod_\gamma D'_\gamma$ .

Since $(A,\sigma)$ is topologically isomorphic to $(A',\sigma')$, $D_\gamma \cong D'_\gamma$ $\forall \gamma \in \Gamma$. Let $\varphi : A \longrightarrow A'$ be a topological isomorphism : $\varphi$ induces a semilinear isomorphism of $W_A$ onto $W_{A'}$. By the topological isomorphisms $(R,\tau) \cong \mathrm{End}(W_A)$, $(R',\tau') \cong \mathrm{End}(W_{A'})$ we obtain that $(R,\tau)$ and $(R',\tau')$ are topologically isomorphic.

<u>7.6</u>. In the situation 7.1, let $(B,\beta)$ be the cobasic ring of $(A,\sigma)$. Then $(B,\beta)$ is left l.c. and $\beta = \beta_*$. Clearly :

$$B/J(B) \cong \prod_\gamma D_\gamma \cong A/J(A) .$$

The ring $(B,\beta)$ will be called the <u>basic ring</u> of $(R,\tau)$. It can be shown that there exists an idempotent $e \in R$ such that $B$ may be identified with $eRe$ and $\beta$ with the relative topology of $\tau_*$. For more informations about $(B,\beta)$ see [DO].

<u>7.7</u>. <u>Remark</u>. The above results show a strong analogy between l.c. rings and semiperfect abstract rings (note that if $(R,\tau)$ is left l.c. then the idempotents of $R/J(R)$ can be lifted and $R/J(R)$ is a generalized Artin-Wedderburn ring). We think this analogy could be a starting point for further research.

## <u>R E F E R E N C E S</u>

1. Pham Ngoc Ánh , "<u>Duality of Modules over topological rings</u>", J. of Algebra <u>75</u> (1982) 395-425.

2. D. Dikranjan-A. Orsatti , "<u>Sugli anelli linearmente compatti</u>", to appear in Atti del Convegno di Topologia at L'Aquila 1983 in Rendiconti del Circolo Matematico di Palermo.

3. D. Dikranjan-W Wieslaw , "<u>Rings with only ideal topologies</u>", Com. Univ. San Pauli, <u>29</u>, n.2(1980) 157-167.

4. C. Faith , "<u>Algebra II. Ring Theory</u>" , Berlin , 1976.

5. P. Gabriel, "Des categories abeliennes",Bull. Soc. Math. France, 90, (1962) 323-448.

6. H. Leptin , "Linear Kompakten Ringe und Moduln", Math. Z. 62 (1955) 241-267.

7. H. Leptin , "Linear Kompakten Ringe und Moduln II", Math. Z. 66 (1957) 289-327.

8. C. Menini , "Linearly Compact Rings and Strongly Quasi-Injective Modules", Rend. Sem. Mat. Univ. Padova 65 (1980) 251-262.

9. C. Menini , "Linearly Compact Rings and Selfcogenerators", to appear in Rend. Sem. Mat. Univ. Padova.

10. C. Menini-A. Orsatti , "Good Dualities and Strongly Quasi-Injective Modules", Ann. Mat. Pura e Appl. 127 (1981) , 182-230.

11. C. Menini-A. Orsatti , "Dualities between Categories of Topological Modules" , Communications in Algebra, 11 (1) , (1983) , 21-66.

12. C. Menini-A. Orsatti , "Topologically Left Artinian Rings", to appear in J. of Algebra.

13. B. Müller , "Linear Compactness and Morita Duality" , J. of Algebra 16 , (1970) , 60-66.

14. U. Oberst , "Duality Theory for Grothendieck categories and Linearly compact rings", J. of Algebra 15 (1970) 473-542.

15. A. Orsatti , "Dualità per alcune classi di moduli E-compatti", Ann. Mat. Pura e Appl. 113 (1977) 211-235.

16. A. Orsatti, "Anelli linearmente compatti e teoremi di Leptin", Bollettino UMI (6) 1-A (1982) 331-357.

17. A. Orsatti-V. Roselli , "A Characterization of Discrete Linearly Compact Rings by Means of a Duality", Rend. Sem. Mat. Univ. Padova 64 (1981) 219-234.

18. S. Warner , "Linearly Compact Rings and Modules", Math. Ann. 197 , (1972) 29-43.

19. D. Zelinsky , "Linearly Compact Modules and Rings" , Am. J. Math. 75 (1953) 79-90.

# DIRECT SUM CANCELLATION OVER NOETHERIAN RINGS

ROGER WIEGAND
UNIVERSITY OF NEBRASKA

## INTRODUCTION

Suppose X, Y and Z are modules over a commutative ring R and that $X \oplus Z \cong Y \oplus Z$. This cancellation problem is to determine conditions under which one can conclude that $X \cong Y$. There are many counterexamples, some interesting, some boring, involving modules that aren't finitely generated. (For an example of the second sort, take $X = 0$, $Y = R$, $Z$ = free module of infinite rank.) In this paper we will assume that <u>all modules are finitely generated</u>. Further, <u>all rings are assumed to be commutative and Noetherian</u>.

Under these restrictions there are no boring counterexamples to cancellation. The best known example comes from topology. Take $R = \mathbb{R}[x,y,z]$, with $x^2 + y^2 + z^2 = 1$, the coordinate ring of the real 2-sphere, and let $M$ be the kernel of the map $\phi = [x,y,z] : R^3 \to R$. Then $\phi$ is split by $\begin{bmatrix} x \\ y \\ z \end{bmatrix}$, so $M \oplus R = R^{(3)}$. On the other hand, $M \not\cong R^{(2)}$, since a basis vector $\begin{bmatrix} f \\ g \\ h \end{bmatrix}$ would be non-vanishing and everywhere orthogonal to $\begin{bmatrix} x \\ y \\ z \end{bmatrix}$, that is, tangent to the sphere. It is a standard (but by no means easy) theorem of algebraic topology that the sphere has no non-vanishing tangent vector fields.

Most research on the cancellation problem has dealt with projective modules. The original impetus for this research was Serre's 1957 paper, [Se], in which he raised the question ( often referred to as Serre's conjecture): If $R$ is the polynomial ring in $n$ variables over a field, is every projective R-module free? Serre proved that every projective module (in this case) is stably free, that is, $P \oplus \text{free} = \text{free}$, so Serre's problem amounts to a cancellation problem. Three decades of research have produced lots of important results on projective cancellation. One such result, obtained shortly before the solution of Serre's problem in 1976, will play a role later in this paper.

0.1 Theorem: (Murthy & Swan [MS]): Let $R$ be an integral domain of Krull dimension 2, finitely generated as an algebra over an algebraically closed field. If $X$, $Y$, $Z$ are projective modules such that $X \oplus Z \cong Y \oplus Z$, then $X \cong Y$.

Another well-known result we'll need is local cancellation.

0.2 Theorem: Let $X$, $Y$, $Z$ be modules over a semilocal ring, and suppose $X \oplus Z \cong Y \oplus Z$. Then $X \cong Y$.

The crux of the proof is a theorem due to Grothendieck, [EGA,2.5.8], which states that if X and Y are modules over a semilocal ring R, and $S\otimes_R X \cong S\otimes_R Y$ for some faithfully flat ring extension $R \to S$, then $X \cong Y$. For a semilocal ring R with maximal ideals $M_1,\ldots,M_t$ one applies Grothendieck's result to the extensions $R \to R_{M_1} \times \cdots \times R_{M_t}$, and $R_{M_i} \to (R_{M_i})^\wedge$, thus reducing the proof of (0.2) to the case of a complete local ring. But in this case the Krull-Schmidt Theorem is valid [Swl, Part II,2.22], and cancellation follows easily. This approach was used by Vasconcelos, [V]. An alternate method, due to Evans, [E], utilizes stable range conditions on the endomorphism ring of Z. This method was used by Goodearl and Warfield, [GW], to get cancellation results over non-Noetherian rings.

## DIMENSION ONE

In this section R is a one-dimensional ring with no non-zero nilpotent elements. Serre proved in [Se] that every projective module M of rank n is of the form $R^{(n-1)} \oplus I$, where I is an invertible ideal. Moreover, I is determined, up to isomorphism, by M. It follows easily that cancellation is valid for projective modules. The question of whether cancellation holds in general was probably first raised by Eisenbud and Evans in [EE]. My guess is that for some time the answer has been known but considered uninteresting by the algebraic number-theorists. For algebraists, the opposite appears to be true.

Before we get to our main results, we need to establish some notation and conventions. The total quotient ring K is a finite product of fields, and the integral closure $\tilde{R}$ of R in K is a finite product of Dedekind domains. We always assume that $\tilde{R}$ *is finitely generated as an R-module.* This means that the conductor ideal $\underline{c} = \{r \in R \mid r\tilde{R} \subseteq R\}$ contains a non-zerodivisor. Since R is one-dimensional $R/\underline{c}$ is Artinian, thus a finite product of Artinian local rings. We have a Cartesian square (pullback diagram)

(1.1)
$$\begin{array}{ccc} R & \hookrightarrow & \tilde{R} \\ \downarrow & & \downarrow \\ R/\underline{c} & \hookrightarrow & \tilde{R}/\underline{c} \end{array}$$

Now cancellation holds over $\tilde{R}$, $R/\underline{c}$, and $\tilde{R}/\underline{c}$, and it seems reasonable to try to use the pullback diagram to produce cancellation over R. The first difficulty is that (1.1) doesn't yield a Cartesian square when it's tensored with an arbitrary R-module. For this reason we restrict our attention to <u>torsionfree</u> R-modules. (R-module A is torsionfree provided $ra = 0 \Rightarrow a = 0$ or $r$ is a zerodivisor in R.) Even then, $\tilde{R}\otimes_R A$ may have torsion, so we work instead with $\tilde{R}A = (\tilde{R}\otimes_R A)/\text{torsion}$. Then we obtain, from (1.1), a Cartesian square (called the <u>standard pullback</u> for A)

(1.2)
$$\begin{array}{ccc} A & \hookrightarrow & \tilde{R}A \\ \downarrow & & \downarrow \pi \\ A/\underline{c}A & \underset{j}{\hookrightarrow} & \dfrac{\tilde{R}A}{\underline{c}A} \end{array}$$

In [W] I approached the cancellation problem obliquely, by perturbing A as follows: Let $\phi$ be an automorphism of $RA/\underline{c}A$, and form a new pullback

(1.3)
$$\begin{array}{ccccc} B & & \hookrightarrow & & \tilde{R}A \\ \downarrow & & & & \downarrow \pi \\ \dfrac{A}{\underline{c}A} & \underset{j}{\hookrightarrow} & \dfrac{\tilde{R}A}{\underline{c}A} & \underset{\phi}{\longrightarrow} & \dfrac{\tilde{R}A}{\underline{c}A} \end{array}$$

Then B is a torsionfree $\tilde{R}$-module, and (1.3) is naturally isomorphic to the standard pullback for B. Now the important point is that <u>if det $\phi$ lifts to a unit of $\tilde{R}$ we can lift $\phi$ to an automorphism of $\tilde{R}A$, and in this case</u> $B \cong A$. Thus we get a well-defined action of the group

$$E(R) = \text{coker}\,(\tilde{R}^* \to \tilde{R}/\underline{c}^*),\ (\ )^* = \text{group of units}$$

on {isomorphism classes of torsion-free modules}. The following theorem from [W] shows the relevance of this action to the cancellation problem.

<u>1.4 Theorem:</u> These four conditions are equivalent, for torsionfree modules A and B :

a) $A^x \cong B$ for some $x \in E(R)$.

b) $\tilde{R}A \cong \tilde{R}B$ and $A_M \cong B_M$ for every maximal ideal $M$ of R.

c) $A \oplus C \cong B \oplus C$ for some R-module C.

d) $A \oplus \tilde{R} \cong B \oplus \tilde{R}$.

What's missing from the theorem is a criterion for deciding whether or not $A^x \cong A$. When A is a rank-one projective, the answer lies in the Mayer-Vietoris exact sequence, [Mi], associated to (1.1):

$$0 \to R^* \to R/\underline{c}^* \times \tilde{R}^* \xrightarrow{\alpha} \tilde{R}/\underline{c}^* \xrightarrow{\beta} \text{Pic}R \xrightarrow{\gamma} \text{Pic}\tilde{R} \to 0$$

Here $\alpha$ is the multiplication map $(u,v) \to (\text{image of } u^{-1})(\text{image of } v)$. The map $\beta$ can be interpreted in terms of our group action: If

$y \in \tilde{R}/\underline{c}^*$ and $x$ is the image of $y$ in $E(R)$, then $\beta(y) = R^x$. Thus $R^x \cong R$ if and only if $y$ is in the image of $\alpha$. Letting

$$D(R) = \ker(\mathrm{Pic}R \to \mathrm{Pic}\tilde{R}) = \mathrm{coker}\ \alpha ,$$

we have

1.5 <u>Corollary</u>: (a) If $E(R) = 0$, then $R$ has torsionfree cancellation (i.e., if $X$ and $Y$ are torsionfree $R$-modules and $X \oplus Z \cong Y \oplus Z$, then $X \cong Y$).

(b) If $D(R) \neq 0$, there is an invertible ideal $I$ of $R$ such that $I \oplus \tilde{R} \cong R \oplus \tilde{R}$ but $I \not\cong R$.

This provides lots of examples of failure of cancellation. Take, for example, $R = k[t^2,t^3]$, where $k$ is a field. Then $\mathrm{Pic}R$ is isomorphic to the additive group of the field; but $\tilde{R} = k[t]$, so $\mathrm{Pic}\tilde{R} = 0$. Therefore $D(R) \neq 0$, and torsionfree cancellation fails for $R$-modules. This example is typical rather than exceptional, as shown by the following sharpening of [W, Theorem 3.2 ]:

1.6 <u>Theorem</u>: Let $R$ be a one-dimensional integral domain, finitely generated as an algebra over an infinite field $k$, let $K$ be the quotient field of $R$, and let $\bar{k}$ be the algebraic closure of $k$ in $K$. If $\bar{k}R$ is properly contained in $\tilde{R}$, then $D(R)$ is not finitely generated; hence, torsionfree cancellation fails over $R$.

It follows easily from the Mayer-Vietoris sequence associated to the conductor square for $R \hookrightarrow \bar{k}R$, that $D(R)$ maps onto $D(\bar{k}R)$. Therefore we may assume that $k = \bar{k}$. With this reduction, the proof follows immediately from two lemmas, which say (1) that the cokernel of the horizontal map $R/\underline{c}^* \to \tilde{R}/\underline{c}^*$ is not finitely generated, and (2) that $\tilde{R}^*/k^*$ is finitely generated. The first lemma and its proof are essentially taken from [W]; while the second lemma, obtained in conversations with Bill Heinzer, replaces the geometric part of the proof of [W, Theorem 3.2].

1.7 Lemma: Let $A \subseteq B$ be an integral extension of Artinian rings, and assume each local ring of $A$ is infinite. If $B^*/A^*$ is a finitely generated group, then $A = B$.

Proof. We may assume $A$ is local, with maximal ideal $M$, and that $B$ is a finitely generated $A$-module. Suppose first that $M$ is equal to the Jacobson radical of $B$. Then $B^*/A^* = (B^*/(1+M))/(A^*/(1+M)) = (B/M)^*/(A/M)^*$. Let $K = A/M$, an infinite field, and $E = B/M$, a semisimple $K$-algebra. If $A \neq B$, then either $E$ is a product of $t \geq 2$ fields, each containing $K$, or else $E$ is a proper finite-dimensional field extension of $K$. In the first case $E^*/K^*$ has a subgroup isomorphic to $K^{*(t-1)}$, which is not finitely generated. In the second case we know, [DM, (3.6)], that $E^*/K^*$ either has infinite rank or is a torsion group. But if $E^*/K^*$ is torsion, it can't be finitely generated, since it's infinite: If $x \in E-K$ the elements $x + \alpha$, $\alpha \in K$, are all in distinct cosets modulo $K^*$.

In the other case ($\mathcal{M} \neq J$ = Jacobson radical of B), there is an element $f \in J$ such that $f \notin \underline{c}$ (the conductor ideal) but $f^2 \in \underline{c}$. There is an epimorphism

$$\frac{1+Bf}{(1+Bf)\cap A^*} \twoheadrightarrow \frac{Bf}{Bf\cap A}$$

taking the coset of $1 + bf$ to the coset of $bf$. Now $Bf/(Bf\cap A)$ is a non-zero A-module, so it maps onto $A/\mathcal{M}$, whose additive group is not finitely generated.

1.8 Lemma: Let k be a field, let D be a Dedekind domain finitely generated as a k-algebra, and let K be the quotient field of D. If k is algebraically closed in K, then $D^*/k^*$ is a finitely generated group.

Proof: Let $\mathcal{V}$ be the set of all discrete valuation rings between k and K. Each element of K is in all but finitely many members of $\mathcal{V}$, and since D is finitely generated, D is contained in all but finitely many members $V_1,\ldots,V_n$ in $\mathcal{V}$. Since k is algebraically closed in K we know that $k = D \cap V_1 \cap \ldots \cap V_n$, so it will suffice to show that $E^*/(E\cap V)^*$ is cyclic, when E is a ring between k and K and $V \in \mathcal{V}$. Let v be the valuation of K with valuation ring V. If $v(x) = 0$ for all $x \in E^*$ then $E^* = (E\cap V)^*$. Otherwise, choose an element $x \in E^*$ of smallest positive value. One checks directly that $E^*/(E\cap V)^*$ is generated by the coset of x.

For general one-dimensional rings, the Picard group doesn't tell the whole story. There is an example in [W] of a one-dimensional ring R over which every projective module is free, but for which torsionfree cancellation fails. In fact, R has two non-isomorphic modules A, B, of rank 2, such that $A \oplus R \cong B \oplus R$. The details are carried out in [W,2.8] and [LW,6.8], but this is the general idea: By Theorem 1.4, B is obtained from A by a pullback like (1.3). Now the <u>determinant</u> of $\phi$ is a product of a unit that lifts to $\tilde{R}^*$ and one that lifts to $\tilde{R}/\underline{c}^*$, but $\phi$ itself can't be factored into two liftable automorphisms. The problem is that $A/\underline{c}A$ doesn't have enough automorphisms.

The difficulty encountered above doesn't occur if A has a faithful ideal as a direct summand. This leads to consideration of those rings for which every faithful torsionfree module has a direct summand isomorphic to a faithful ideal. These rings were shown in [W] to be just the 2-generator rings studied by Bass in [B].

1.9 <u>Theorem</u> ([B],[W]): Let R be a one-dimensional ring with no nonzero nilpotents, and assume $\tilde{R}$ is finitely generated over R. These conditions are equivalent:

(a) Every faithful torsionfree R-module has a faithful ideal as a direct summand.

(b) Every ideal of R can be generated by 2 elements.

(c) Every local ring of R has multiplicity $\leq 2$.

The rings satisfying these conditions were termed <u>Bass rings</u> in

[LW]. For Bass rings, we have the following results on cancellation ([W,2.7] and [LW,6.2]):

1.10 <u>Theorem</u>: Let R be a Bass ring, and let X, Y, and Z be torsionfree R-modules such that $X \oplus Z \cong Y \oplus Z$. If either (i) $D(R) = 0$ or (ii) Z is projective, then $X \cong Y$.

Quadratic orders over $\mathbb{Z}$ provide interesting examples of Bass rings. These are the rings $R(d,n) = \mathbb{Z}[n\omega]$, where

d is a squarefree integer $\neq 1$,

n is a positive integer, and

$$\omega = \begin{cases} \sqrt{d}, & \text{if } d \equiv 2 \text{ or } 3 \pmod 4 \\ \frac{1}{2}(1+\sqrt{d}), & \text{if } d \equiv 1 \pmod 4 \end{cases}$$

We know, by (1.5) and (1.10), that R(d,n) has torsionfree cancellation if and only if $D(R(d,n)) = 0$. In [W] I developed a systematic approach for determining which quadratic orders have torsionfree cancellation. When $d < 0$, the answer is rather simple ([W,4.5]):

1.11 <u>Theorem</u>: Let d be a square-free negative integer, and let n be an integer greater than 1 (so that $R(d,n) \neq \widetilde{R(d,n)}$). Then R(d,n) has torsionfree cancellation if and only if one of the following holds:

(a) $n = 2$ and $d \equiv 1 \pmod 8$, i.e., $R = \mathbb{Z}[\sqrt{-7}], \mathbb{Z}[\sqrt{-15}], \ldots$,

(b) $n = 2$ and $d = -1$, i.e., $R = \mathbb{Z}[2\sqrt{-1}]$,

(c) $n = 2$ and $d = -3$, i.e., $R = \mathbb{Z}[\sqrt{-3}]$,

(d) $n = 3$ and $d = -3$, i.e., $R = \mathbb{Z}[\frac{3}{2}(1+\sqrt{-3})]$.

From now on we assume $d$ is a squarefree positive integer. I'll indicate the general approach used in [W], and tighten up a couple of the conclusions. The pullback (1.1) looks like this:

(1.12)
$$\begin{array}{ccc} R(d,n) & \hookrightarrow & \mathbb{Z}[\omega] \\ \downarrow & & \downarrow \\ \mathbb{Z}/(n) & \hookrightarrow & (\mathbb{Z}/(n))[\omega] \end{array}$$

The ring in the lower right corner is actually $(\mathbb{Z}/(n))[X]/(\psi_d)$, where $\psi_d$ is the minimal polynomial for $\omega$, reduced modulo $n$. Let $U(d,n)$ denote the "horizontal cokernel":

$$U(d,n) = (\mathbb{Z}/(n))[\omega])^*/(\mathbb{Z}/(n))^*$$

Next, let $\varepsilon = a + b\omega$ be the fundamental unit (the smallest unit of $\mathbb{Z}[\omega]$ greater than 1). The Dirichlet Unit Theorem says that $\mathbb{Z}[\omega]^* = \{\pm \varepsilon^n | n \varepsilon \mathbb{Z}\}$, so the multiplication map $\alpha$ of (1.5) is surjective if and only if $U(d,n)$ is generated by the image of $\varepsilon$.

1.13 Proposition: The real quadratic order $R(d,n)$ has torsionfree cancellation if and only if $U(d,n)$ is a cyclic group generated by the image of the fundamental unit.

1.14 Corollary: Let $n = n_1n_2$, where $n_1$ and $n_2$ are relatively prime positive integers. Then $R(d,n)$ has cancellation if and only if

(a) $R(d,n_1)$ and $R(d,n_2)$ both have torsionfree cancellation, and

(b) $U(d,n_1)$ and $U(d,n_2)$ have relatively prime orders.

Now (1.14) reduces our problem to the case $n$ = prime power (provided we keep track of the order of $U(d,p^e)$). As expected, the prime $p = 2$ has to be dealt with separately. If $d \equiv 1 \pmod 8$, then $U(d,2)$ is trivial, so $R(d,2)$ has torsionfree cancellation. (The vertical map isn't needed.) If $d \not\equiv 1 \pmod 8$. Then $(U(d,2))$ is either 2 or 3; so torsionfree cancellation holds if and only if the image in $U(d,n)$ of $\varepsilon$ is non-trivial. In summary:

1.15 Proposition: Let $d$ be a squarefree integer greater than 1, and let $\varepsilon = a + b\omega$ be the fundamental unit of $\mathbb{Z}[\omega]$. Then $R(d,2)$ has torsionfree cancellation if and only if either $d \equiv 1 \pmod 8$ or $b$ is odd.

The case $n = 4$ is worked out in Lemma 4.6 of [W]. It turns out that $U(d,4)$ has order 2, 4 or 6, depending on the congruence class of $d \pmod{16}$. When $d \equiv 3 \pmod 4$, $U(d,4)$ isn't cyclic, so, for example, $\mathbb{Z}[4\sqrt{3}]$ does not have torsionfree cancellation. If $d \equiv 5 \pmod{16}$, $U(d,4)$ is cyclic of order 6 with $\omega$ and $\omega - 1$ as generators. Thus, writing $\varepsilon = a + b\omega$ for the fundamental unit of

$\mathbb{Z}[\omega]$, we see that $\mathbb{Z}[4\omega]$ has torsionfree cancellation if and only if either $a \equiv 0 \pmod 4$ or $a + b \equiv 0 \pmod 4$. The other cases mentioned in [W,4.6] are similar.

Intimidated by this case analysis, I quit worrying about $R(d,2^e)$, one step too soon as it turns out. Recently I have discovered that $U(d,2^e)$ is <u>never</u> cyclic if $e \geq 3$, so torsionfree cancellation always fails. (The proof is rather easy, and the result would no doubt have been anticipated by most number theorists.) Thus we have a complete classification, in terms of the fundamental unit, of those rings $\mathbb{Z}[2^e\omega]$ that have torsionfree cancellation.

The case of an odd prime $p$ involves some substantial computations, but the end results are fairly nice. First of all, $U(d,p^e)$ is almost always cyclic. We use the Legendre symbol $(\frac{d}{p})$, and define $(\frac{d}{p}) = 0$ if $d \equiv 0 \pmod p$. Thus $(\frac{d}{p})$ is one less than the number of distinct solutions to $x^2 = d$ in $\mathbb{Z}/(p)$.

1.16 <u>Theorem</u> ([W,4.9 and 4.12]): Let $p$ be an odd prime, and let $e \geq 1$.

(a) $U(d,p^e)$ has order $p^{e-1}(p-(\frac{d}{p}))$.

(b) $U(d,p^e)$ fails to be cyclic if and only if $p = 3$, $e \geq 2$ and $d \equiv 6 \pmod 9$.

Even when $U(d,p^e)$ is cyclic (i.e., we're not in the case described in (b) above) we still need to know whether the image of the fundamental unit generates the group. When $d \equiv 0 \pmod p$, the answer is very simple (though the proof, in [W], is rather computational).

1.17 Theorem: Let $p$ be an odd prime, and assume that $U(d,p^e)$ is cyclic and that $d \equiv 0 \pmod{p}$. Then $R(d,p^e)$ has torsionfree cancellation if and only if $b \not\equiv 0 \pmod{p}$, where $\varepsilon = a + b\omega$ is the fundamental unit.

Fortunately, the case $d \equiv 0 \pmod{p}$ is forced by torsionfree cancellation "three-fourths of the time".

1.18 Theorem: Let $p$ be an odd prime, let $e \geq 1$, and let $d$ be a squarefree integer greater than 1. Assume either that $p \equiv 1 \pmod{4}$ or that the fundamental unit of $\mathbb{Z}[\omega]$ has norm 1. If $R(d,p^e)$ has torsionfree cancellation then $d \equiv 0 \pmod{p}$.

Using these two theorems, we can often resolve the cancellation problem very quickly.

1.19 Example: Torsionfree cancellation fails over $R(46,m)$ for every $m \geq 2$. To see this we may assume $m$ is prime, because $R(d,m_1)$ has torsionfree cancellation if $R(d,m_1m_2)$ does ([W,4.3]). The fundamental unit of $\mathbb{Z}[\sqrt{46}]$ is $24335 + 3588\omega$, with norm = 1. If $m = 2$, we can apply (1.15). Since $3588 \equiv 0 \pmod{23}$, (1.17) and (1.18) show that torsionfree cancellation fails for $m$ odd. Incidentally, 46 is the smallest example of this sort. For every $d$, $1 < d < 46$, there is a suitable $m \geq 2$ for which torsionfree cancellation holds. Take

$$m = \begin{cases} d & \text{if } d = 2 \text{ or } d \text{ is odd} \\ d/2 & \text{otherwise.} \end{cases}$$

One can verify easily that $R(d,m)$ has torsionfree cancellation, by using (1.17) and (1.18) together with a table of fundamental units, [BS, Table 1].

The case not covered by (1.17) and (1.18), that is, norm = 1 and $p \equiv 3 \pmod 4$ appears to be difficult, and I don't expect to find easily stated criteria. To compound the difficulty, there seem to be sporadic cases in which $R(d,p)$ has torsionfree cancellation but $R(d,p^2)$ does not (in contrast to the situation described by (1.17) and (1.18)). It is interesting to see how this can happen.

1.20 <u>Example</u>: Let $d = 37$ and $p = 7$. Then $U(d,p) = 6$ and $U(d,p^2) = 42$. We have $\varepsilon = 6 + \sqrt{37}$, $\varepsilon^2 = 73 + 12\sqrt{37}$ and $\varepsilon^3 = 882 + 145\sqrt{37}$. Since neither 12 nor 145 is a multiple of 7, $R(d,p)$ has torsionfree cancellation. But $882 \equiv 0 \pmod{49}$, so the coefficient of $\sqrt{37}$ in $\varepsilon^6$ is a multiple of 49. Therefore $R(d,p^2)$ does not have torsionfree cancellation

Fortunately, when the conductor goes beyond $p^2$, no further problems arise.

1.21 <u>Proposition</u>: Let p be an odd prime. If $R(d,p^2)$ has torsionfree cancellation, then so does $R(d,p^e)$ for all e.

I am indebted to Leo Chouinard for this proposition. Note how this contrasts sharply with the case $p = 2$.

In [L] Levy discusses a special subclass of Bass rings, called "Dedekind-like" rings. The neatest characterization of Dedekind-like rings R among Bass rings in that each singular maximal ideal of R has exactly two maximal ideals of $\tilde{R}$ lying over it. This means there's no ramification or residue field extension. Thus, for example, $\mathbb{Z}[5\sqrt{-1}]$ is Dedekind-like, but $\mathbb{Z}[2\sqrt{-1}]$ and $\mathbb{Z}[3\sqrt{-1}]$ are not. (Levy also assumes that R has no maximal ideals of height 0, but this is presumably unimportant.) Over Dedekind-like rings, Levy is able to deal with the torsion parts of modules, something that has so far not worked for general Bass rings. One of the results from [L] runs as follows.

1.22 <u>Theorem</u> ([L,13.7]): Let R be a Dedekind-like ring. The following conditions are equivalent:

a) $X\oplus Z \cong Y\oplus Z \implies X \cong Y$ for R-modules X, Y, Z.

b) $D(R) = 0$.

We conclude this section with a brief discussion of integral group rings. If G is a finite abelian group, then $\mathbb{Z}G$ is always one-dimensional with finite integral closure. But $\mathbb{Z}G$ is a Bass ring $\iff$ $\mathbb{Z}G$ is Dedekind-like $\iff$ $|G|$ is squarefree. Now the finite groups

$G$ for which $\mathrm{Pic}(\mathbb{Z}G) = 0$ were catalogued completely in [C-N]. They are the cyclic groups of prime order, both groups of order 4, and the cyclic groups of order 1, 6, 8, 9, 10 and 14. If $G$ has prime order, or order 1, 6, 10 or 14, Levy's theorem (1.22) says that $\mathbb{Z}G$ has general cancellation (for finitely generated modules). If $|G| = 4$, an easy computation shows that $E(\mathbb{Z}G) = 0$, so $\mathbb{Z}G$ has torsionfree cancellation, by (1.5). For the cyclic groups of order 8 and 9, $E(\mathbb{Z}G) \neq 0$ but $D(\mathbb{Z}G) = 0$, and I have been unable to decide whether torsionfree cancellation holds. (My guess: It fails for order 8 but holds for order 9.) For all other finite abelian groups, $D(\mathbb{Z}G) \neq 0$, so torsion-free cancellation fails.

## DIMENSION TWO

The first results on cancellation over 2-dimensional rings appeared in Chase's 1962 paper, [C]. He proved essentially the following:

2.1 Theorem: Suppose R is a two-dimensional integral domain, finitely generated as an algebra over an algebraically closed field k. Assume R is regular (that is, $R_{\mathcal{M}}$ is a regular local ring for every maximal ideal $\mathcal{M}$) and that every projective R-module is free. If X, Y and Z are R-modules such that $X \oplus Z \cong Y \oplus Z$, and X and Y are torsionfree of rank not a multiple of the characteristic of k, then $X \cong Y$.

Chase applied this to the polynomial ring $k[T_1, T_2]$, and used Seshadri's theorem [Se] that every projective module is free over the polynomial ring in one variable over a principal ideal domain. He also produced non-isomorphic torsionfree modules X and Y over $R = \mathbb{R}[T_1, T_2]$ such that $X \oplus R \cong Y \oplus R$.

In [W] I was able to prove (2.1) without the requirement that projective R-modules are free. The key ideas were (1) a lifting theorem for automorphisms of projective modules (rather than just free modules as in [C]) and (2) the projective cancellation theorem of Murthy and Swan (0.1). I've not been able to determine whether or not the condition on the characteristic can be dropped.

## DIMENSION THREE

In this section I'll show that torsionfree cancellation fails in dimension 3, even for well-behaved rings. I am indebted to David Eisenbud for the key idea of using syzygies and duality to get torsion-free examples.

3.1 Theorem: Let $R$ be an integral domain possessing a prime ideal $P$ such that

a) $\dim R/P = 1$,

b) $(R/P)^{\sim}$ is finitely generated over $R/P$,

c) $D(R/P) = \ker(\mathrm{Pic}(R/P) \to (\mathrm{Pic}(R/P)^{\sim}) \neq 0$, and

d) all the rings $R_{\mathcal{M}}$, $\mathcal{M} \supset P$ are regular local rings of the same dimension $n \geq 3$. Then torsionfree cancellation fails for $R$-modules.

Proof: Let $M$ be an arbitrary torsionfree $(R/P)$-module. We will determine the homological properties of $M$ as an $R$-module. (See [Ma, §§7,15-17] for a general reference.) First, notice that $\mathrm{Ass}_R(M) = \{P\}$, so $\mathrm{depth}_{R_{\mathcal{M}}} M_{\mathcal{M}} = 1$ for all $\mathcal{M} \supset P$. Also, $M$ has finite projective dimension (check locally), so $\mathrm{proj.\ dim}_R M = n-1$ by [Ma, Ex. 4, p.113]. Choose a projective resolution

$$0 \to E_{n-1} \to \dots \to E_0 \to M \to 0$$

and let $V = \operatorname{Im}(E_{n-2} \to E_{n-3})$. Applying [Ma, Theorem 26] to each of the rings $R_M$, $M \supset P$, we learn that $\operatorname{Ext}^i_R(M,R) = 0$ for $i < n-1$. Letting $(\ )^* = \operatorname{Hom}_R(\_,R)$, we see that the complex

$$0 \to M^* \to E_0^* \to \dots \to E^*_{n-1} \to 0$$

is exact except at $E^*_{n-1}$, and

$$\operatorname{coker}(E^*_{n-2} \to E^*_{n-1}) = \operatorname{Ext}^{n-1}_R(M,R) = \operatorname{Ext}^1_R(V,R).$$

Noting that $M^* = 0$, we have a projective resolution

$$0 \to E_0^* \to \dots \to E^*_{n-1} \to \operatorname{Ext}^1_R(V,R) \to 0$$

Dualizing again, we have

$$\operatorname{Ext}^{n-1}_R(\operatorname{Ext}^1_R(V,R),R) = \operatorname{coker}(E_1^{**} \to E_0^{**}) = \operatorname{coker}(E_1 \to E_0) = M.$$

In summary: If $M$ is any torsionfree $R/P$-module, then $M \cong \operatorname{Ext}^{n-1}_R(\operatorname{Ext}^1_R(V,R),R)$, where $V$ is any $(n-3)$rd syzygy of $M$.

By (1.6) there are torsionfree $R/P$-modules $X$, $Y$, $Z$ such that $X \oplus Z \cong Y \oplus Z$, but $X \not\cong Y$. Let $U$, $V$, $W$ be their respective $(n-3)$rd syzygies. Then $U \oplus W$ is an $(n-3)^{rd}$ syzygy of $X \oplus Z$, and $V \oplus W$ is an

$(n-3)^{rd}$ syzygy of $Y \oplus Z$. By Schanuel's lemma there are projective modules $E$ and $F$ such that $U \oplus W \oplus E \cong V \oplus W \oplus F$. If torsionfree cancellation were valid over $R$ we'd have $U \oplus E \cong V \oplus F$. Applying $\mathrm{Ext}_R^{n-1}(\mathrm{Ext}_R^1(\_,R),R)$ to both sides, we'd get $X \cong Y$, the desired contradiction.

I believe this theorem ought to apply to any integral domain that's finitely generated as an algebra over a field, as long as the dimension is at least 3. Also, one can probably find the requiste prime ideal in any finitely generated (over $\mathbb{Z}$) domain of dimension $\geq 3$, but this is likely to be harder. Here are two cases that are easy:

3.2 Corollary: Let $R$ be either $k[T_1,T_2,T_3]$, $k$ a field, or $\mathbb{Z}[T_1,T_2]$. Then torsionfree cancellation fails for $R$-modules.

Proof: Apply (3.1), with $P = (T_1{}^2-T_2{}^3,T_3)$ in the first case and $P = (2,T_1{}^2-T_2{}^3)$ in the second case.

3.3 Corollary: Let $R$ be an integral domain, finitely generated over an infinite perfect field $k$. If $\dim R = n \geq 3$, then torsion-free cancellation fails for $R$-modules.

Proof: Choose a non-zero non-unit $f \in R$ such that $R_{\mathcal{M}}$ is regular local for every $\mathcal{M}$ not containing $f$. The Jacobson radical of $R$ is $0$, so there is an element $g \in R$ such that $1 + fg$ isn't a unit. Let $Q$ be any prime of height $n-2$ containing $1+fg$. For every maximal ideal $\mathcal{M}$ of $R$ containing $Q$, $R_{\mathcal{M}}$ is regular, so it will suffice to prove that $R/Q$ possess a height-one prime $P/Q$ such that

$D(R/P) \neq 0$. Letting $S = R/Q$, a 2-dimensional, finitely generated $k$-domain, we want to prove that $S$ has a height-one prime $P$ for which $D(S/P) \neq 0$.

Choose a maximal ideal $M$ of $S$ such that $S_M$ is regular, and choose $x,y \in M$ such that $(x,y)S_M = MS_M$. Let $P$ be any minimal prime ideal of $(y^2-x^3)S$. We will use (1.6) to show that $D(S/P) \neq 0$. Letting $A = S/P$ and $\bar{k}$ = algebraic closure of $k$ in the quotient field $K$ of $A$, we need only check that $\bar{k}A$ is properly contained in $\tilde{A}$. In fact, it will suffice to check that $\bar{k}B \neq \tilde{B}$, where $B$ is the local ring of $A$ at the maximal ideal $N = M/P$.

Let $z = y/x \in K$, and note that $z^2 = x$, $z^3 = y$, but $z \notin B$. Thus $B$ is not seminormal. (See [Sw2] for an excellent discussion of seminormality.) Note that $B/N \cong B[z]/zB[z]$, so $zB[z]$ is a maximal ideal of $B[z]$. Moreover, if $M$ is any maximal ideal of $B[z]$, we have $z^2 \in N = M \cap B$, so $M = zB[z]$. Thus $B[z]$ is a discrete valuation ring, whence $B[z] = \tilde{B}$. It is easy to see that $N\tilde{B} = N$, so that $N$ is the conductor of $B$ in $\tilde{B}$. If, now $\bar{k}B = \tilde{B}$, we would have $\bar{k}(B/N) = \tilde{B}/N$. But $\bar{k}(B/N)$ is reduced (has no non-zero nilpotents) because $k$ is perfect, [Ma,(27.E)], whereas $\tilde{B}/N$ is <u>not</u> reduced because $B$ is not seminormal. This contradiction completes the proof.

## REFERENCES

[B] H. Bass, "On the ubiquity of Gorenstein rings", Math Z. 82 (1963), 8-28.

[BS] Z.I. Borevich and I.R. Shafarevich, Number Theory, Academic Press, New York, 1966.

[C] S.U. Chase, "Torsion-free modules over K[x,y]", Pacific J. Math. 12(1962), 437-447.

[C-N] P. Cassou-Noguès, "Classes d'ideaux de l'algèbre d'un groupe abélien", C.R. Acad. Sci. Paris 276(1973, Ser. A., 973-975.

[DM] E.D. Davis and P. Maroscia, "Affine curves on which every point is a set-theoretic complete intersection", preprint.

[E] E.G. Evans, Jr., "Krull-Schmidt and cancellation over local rings", Pacific J. Math. 46(1973), 115-121.

[EE] D. Eisenbud and E.G. Evans, Jr., "Generating modules efficiently: theorems from algebraic K-theory", J. Algebra 27(1973), 278-305.

[EGA] A. Grothendieck and J. Dieudonné, Eléments de Géométrie Algébrique IV Partie 2, Publ. Math. IHES 24(1967).

[GW] K.R. Goodearl and R.B. Warfield, "Algebras over zero-dimensional rings", Math. Ann. 223(1976), 157-168.

[L] L.S. Levy, "Modules over Dedekind-like rings", J. Algebra (to appear)

[LW] L.S. Levy and R. Wiegand, "Dedekind-like behavior of rings with 2-generated ideals", preprint.

[Ma] H. Matsumura, Commutative Algebra, Benjamin, New York, 1970.

[Mi] J. Milnor, Introduction to Algebraic K-Theory, Princeton University Press, Princeton, 1971.

[MS] M.-P. Murthy and R.G. Swan, "Vector bundles over affine surfaces", Invent. Math. 36(1976), 125-165.

[Se] J.-P. Serre, "Modules projectifs et espaces fibrés à fibre vectorielle", Sém. P. Dubreil, 1957/1958.

[SW1] R.G. Swan, Algebraic K-Theory, Lecture Notes in Math. 76(1968).

[SW2] R.G. Swan, "On seminormality", preprint.

[V] W. Vasconcelos, "On local and stable cancellation", An. Acad. Brasil Ci. 37(1965).

[W] R. Wiegand, "Cancellation over commutative rings of dimension one and two", J. Algebra (to appear).

# ON COMPOSITION SERIES OF A MODULE WITH RESPECT TO A SET OF GABRIEL TOPOLOGIES

TOMA ALBU
UNIVERSITY OF BUCHAREST

Composition series for modules with respect to a Gabriel topology (or equivalently, with respect to a hereditary torsion theory) were introduced by Goldman [1] in 1975, but only for torsion-free modules, and have been further studied, among others, by Beachy, Golan,etc.

A series of recent papers of Boyle and Feller [1], [2], [3] consider the semicritical socle series as well as the Krull critical composition series of a module over a right noetherian ring.

Teply [1] announced in 1982 the definition of the notion of composition series of a module for sets of torsion theories, which works in particular for the chain of torsion theories for Krull dimension over a right noetherian ring.

The aim of this paper is to place all these notions in a latticial setting. This allows to obtain simplified proofs in a slight generalized context of some previous results and to enlighten the machinery involved in their proofs.

## 0. *PRELIMINARIES*

Throughout this section L will denote a modular lattice with a least element 0 and with a greatest element 1. The notation and terminology will follow Stenström [1].

0.1. *Lemma*. Let $(x_i)_{0\le i\le n}$ be a finite family of elements of L such that

$0=x_o<x_1<\ldots<x_n=1$.

If $y\in L$, $y\ne 0$, then there exists j, $1\le j\le n$, $z\in]x_{j-1},x_j]$, and $u\in]0,y]$ such that the intervals $[0,u]$ and $[x_{j-1},z]$ are similar.

*Proof*. Denote by j the least integer $\le n$ such that $y\wedge x_j\ne 0$. Clearly $j>0$; then $x_{j-1}<(y\wedge x_j)\vee x_{j-1}$, for otherwise $y\wedge x_j\le x_{j-1}$, and so $y\wedge x_j=(y\wedge x_j)\wedge x_{j-1}=y\wedge x_{j-1}=0$, a contradiction. Denote $z=(y\wedge x_j)\vee x_{j-1}$. Then $z\in]x_{j-1},x_j]$ and

$[x_{j-1},z]\simeq[x_{j-1}\wedge(y\wedge x_j),y\wedge x_j]=[y\wedge x_{j-1},y\wedge x_j]=[0,u]$, where $u=y\wedge x_j$.

Throughout the remainder of this section L will be supposed to be an upper continuous and modular lattice.

We shall recall now some definitions and properties. An *atom* of L is a nonzero element $a\in L$ such that whenever $b\in L$ and $b<a$, then $b=0$, i.e.,the interval $[0,a]$ has exactly two elements, 0 and a. If $x,y\in L$ and $x<y$, then the interval $[x,y]$ is said to be *simple* if y is an atom in the sublattice $[x,y]$ of L. The lattice L is called *semi-atomic* if 1 is a join of atoms, and L is called *semi-artinian* if for every $x\in L$, $x\ne 1$, the sublattice $[x,1]$ of L contains an atom. If L is semi-atomic, then for every $x\le y$ in L,the interval $[x,y]$ of L is a semi-atomic lattice.

The *(ascending) Loewy series* of L

$$s_o(L)<s_1(L)<\ldots<s_{\lambda(L)}(L) \qquad (*)$$

is defined inductively as follows: $s_o(L)=0$, $s_1(L)$ is the *socle* So(L) of L (i.e., the join of all atoms of L), and if the elements $s_\beta(L)$ of L have been defined for all ordinals $\beta<\alpha$, then $s_\alpha(L)=\bigvee_{\beta<\alpha} s_\beta(L)$ if $\alpha$ is a limit ordinal, and $s_\alpha(L)=\mathrm{So}([s_\gamma(L),1])$ if $\alpha=\gamma+1$; $\lambda(L)$ is the least ordinal $\lambda$ such that $s_\lambda(L)=s_{\lambda+1}(L)$, and is called the *Loewy length* of L. The intervals $[s_\alpha(L),s_{\alpha+1}(L)]$ are called the *factors* of the series (*), and they are for each ordinal $\alpha,\alpha<\lambda(L)$ semi-atomic lattices. The lattice L is semi-artinian if and only if $s_{\lambda(L)}(L)=1$. Note also that if x,y and z are elements of L such that $x<y<z$, then $[x,z]$ is a semi-artinian lattice if and only if $[x,y]$ and $[y,z]$ are both semi-artinian.

For all these summarized facts concerning Loewy series of a lattice the reader is referred to Năstăsescu [1].

0.2. *Lemma*. Let $(x_i)_{0\le i\le n}$ be a finite family of elements of L $(n\ge 1)$, such that

$$0=x_o<x_1<\ldots<x_n$$

and $[x_{i-1},x_i]$ are simple intervals for each i, $1\le i\le n$. If the interval $[x_n,1]$ contains no atom, then $\lambda(L)\le n$ and $x_n=s_{\lambda(L)}(L)$.

*Proof*. We shall prove by induction that $x_i\le s_i(L)$ for each $i, 0\le i\le n$. Trivially this holds if i=0 and i=1, so assume it holds for i<n and prove it for i+1. By modularity, one has

$$[s_i(L),x_{i+1}\vee s_i(L)]=[x_i\vee s_i(L),(x_i\vee s_i(L))\vee x_{i+i}]\simeq[(x_i\vee s_i(L))\wedge x_{i+1},x_{i+i}]=$$
$$=[x_i\vee(s_i(L)\wedge x_{i+1}),x_{i+1}]\subseteq[x_i,x_{i+1}].$$ But $[x_i,x_{i+1}]$ is a simple interval, so $[s_i(L),x_{i+1}\vee s_i(L)]$ is a simple interval or is reduced to a single element, and therefore $x_{i+1}\vee s_i(L)\le s_{i+1}(L)$. In particular $x_{i+1}\le s_{i+1}(L)$.

Let now $k\ge 1$ be the least natural number such that $x_n\le s_k(L)$ and $x_n\not\le s_{k-1}(L)$. Since each interval $[s_j(L),s_{j+1}(L)]$ is a semi-atomic lattice, hence a semi-artinian lattice, it follows that $[0,s_k(L)]$ is a semi-artinian lattice. Hence $x_n=s_k(L)$, for otherwise, the interval $[x_n,s_k(L)]$ would contain an atom, a contradiction. By the same argument, $s_k(L)=s_{k+1}(L)$; so $\lambda(L)=k$, and consequently $x_n=s_{\lambda(L)}(L)$.

## 1. *COMPOSITION SERIES OF A MODULE WITH RESPECT TO A GABRIEL TOPOLOGY*

Let L be a modular lattice with elements 0 and 1, $0\ne 1$. Recall that a *(Jordan-Hölder) composition series* of L is a chain

$$0=a_o<a_1<\ldots<a_n=1$$

such that each interval $[a_{i-1},a_i]$, $1\le i\le n$ is a simple interval. The lattice L is said to be *of finite length* if L has a composition series; in this case, a well-known result asserts that any two composition series of L are equivalent (see e.g. Stenström [1]). Note that a modular lattice with 0 and 1 is of finite length if and only if it is both artinian and noetherian.

Let now R be an associative, unitary and nonzero ring, and Mod-R the category of unitary right R-modules. If M is a right R-module then $L(M)$ will denote the lattice of all submodules of M.

The set of all right Gabriel topologies on R will be denoted by $Gab(R)$. If $F \in Gab(R)$, then $(T_F, F_F)$ will denote the corresponding hereditary torsion theory on Mod-R, and $t_F$ the torsion radical associated to $(T_F, F_F)$. If $M \in$ Mod-R, we shall use the following notation

$C_F(M) = \{N \in L(M) \mid M/N \in F_F\}$.

For each $P \in L(M)$, $\tilde{P}$ will denote the F-saturation of P in M, i.e. $\tilde{P}/P = t_F(M/P)$. Thus $P \in C_F(M)$ iff $P = \tilde{P}$, i.e., P is F-saturated. If $(N_i)_{i \in I}$ is a family of elements of $C_F(M)$, then $\bigvee_{i \in I} N_i = (\sum_{i \in I} N_i)^{\sim}$ and $\bigwedge_{i \in I} N_i = \bigcap_{i \in I} N_i$ are elements of $C_F(M)$. Moreover, $C_F(M)$ is an upper continuous and modular lattice with respect to the partial ordering given by "$\subseteq$" (inclusion) and with respect to the operations "$\vee$" and "$\wedge$", having the least element $t_F(M)$ and the greatest element M.

Recall that $M \in$ Mod-R is said to be F-*cocritical* if $M \neq 0$, $M \in F_F$ and $M/M' \in T_F$ for every nonzero submodule M' of M, or equivalently, if $M \neq 0$ and $C_F(M) = \{0, M\}$, or equivalently, if $M \in F_F$ and $T_F(M)$ is a simple object in the quotient category Mod-R/$T_F$, where

$T_F$: Mod-R $\to$ Mod-R/$T_F$

is the canonical functor.

An F-*composition series* of $M \in$ Mod-R is a chain

$t_F(M) = M_0 < M_1 < \ldots < M_n = M$

of submodules of M such that $M_i/M_{i-1}$ is an F-cocritical module for each i, $0 < i \leq n$.

1.1. *Proposition*. Let $M \in$ Mod-R and

$$M_o < M_1 < \ldots < M_n = M \qquad (*)$$

be a chain of submodules of M. Then (*) is an F-composition series of M if and only if (*) is a composition series of the lattice $C_F(M)$.

*Proof.* Suppose that (*) is an F-composition series of M. The exact sequence of R-modules

$$0 \to M_{n-1}/M_{n-2} \to M/M_{n-2} \to M/M_{n-1} \to 0$$

with $M_{n-1}/M_{n-2} \varepsilon F_F$ and $M/M_{n-1} \varepsilon F_F$ yields $M/M_{n-2} \varepsilon F_F$. Then, from the exact sequence of R-modules

$$0 \to M_{n-2}/M_{n-3} \to M/M_{n-3} \to M/M_{n-2} \to 0$$

with $M_{n-2}/M_{n-3} \varepsilon F_F$ and $M/M_{n-2} \varepsilon F_F$ one gets $M/M_{n-3} \varepsilon F_F$, and so on, $M/M_i \varepsilon F_F$ for all i, $0 \le i < n$, i.e., $M_i \varepsilon C_F(M)$.

Let now $X \varepsilon C_F(M)$ with $M_{i-1} < X \le M_i$. Then $M/X \varepsilon F_F$, hence $M_i/X \varepsilon F_F$. But $(M_i/M_{i-1})/(X/M_{i-1}) \simeq M_i/X \varepsilon T_F$ because $M_i/M_{i-1}$ is F-cocritical, hence $M_i/X \varepsilon T_F \cap F_F = \{0\}$, i.e., $X = M_i$, and so $[M_{i-1}, M_i]$ is a simple interval in $C_F(M)$, i.e., (*) is a composition series of the lattice $C_F(M)$.

Conversely, suppose that (*) is a composition series of the lattice $C_F(M)$. Then necessarily $M_o = t_F(M)$. We have to prove that $M_i/M_{i-1}$ is F-cocritical for each i, $0 < i \le n$. First of all, $M_i/M_{i-1} \varepsilon F_F$ because $M/M_{i-1} \varepsilon F_F$ . Let $Y/M_{i-1} \varepsilon C_F(M_i/M_{i-1})$ with $Y \ne M_{i-1}$. Then $(M_i/M_{i-1})/(Y/M_{i-1}) \simeq$ $\simeq M_i/Y \varepsilon F_F$. The exact sequence of R-modules

$$0 \to M_i/Y \to M/Y \to M/M_i \to 0$$

with $M_i/Y \varepsilon F_F$ and $M/M_i \varepsilon F_F$ yields $M/Y \varepsilon F_F$ , i.e., $Y \varepsilon C_F(M)$. But $M_{i-1} < Y \le M_i$, hence $Y = M_i$ because $[M_{i-1}, M_i]$ is a simple interval in $C_F(M)$. Consequently $M_i/M_{i-1}$ is an F-cocritical module for each i, $0 < i \le n$.

1.2. *Corollary.* The following assertions are equivalent for a right R-module M:

(1) M has an F-composition series.

(2) $C_F(M)$ is a lattice of finite length.

(3) $T_F(M)$ is an object of finite length in Mod-$R/\mathcal{T}_F$. Moreover, if

$t_F(M)=M_o<M_1<\ldots<M_n=M$

is a composition series of the lattice $C_F(M)$, then the injective hulls $E_R(M_i/M_{i-1})$ of the modules $M_i/M_{i-1}$ are unique up to order and isomorphism.

*Proof*. Since the lattice $C_F(M)$ is isomorphic to the lattice $S(T_F(M))$ of all subobjects of the object $T_F(M)$ (see e.g. Albu and Năstăsescu [1]), it follows immediately the equivalence of the assertions (1), (2) and (3).

Applying the exact functor $T_F$ to the chain

$t_F(M)=M_o<M_1<\ldots<M_n=M$

one obtains a composition series

$0=T_F(M_o)\subset T_F(M_1)\subset\ldots\subset T_F(M_n)=T_F(M)$

of $T_F(M)$ in Mod-$R/\mathcal{T}_F$. By the Jordan-Hölder theorem we have only to prove that if $X_1$ and $X_2$ are two F-cocritical R-modules with $T_F(X_1)\simeq T_F(X_2)$, then $E_R(X_1)\simeq E_R(X_2)$. Since $T_F(X_1)\simeq T_F(X_2)$ it follows that the injective hull $E(T_F(X_1))$ of $T_F(X_1)$ is isomorphic to the injective hull $E(T_F(X_2))$ of $T_F(X_2)$. But $X_1,X_2\varepsilon\mathcal{F}_F$, hence

$E(T_F(X_1))\simeq T_F(E_R(X_1))$ and $E(T_F(X_2))\simeq T_F(E_R(X_2))$

(see Gabriel [1]), and so, $S_FT_F(E_R(X_1))\simeq S_FT_F(E_R(X_2))$, where $S_F$:Mod-$R/\mathcal{T}_F\to$Mod-R is the right adjoint of the functor $T_F$. Since $E_R(X_1)$ and $E_R(X_2)$ are injective modules, hence F-closed, it follows that

$E_R(X_1)\simeq S_FT_F(E_R(X_1))\simeq S_FT_F(E_R(X_2))\simeq E_R(X_2)$.

## 2. RELATIVE X-COMPOSITION SERIES OF A MODULE

*Definition* (Teply [1],[2]). Let $\mathcal{X}$ be a nonempty set of (right) Gabriel topologies on R. A right R-module M is said to have an $\mathcal{X}$-composition series if there exists a chain of submodules of M

$M_o<M_1<\ldots<M_n=M$ (*)

such that

(1) Each $M_i/M_{i-1}$ is $F_i$-cocritical for some $F_i\varepsilon\mathcal{X}$.

(2) If $M_i/M_{i-1}$ contains an F-cocritical submodule for some $F\varepsilon\mathcal{X}$, then $F=F_i$.

(3) $F_1 \leq F_2 \leq \ldots \leq F_n$.

(4) $M_o = t_{F_1}(M)$.

Note that the original definition of Teply requires $M_o=0$, i.e., $M$ is $F_1$-torsion-free. If $X$ has only one element $F$ one obtains the definition of an $F$-composition series.

If $M$ has an $X$-composition series (*), then by renumbering the Gabriel topologies $F_i$ one gets for $M$ a so called $X$-composition series of type

$(F_1,n_1;F_2,n_2; \ldots ; F_k,n_k)$,

where $k \geq 1$, $n_1, n_2, \ldots, n_k$ are natural numbers $\geq 1$, and $F_1,F_2,\ldots,F_k \in X$ such that

(i) $F_1 < F_2 < \ldots < F_k$.

(ii) The first $n_1$ factors of the series (*) are $F_1$-cocritical, the next $n_2$ factors are $F_2$-cocritical, and so on, the last $n_k$ factors of (*) are $F_k$-cocritical.

2.1. *Lemma.* Let $F \in Gab(R)$ and $X \in$ Mod-R. Then $C_F(X)=\{X\}$ if and only if $X \in T_F$.

*Proof.* If $C_F(X)=\{X\}$, then $t_F(X)=X$ because $t_F(X) \in C_F(X)$, i.e., $X \in T_F$. Conversely, if $X \in T_F$ and $Y \in C_F(X)$, then $X/Y \in T_F \cap F_F = \{0\}$, hence $Y=X$.

2.2. *Proposition.* Let $X \in$ Mod-R and $F_1,F_2 \in Gab(R)$ such that $F_1 \leq F_2$. If $X$ is $F_1$-cocritical, then either $X$ is $F_2$-cocritical or $X \in T_{F_2}$.

*Proof.* Since $F_1 \leq F_2$, it follows that $F_{F_2} \subseteq F_{F_1}$, hence $C_{F_2}(X) \subseteq C_{F_1}(X) = \{0,X\}$, and so, $C_{F_2}(X)=\{0,X\}$ or $C_{F_2}(X)=\{X\}$, i.e., $X$ is $F_2$-cocritical or $X \in T_{F_2}$ by 2.1.

2.3. *Theorem* (Teply [1], [2]). Let $M \in$ Mod-R and $\emptyset \neq X \subseteq Gab(R)$. If $M$ has an $X$-composition series $M_o < M_1 < \ldots < M_n = M$ of type $(F_1,n_1;F_2,n_2; \ldots ;F_k,n_k)$, then any other $X$-composition series of M beginning with $M_o$ is of the same type, and the injective hulls of the factors $M_i/M_{i-1}$ are unique up to order and isomorphism. Moreover,

$$t_{F_{j+1}}(M)=M_{n_1+n_2+\ldots+n_j}$$

for each j, $1\leq j\leq k-1$.

*Proof.* Let $M_o=N_o<N_1<\ldots<N_m=M$ be another $X$-composition series of M of type $(G_1,m_1;G_2,m_2;\ \ldots\ ;\ G_s,m_s)$. We have to prove that s=k, $G_i=F_i$ and $m_i=n_i$ for each i, $1\leq i\leq k$.

First of all, $M/M_{n-1}\varepsilon F_{F_k}\subseteq F_{F_1}$, and, as in the proof of 1.1 one deduces that $M/M_i\varepsilon F_{F_1}$ for all i, i.e., the given chain of submodules of M is a chain in the lattice $C_{F_1}(M)$. The same is true also for the other above considered $X$-composition series of M. Note that by the same argument, if $k\geq 2$, then $M_{n_1}\varepsilon C_{F_2}(M)$.

By 0.1, applied to the interval $[M_o,M]$ considered in the lattice $L(M)$, one deduces that there exists j, $1\leq j\leq n$, $Z\varepsilon L(M)$, $M_{j-1}<Z\leq M_j$, and $U\varepsilon L(N_1)$, $N_o<U$ such that $U/N_o\simeq Z/M_{j-1}$. But $U/N_o$ is $G_1$-cocritical, hence $G_1=F_p\geq F_1$, where $M_j/M_{j-1}$ is $F_p$-cocritical for some $p\geq 1$. By symmetry, $F_1\geq G_1$, and so $F_1=G_1$.

If $M_{n_1}\neq M$, then applying again 0.1 to the interval $[M_{n_1},M]$ considered in the lattice $L(M)$, one deduces that $[M_{n_1},M]$, considered now as an interval in the lattice $C_{F_1}(M)$, contains no atom, for otherwise the R-module $M_j/M_{j-1}$ would contain a submodule which is $F_1$-cocritical for some $j>n_1$, a contradiction. By 0.2, $M_{n_1}=s_\lambda(C_{F_1}(M))=s_\lambda(C_{G_1}(M))=N_{m_1}$, where $\lambda=\lambda(C_{F_1}(M))=$ $=\lambda(C_{G_1}(M))$. The same is also true if $M_{n_1}=M$.

By 1.2, $n_1=m_1$ and $E_R(M_i/M_{i-1})$ are unique up to order and isomorphism for all i, $i \leq n_1$. Applying now again 0.1, 0.2 and 1.2 to the chains

$$M_{n_1} < M_{n_1+1} < \ldots < M_n = M$$

$$M_{n_1} = N_{n_1} < N_{n_1+1} < \ldots < N_m = M$$

of elements of $C_{F_2}(M)$ one gets $n_2 = m_2$, etc.

We shall prove the last assertion of the theorem only for j=1. By 2.2, $M_1/M_o \varepsilon \mathcal{T}_{F_2}$; on the other hand, $M_o \varepsilon \mathcal{T}_{F_1} \subseteq \mathcal{T}_{F_2}$, hence $M_1 \varepsilon \mathcal{T}_{F_2}$, and so on, $M_i \varepsilon \mathcal{T}_{F_2}$ for all $i \leq n_1$. It follows that $M_{n_1} = t_{F_2}(M_{n_1}) \leq t_{F_2}(M)$. But we have seen that $M_{n_1} \varepsilon C_{F_2}(M)$, hence $t_{F_2}(M) \leq M_{n_1}$, and thus $t_{F_2}(M) = M_{n_1}$.

2.4. *Remark.* It is easy to see that the notion of an F-*semicritical module*, as it appears in Teply [2], can be reformulated in a latticial setting as follows: an R-module M is F-semicritical iff $M \varepsilon F_F$ and $C_F(M)$ is a semi-atomic lattice of finite length. Note also that the F-*semicritical socle series* of M is exactly the Loewy series of the lattice $C_F(M)$.

## REFERENCES

Albu, T.

[1] Certain artinian lattices are noetherian. Applications to the relative Hopkins-Levitzki theorem, in *Methods in Ring Theory*, Van Oystaeyen, F., Editor, D. Reidel Publishing Company, Dordrecht-Holland, 1984.

Albu, T. and Năstăsescu, C.

[1] *Relative Finiteness in Module Theory*, Marcel Dekker, Inc., New York and Basel, 1984.

Boyle, A.K. and Feller, E.H.

[1] Semicritical modules and k-primitive rings, in *Module Theory*, Lecture Notes in Mathematics 700, Springer-Verlag, Berlin Heidelberg New York, 1979.

[2] The endomorphism ring of a Δ-module over a right noetherian ring, *Israel J.Math.*, 45, 313-328, 1983.

[3] α-Injectives and the semicritical socle series, *Comm.Algebra*, 11, 1643-1674, 1983.

Gabriel, P.

[1] Des catégories abéliennes, *Bull.Soc.Math.France*, 90, 323-448, 1962.

Goldman, O.

[1] Elements of noncommutative arithmetic I, *J.Algebra*, 35, 308-341, 1975.

Năstăsescu, C.

[1] *Teoria Dimensiunii în Algebra Necomutativă*, Editura Academiei, București, 1983.

Teply, M.L.

[1] Composition series for sets of torsion theories, *Abstracts Amer. Math.Soc.*, 3, 274, 1982.

[2] Torsionfree modules and the semicritical socle series, preprint, 1983.

Stenström, B.

[1] *Rings of Quotients*, Springer-Verlag, Berlin Heidelberg New York, 1975

A REMARK ON LEFT PSEUDO-ARTINIAN RINGS (*)

GABRIELLA D'ESTE
Università di Padova

CLAUDIA MENINI
Università di Ferrara

Problem 2 of [2] asks whether or not any left pseudo-artinian ring is also a strongly left pseudo-artinian ring. Let us recall that a ring R is called left pseudo-artinian (resp. strongly left pseudo-artinian) if any finitely generated submodule of the minimal cogenerator $_R E$ of the category $_R\mathcal{M}$ of all left R-modules is artinian (resp. of finite length).

We have obtained, independently from DIKRANIAN and ORSATTI [1] , some examples of left pseudo-artinian rings which are not strongly left pseudo-artinian.

In section 1 of this paper, we construct one of these rings, which is very easy to describe. In fact such a ring, say $\bar{B}$ , is naturally obtained from a one-point extension of the path algebra of the quiver 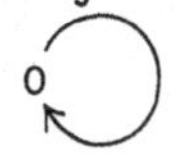 .

(*) This paper was written while the authors were members of the G.N.S.A.G.A. of C.N.R. with a partial financial support by Ministero della Pubblica Istruzione.

Moreover $\bar{B}$ is a semilocal ring (with exactly two non isomrphic simple modules) and from $\bar{B}$ one easily gets another semilocal left pseudo-artinian ring $\hat{B}$ which is not strongly left pseudo-artinian and which is linearly compact discrete (and hence - in the terminology of [1]' - a ring which is s.l.c. but not a.l.c. in its (left) cofinite topology).

In section 2, we construct a local ring $\bar{C}$ which is naturally obtained from the path algebra of a quiver. Again, using similar techniques as in §1., we get from $\bar{C}$ another ring $\hat{C}$ which is even linearly compact discrete. It can be shown that $\hat{C}$ is one of the rings obtained in [1] with different methods.

Finally, we stress the fact that the starting point for our construction was the nice example of a cyclic artinian but not noetherian module, say ${}_R V$, given by B. OSOFSKY [3], where $V$, a vector space over the field $k$ with basis $\{v_n : n \in N\}$, is regarded as a module over the subalgebra $R$ of $\mathrm{End}_k(V)$ generated by the maps $\{f_n : n \in N\}$ defined by the formulas

$$f_0(v_n) = \begin{cases} 0 & \text{if } n = 0,1 \\ v_{n-1} & \text{if } n > 1 \end{cases}$$

and

$$f_i(v_n) = \begin{cases} v_i & \text{if } n = 0 \\ 0 & \text{if } n > 0 \end{cases}$$

for any $i > 0$, as indicated in the following picture.

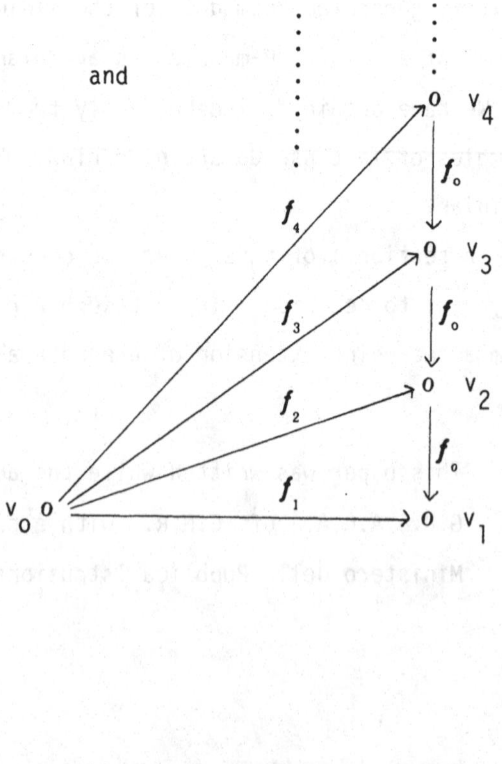

For the definitions of quiver, path algebra of a quiver and representation of a quiver, we refer to the first pages of [4] .

Throughout the paper, we denote by N the set of non-negative integers. If R is a ring, ${}_R\mathcal{M}$ will denote the category of left R-modules. By R-module, we will always mean left R-module.

The first author would like to thank Prof. C.M. RINGEL , for introducing her to the use of quivers and for his suggestions concerning this paper.

§1. Let k be an algebraically closed field and let A be the path algebra $k\Gamma$ , where $\Gamma$ is the quiver $\underset{a}{\circ}\circlearrowleft\alpha$. Let $\bar{A}$ denote the localization of A with respect to the maximal ideal $A\alpha$ , and let B denote the one-point extension of A given by the path algebra $k\Delta$ , where $\Delta$ is the quiver

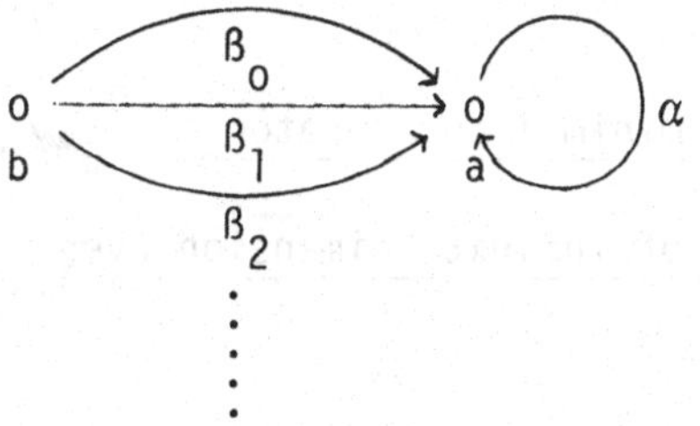


with relations

$$\alpha\beta_n = \begin{cases} 0 & \text{if } n = 0 \\ \beta_{n-1} & \text{if } n > 0 \end{cases}$$

Finally, let $\bar{B}$ be the algebra given by the following pushout diagram, where the solid arrows are the canonical inclusions of A into $\bar{A}$ and B respectively.

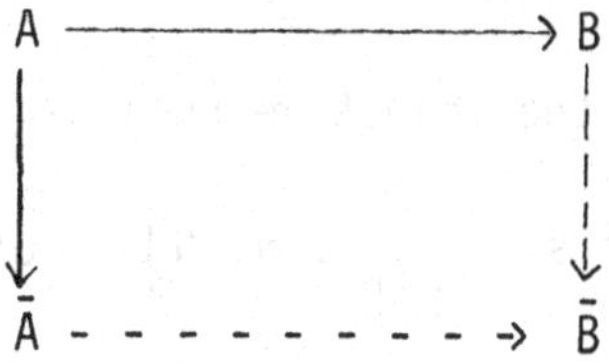

With these notations, we shall prove the following

*Proposition* *$\bar{B}$ is a semilocal left pseudo-artinian ring which is not strongly left pseudo-artinian.*

In the following, we denote by $(U,V;\alpha,\beta_0,\beta_1,\ldots\ldots)$ a representation of $\Delta$ of the form where $U$ and $V$ are $k$-vector spaces and $\alpha$ and the $\beta_i$'s are $k$-linear maps.

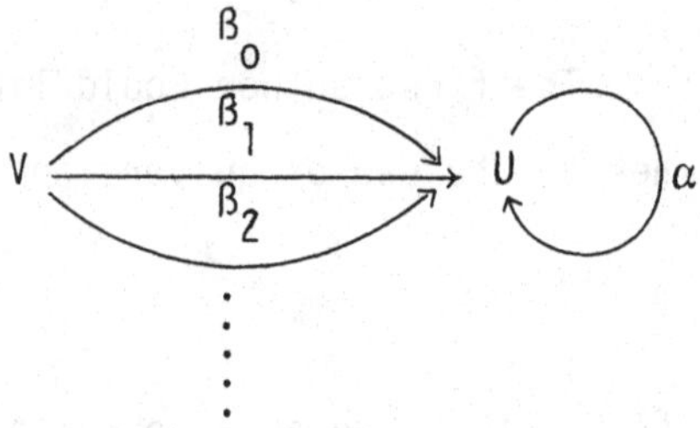


Finally, $I(a)$ and $I(b)$ will be the injective hulls in ${}_B\mathcal{M}$ of the simple left $B$-modules $S(a) = (k,0;0,0,0,\ldots\ldots)$ and $S(b) = (0,k;0,0,0,\ldots\ldots)$ respectively.

The proof of the proposition is based on the following facts.

(i) *$I(a)$ and $I(b)$ can be regarded, in a natural way, as left $\bar{B}$-modules and, in this way, they are the injective hulls in ${}_{\bar{B}}\mathcal{M}$ of the simple $\bar{B}$-modules $S(a)$ and $S(b)$.*

(ii) *The $\bar{B}$-module $I(a) \oplus I(b)$ is the minimal cogenerator of ${}_{\bar{B}}\mathcal{M}$.*

(iii) *Any cyclic $\bar{B}$-submodule of $I(a)$ of infinite dimension over $k$ is artinian but not of finite length.*

*Proof* (i) Evidently $I(b) = S(b)$ and $I(b)$, regarded as a $\bar{B}$-module, is injective. Let $I(a) = (V_a,V_b;\alpha,\beta_0,\beta_1,\ldots\ldots)$ ; then $V_a \circlearrowleft \alpha$ is is the injective hull of $k \circlearrowleft 0$ in ${}_{\bar{A}}\mathcal{M}$. Hence, we can fix a basis $\{y_n : n \in N\}$ of $V_a$ satisfying

$$\alpha(y_n) = \begin{cases} 0 & \text{if } n = 0 \\ y_{n-1} & \text{if } n > 0 \end{cases}$$

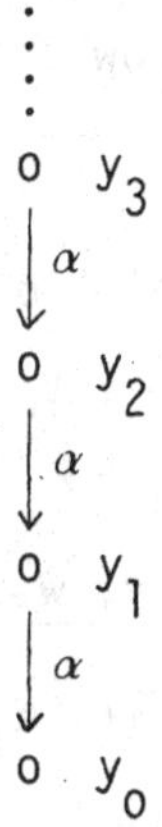


as indicated in the picture.

Consequently, $V_a \circlearrowleft \alpha$ is an $\bar{A}$-module in an obvious way, and so $I(a)$ is a $\bar{B}$-module.

To see that $I(a)$ is the injective hull of $S(a)$ in ${}_{\bar{B}}\mathcal{M}$, let $f: X \longrightarrow I(a)$ and $g: X \longrightarrow Y$ be morphisms of $\bar{B}$-modules with $g$ mono. Then there exists a B-morphism $h: Y \longrightarrow I(a)$ making the following diagram commutative.

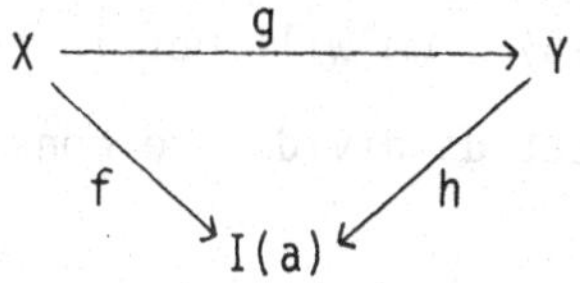


Let $e_a$ and $e_b$ denote the primitive ortogonal idempotents of B, corresponding to the paths of length 0 around a and b respectively. Then any element $x \in \bar{B}$ can be written in the form

$$x = q^{-1}pe_a + \sum_{i=0}^{n} q_i^{-1}p_i\beta_i + ue_b$$

where $n \in N$, $u \in k$, $p, p_i \in A$ and $q_i^{-1}, q^{-1}$ are the inverses in $\bar{A}$ of the elements $q, q_i \in A \setminus A\alpha$ for $i = 0,\dots,n$. Take $q \in A \setminus A\alpha$ and $y \in Y_a$, where $Y = (Y_a, Y_b; \alpha, \beta_0, \beta_1, \dots\dots)$ ; since $h(y) = h(qq^{-1}y) = qh(q^{-1}y) \in V_a$, we have $q^{-1}h(y) = h(q^{-1}y)$ .

Consequently, h is $\bar{B}$-linear, and so $I(a)$ is the injective hull of $S(a)$ in ${}_{\bar{B}}\mathcal{M}$, as claimed.

(ii) Let S be a simple $\bar{B}$-module. Assume first that $e_aS \neq 0$ . Then $e_aS = S$ and so S is a simple $\bar{A}$-module; hence $S \cong S(a)$ .

Suppose now that $e_a S = 0$ ; then $e_b S = S$ , and so $S \cong S(b)$ . Therefore, by (i) , $I(a) \oplus I(b)$ is the minimal cogenerator of $_{\bar{B}}\mathcal{M}$ .

(iii) To determine $I(a)$ , it suffices to describe $V_b$ and all the maps $\beta_i : V_b \longrightarrow V_a$ . To this end, we first note that

(*) <u>If $v, w \in V_b$ and $\beta_i(v) = \beta_i(w)$ for any $i \in N$ , then $v = w$ .</u>

In fact, if $\beta_i(v) = \beta_i(w)$ for any $i$ , then $\alpha^m \beta_i(v) = \alpha^m \beta_i(w)$ for all $m, i \in N$ . Hence $B(v - w) \cap S(a) = 0$ , and so $v = w$ , as claimed.

Using this observation and "glueing" together some cyclic submodules of $I(a)$ , we will obtain the whole module $I(a)$ .

Let us divide the construction into three steps.

<u>Step 1</u> For any $n \in N$ , let $M_n$ denote the representation of $\Delta$ described in the following picture, that is $M_n = ((M_n)_a, (M_n)_b; \alpha, \beta_0, \beta_1, \ldots\ldots)$ , where $((M_n)_a, \alpha) = (V_a, \alpha)$ and $(M_n)_b$ is a one-dimensional k-vector space with basis $\{x_n\}$ and, for all $i \in N$ , the maps $\beta_i$ are defined as follows:

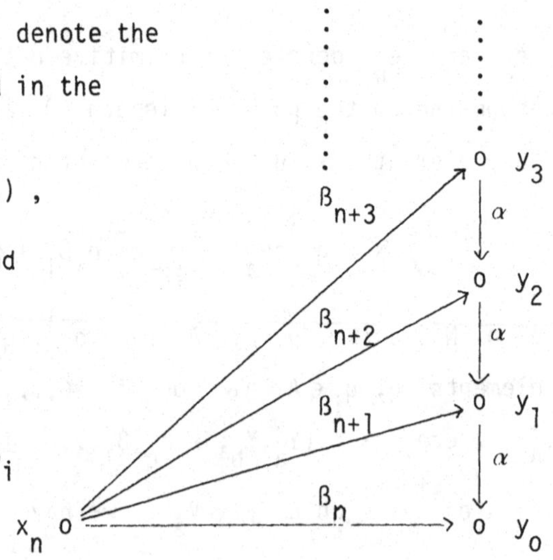


$$\beta_i(x_n) = \begin{cases} 0 & \text{if } i < n \\ y_{i-n} & \text{if } i \geq n \end{cases}$$

Since $S(a)$ is essential in $M_n$, we may assume that $M_n \subseteq I(a)$. Moreover, using the definition of the $M_n$'s, one immediately sees that $\{x_n : n \in N\}$ is a set of linearly independent vectors of $V_b$.

(Otherwise, we can choose some $i$ such that $x_{i+1} = \sum_{n=0}^{i} s_n x_n$, with the $s_n$ belonging to $k$ and not all equal to zero. But then, to find a contradiction, it is enough to note that $\beta_i(x_{i+1}) = 0$, while $\beta_i\left(\sum_{n=0}^{i} s_n x_n\right) \neq 0$.)

Therefore, $\bigoplus_{n\in N} \langle x_n \rangle \subseteq V_b$ and from the definition of $\beta_i$ we get

$\bigoplus_{n>i} \langle x_n \rangle \subseteq \operatorname{Ker} \beta_i$ for any $i \in N$.

Step 2 Now consider the representation $W = (W_a, W_b; \alpha, \beta_0, \beta_1, \ldots\ldots)$ uniquely defined by the following conditions: $(W_a, \alpha) = (V_a, \alpha)$,

$W_b = \prod_{n\in N} \langle x_n \rangle$ and, for any $i$, the map $\beta_i : W_b \longrightarrow W_a$

acts component-wise, that is if $x = (s_n x_n)_{n\in N} \in W_b$, then

$\beta_i(x) = \sum_{n\in N} \beta_i(s_n x_n)_{n\in N}$, and so, by the previous remark,

$$\beta_i(x) = \sum_{n=0}^{i} \beta_i(s_n x_n)_{n\in N} = s_0 y_i + s_1 y_{i-1} + \ldots + s_{i-1} y_1 + s_i y_0 = \sum_{n=0}^{i} s_{i-n} y_n .$$

Again, since $S(a)$ is essential in $W$, we may assume $W \subseteq I(a)$.

Step 3 To see that $I(a) = W$, take any $v \in V_b$. Then, for any $i \in N$, the relations on $\Delta$ imply that $\alpha^{i+1}\beta_i = 0$, so that we can write

$\beta_i(v) = \sum_{j=0}^{i} s_{ij}y_j$ where $s_{ij} \in k$ for any $j = 0,\dots,i$ . Let $s_{ij} = 0$ for all $i, j \in N$ with $i < j$ and let $\sigma$ denote the lower triangular matrix $[s_{ij}]_{i,j\in N}$ . Then it is easy to check that, for any $i \in N$ , all the entries $\{s_{i+n,n} : n \in N\}$ in the i-th diagonal of $\sigma$ coincide with a fixed element, say $s_i$ . (Indeed, since

$\beta_i = \alpha\beta_{i+1} = \dots\dots = \alpha^n\beta_{i+n} = \dots\dots\dots$ , we have

$\beta_i(v) = \alpha\beta_{i+1}(v) = \dots\dots = \alpha^n\beta_{i+n}(v) = \dots\dots\dots$ . Consequently,

$s_{io} = s_{i+1,1} = \dots\dots = s_{i+n,n} = \dots\dots\dots$ , as claimed.)

Using these elements $s_i \in k$ , we now consider the vector $w = (s_nx_n)_{n\in N} \in W_b$ . Since

$$\beta_i(w) = \sum_{n=0}^{i} s_{i-n}y_n = \sum_{n=0}^{i} s_{i-n,0}y_n = \sum_{n=0}^{i} s_{i,n}y_n = \beta_i(v) \quad \text{for any } i \in N ,$$

we deduce from (*) that $v = w$ . Thus $V_b = W_b$ and therefore $I(a) = W$ . Finally, let $0 \neq x \in V_b$ and write $x = (s_nx_n)_{n\in N}$ with $s_n \in k$ for all $n$ . Let $m = \min\{n \in N : s_n \neq 0\}$. Then $\beta_m(x) = s_my_o$ and so $y_o \in \bar{B}x$ . Assume now, by induction, that $y_o,\dots, y_{i-1} \in \bar{B}x$ for some $i \geq 1$ . Since

$\beta_{m+i}(x) = s_{m+i}y_o + s_{m+i-1}y_1 + \dots + s_{m+1}y_{i-1} + s_my_i$ , it follows that $y_i \in \bar{B}x$ . Hence, $\bar{B}x = (V_a, \langle x\rangle; \alpha, \beta_o, \beta_1, \dots\dots)$ and clearly, there is a $\bar{B}$-isomorphism $\bar{B}x_m \longrightarrow \bar{B}x$ , sending $x_m$ to $x$ . As a consequence, $\bar{B}x$ is artinian, but not of finite length. Since the cyclic submodules of $I(a)$ which are not generated by some $0 \neq x \in V_b$ are finite dimensional k-vector spaces contained in $V_a$ , the proof of (iii) is complete.

Proof of the proposition Let $J(\bar{B})$ be the Jacobson radical of $\bar{B}$. Since $\bar{B}/J(\bar{B})$ is a two-dimensional $k$-vector space generated by the images of $e_a$ and $e_b$, it follows that $\bar{B}$ is a semilocal ring. Moreover, by (i), (ii) and (iii), $\bar{B}$ is a left pseudo-artinian ring which is not strongly left pseudo-artinian.

Remark Let $\hat{A}$ be the completion of $A$ in its $A\alpha$-adic topology and let $\hat{B}$ be the algebra given by the following pushout diagram with obvious maps.

$$\begin{array}{ccc} A & \longrightarrow & B \\ \downarrow & & \downarrow \\ \hat{A} & \dashrightarrow & \hat{B} \end{array}$$

Then $\hat{B}$ is a left linearly compact ring and $I(a) \oplus I(b)$ is also the minimal cogenerator of ${}_{\hat{B}}\mathcal{M}$.

In fact any element $\hat{b} \in \hat{B}$ can be written in the form

$$(**) \qquad \hat{b} = \hat{r}e_a + \sum_{i=0}^{n} u_i \beta_i + ue_b$$

where $n \in N$, $\hat{r} \in \hat{A}$, $u$ and $u_i \in k$, for $i = 0,\ldots,n$. Let $L$ be the left ideal of $\hat{B}$ generated by the $\beta_i$'s, $i \in N$. Then $L$ is an artinian module (and in fact isomorphic to the minimal cogenerator of ${}_{\hat{A}}\mathcal{M}$) and we have the following equalities of $\hat{A}$-modules:

$$\hat{B} = \hat{A}e_a \oplus L \oplus ke_b$$

Thus $\hat{B}$ is a left linearly compact discrete $\hat{A}$-module and hence it is a left linearly compact discrete ring. To prove that $I(a) \oplus I(b)$ is the minimal cogenerator of ${}_{\hat{B}}\mathcal{M}$, it is enough to show that $I(a)$ is injective in ${}_{\hat{B}}\mathcal{M}$. First of all, as every element of $I(a)$ is annihilated by a power of $\alpha$, it is clear that $I(a)$ has a natural $\hat{B}$-module structure. Let $H(a)$ be the injective envelope of $I(a)$ in ${}_{\hat{B}}\mathcal{M}$. Then $e_a H(a)$ is the injective envelope in ${}_{\hat{A}}\mathcal{M}$ of $S(a)$ and hence - as is well known -

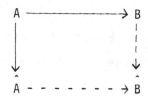

$e_a H(a) = e_a I(a)$ . Let $0 \neq x \in H(a)$ . Then there exists a $\hat{b} \in \hat{B}$ such that $0 \neq \hat{b}x \in S(a)$ . Write $\hat{b}$ in the form (**). Since $x = e_a x + e_b x$ , it is clear that there exists an $m \in N$ such that $\alpha^m x = 0$ . Let $r \in A$ such that

$$\hat{r} - r \in \hat{A}\alpha^m$$

and set

$$c = re_a + \sum_{i=0}^{n} u_i \beta_i + ue_b \quad .$$

Then $c \in \bar{B}$ and $(\hat{b} - c)x = (\hat{r} - r)x = 0$ . Thus $0 \neq \hat{b}x = cx$ and $S(a)$ is an essential $\bar{B}$-submodule of $H(a)$ . Therefore $I(a) = H(a)$, as claimed.

§2. Using the same techniques as in §1. , we can easily construct other examples of left pseudo-artinian rings which are not strongly left pseudo-artinian.

For instance, let Q be the quiver

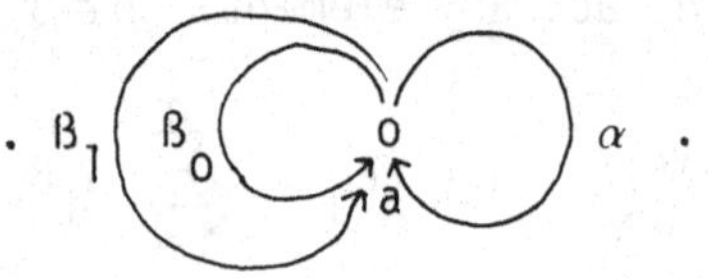


with relations

$$\alpha \beta_i = \begin{cases} 0 & \text{if } i = 0 \\ \beta_{i-1} & \text{if } i > 0 \end{cases} \quad , \quad \beta_i \alpha = 0 \quad \text{and} \quad \beta_i \beta_j = 0 \quad \text{for all } i, j \in N$$

Let A and $\bar{A}$ be as in §1. and let C be the path algebra kQ . Let now $\bar{C}$ be the algebra given by the following pushout diagram, where again the solid arrows are the obvious embeddings of A into $\bar{A}$ and C respectively.

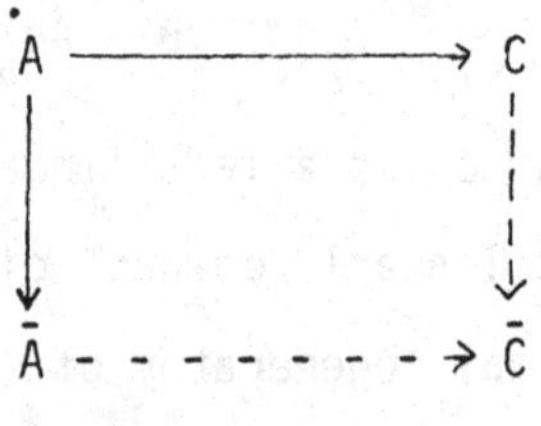


With these hypotheses, we want to prove the following

<u>Proposition</u> <u>$\bar{C}$ is a local left pseudo-artinian ring which is not strongly left pseudo-artinian.</u>

Sketch of the proof  Proceeding as in §1. , one directly verifies that the injective hull $I(a)$ of $S(a)$ in ${}_{C}\mathcal{M}$ is the minimal cogenerator of ${}_{\bar{C}}\mathcal{M}$ . Moreover, $I(a)$ consists of a vector space $V_a$ admitting a family of linearly independent vectors, say $\{x_n , y_n : n \in N\}$ such that

$$V_a = \prod_{n\in N} < x_n > \oplus \bigoplus_{n\in N} < y_n > ,$$ and the maps $\alpha$ and $\beta_i$ , for all $i \in N$ , are defined as follows: if $v = (s_n x_n)_{n\in N} + \sum_{n\in N} t_n y_n$ with $s_n , t_n \in k$ for any $n$ and the $t_n$ 's almost all zero, then

$$\alpha(v) = \sum_{n\in N} t_{n+1} y_n \quad \text{and} \quad \beta_i(v) = \sum_{n=0}^{i} s_{i-n} y_n$$ for all $i$ . Also in this case, any cyclic $\bar{C}$-submodule of $I(a)$ of infinite dimension over $k$ is artinian but not of finite length. In fact, without loss of generality, we may assume that it is generated by an element, say $x$ , of the form

$$0 \neq x = (s_n x_n)_{n\in N} \in \prod_{n\in N} < x_n >.$$

This implies that, if $m = \min \{ n \in N : s_n \neq 0 \}$ and $z_i = \beta_{i+m}(x)$ for any $i$ , then $\bar{C}x$ is the representation of $Q$ described in the following picture. The proof is now complete.

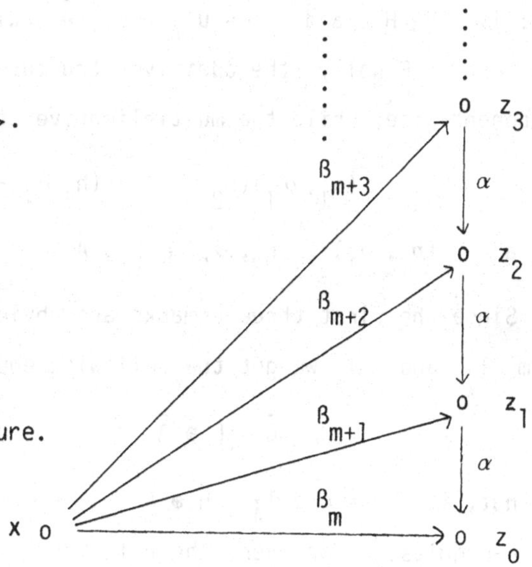


Remark  Let us note the following further properties of $\bar{C}$ .

1) Any element of $\bar{C}$ can be written in the form

$$\sum_{i=0}^{n} u_i \beta_i + \varphi \quad \text{with} \quad \varphi \in \bar{A}\ ,\ n \in N \ \text{and}\ u_i \in k \quad \text{for any}\ i\ .$$

2) We may identify the injective hull of $S(a)$ in ${}_{\bar{A}}\mathcal{M}$, say $H$, with the $\bar{C}$-submodule $(\bigoplus_{n\in N} < y_n >; \alpha, 0, 0, \ldots\ldots)$ of $I(a)$.

3) The assignement $y_n \longmapsto \beta_n$ gives rises to an isomorphism $g$ of $\bar{C}$-modules between ${}_{\bar{C}}H$ and the left ideal $L$ of $\bar{C}$ generated by the $\beta_n$ 's , $n \in N$.

4) There is an algebra isomorphism $H \oplus \bar{A} \longrightarrow \bar{C}$ where $H \oplus \bar{A}$ is the trivial extension of ${}_{\bar{A}}H_{\bar{A}}$ by $\bar{A}$, defined in analogy with the construction in [1] .

More precisely, $H$ is regarded as a right $\bar{A}$-module in the following way: if $h \in H$ and $\varphi = u + \alpha\varphi'\ \bar{A}$ with $u \in k$, $\varphi' \in \bar{A}$, then $h\varphi = hu$. Finally, the additive structure of $H \oplus \bar{A}$ is defined component-wise, while the multiplicative structure satisfies

$$(h_1, \varphi_1)(h_2, \varphi_2) = (h_1\varphi_2 + \varphi_1 h_2\ ,\ \varphi_1\varphi_2)$$

for all $(h_1, \varphi_1)\ ,\ (h_2, \varphi_2) \in H \oplus \bar{A}$.

Since the first three remarks are obvious, we only have to prove 4) . From 1) and 3) we get the following equalities of $\bar{A}$-modules:

$$\bar{C} = L \oplus \bar{A}$$

so that, if $f = g \oplus 1_{\bar{A}} : H \oplus \bar{A} \longrightarrow \bar{C}$, then $f$ is an isomorphism of $\bar{A}$-modules. Moreover, the relations on $Q$ imply that $f$ is a ring homomorphism, as is easily checked.

Nevertheless, note that $\bar{C}$ is not a ring of the type constructed in §6. of [1] . In fact $\bar{A}$ is not complete in its $\bar{A}\alpha$ -adic topology.

However, if $\hat{A}$ denotes the completion of $A$ with respect to its $A\alpha$-adic topology and $\hat{C}$ is given by the following pushout diagram with obvious maps, then, proceeding as in §1. , one proves that $\hat{C}$ is left linearly compact discrete, left pseudo-artinian, but not strongly left pseudo-artinian. Moreover, there is again an isomorphism $H \oplus \hat{A} \longrightarrow \hat{C}$ , and hence $\hat{C}$ has precisely the same form as one of the rings constructed in [1] , §6. .

$$\begin{array}{ccc} A & \longrightarrow & C \\ \downarrow & & \downarrow \\ \hat{A} & \dashrightarrow & \hat{C} \end{array}$$

## REFERENCES


[1] D. DIKRANIAN - A. ORSATTI On the structure of linearly compact rings, Abelian Groups and Modules, Udine 1984 , Springer Verlag Wien .

[2] C. MENINI - A. ORSATTI Topologically left artinian rings, to appear on the Journal of Algebra.

[3] B. OSOFSKY An example of a cyclic artinian module of infinite length, unpublished (communicated by D. EISENBUD to the second author).

[4] C.M. RINGEL Tame Algebras, Springer LMN 831 (1980) , 137-287 .

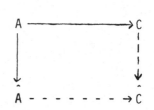

However, if $\hat{A}$ denotes the completion of $A$ with respect to its $\mathfrak{m}$-adic topology and $C$ is given by the following pushout diagram with obvious maps, then, proceeding as in §1, one proves that $C$ is left linearly compact, discrete, left pseudo-artinian, but not semiprimary (left pseudo-artinian). Moreover, there is again a pushout [illegible] and hence $C$ has precisely the same form as the one constructed in [1].

REFERENCES


[1] [illegible] and A. ORSATTI, On the structure of linearly compact rings, Italian Groups and Modules, Udine 1984, Springer, Wien.

[2] [illegible] and A. ORSATTI, Topologically [illegible] and [illegible] rings, to appear in the Journal of Algebra.

[3] B. OSOFSKY, An example of a cyclic artinian module of infinite length (unpublished, communicated by D. Fisher to the second author).

[4] C.M. RINGEL, Tame Algebras, Springer LNM 831 (1980), 137–287.

# LATTICE OF SUBMODULES AND ISOMORPHISM OF SUBQUOTIENTS

Alberto FACCHINI
Istituto di Matematica, Informatica e Sistemistica
Università di Udine
Via Mantica 3 - 33100 Udine - Italy

In this paper we consider the following problem: Let $\mathcal{L}$ be a lattice and $I(\mathcal{L})=\{[A,B] \mid A,B\in\mathcal{L},\ A<B\}$ denote the set of all intervals in $\mathcal{L}$. Let M be a module over an arbitrary ring R, and assume there is an isomorphism $\Phi:\mathcal{L}\to\mathcal{L}(M)$ of $\mathcal{L}$ onto the lattice $\mathcal{L}(M)$ of all submodules of M. We may define an equivalence relation $\sim$ on $I(\mathcal{L})$ by setting $[A,B]\sim[C,D]$ if the R-modules $\Phi(B)/\Phi(A)$ and $\Phi(D)/\Phi(C)$ are isomorphic. The relation $\sim$ depends on R, M and $\Phi$. Characterize the equivalence relations $\sim$ that can be defined in this way.

It is clear that the structure of the lattice $\mathcal{L}(M)$ determines some isomorphisms between certain subquotients of M, for instance projective intervals must correspond to isomorphic subquotients, but a general solution to our problem appears to be very difficult. In this paper we consider two particular cases: uniserial modules and semisimple modules.

In the first section we treat uniserial modules. The lattice of all submodules of the module M will be briefly called "the lattice of M" and denoted $\mathcal{L}(M)$. Thus a module is uniserial if its lattice is totally ordered. Lattices of uniserial modules are easy to characterize: a lattice $\mathcal{L}$ is isomorphic to the lattice of a uniserial module if and only if it is an algebraic totally ordered set, that is, if it is a totally ordered set which is a complete lattice and in which every element is the join of compact elements. Equivalently $\mathcal{L}$ is a totally ordered set in which every subset has an upper and a lower bound, and if $A,B\in\mathcal{L}$ and $A<B$ then there exist $A_1,B_1\in\mathcal{L}$ such that $A\leq A_1<B_1\leq B$ and $B_1$ covers $A_1$. We consider the set $I_f(\mathcal{L})$ of all intervals of $\mathcal{L}$ of finite length and we prove that there is a complete liberty in the choice of the equivalence relation $\sim$ described above, apart from the obvious condition that if two subquotients are isomorphic then their corresponding subquotients must also be isomorphic. Our construction generalizes an idea of L. Fuchs [4, pp. 175-176]. We then consider the same problem for commutative rings and the related topic of rings of triangular matrices.

In the second section we study semisimple modules. We prove that if $M_i$ is a semisimple $R_i$-module, $i=1,2$, and $\Phi:\mathcal{L}(M_1)\to\mathcal{L}(M_2)$ is a lattice isomorphism, then for all submodules $A\leq B$ and $C\leq D$ of $M_1$ the $R_1$-modules $B/A$ and $D/C$ are isomorphic if and only if the $R_2$-modules $\Phi(B)/\Phi(A)$ and $\Phi(D)/\Phi(C)$ are isomorphic. In the above terminology this means that the equivalence relation $\sim$ on $I(\mathcal{L})$ induced by the isomorphism relation among subquotients can be chosen in a unique way.

Therefore these are the two extreme cases: the equivalence relation $\sim$ may be completely arbitrary in the first case (uniserial modules), but it can be chosen in a unique way in the second case (semisimple modules).

The symbol $\mathcal{L}$ will always denote a lattice, and A,B,C... will denote its elements (usually $\mathcal{L}$ will be the lattice $\mathcal{L}(M)$ of a module M, and A,B,C... will be submodules of M.) A subset S of $\mathcal{L}$ is convex if $A,B\in S$, $C\in\mathcal{L}$ and $A\leq C\leq B$ imply that $C\in S$. An element B covers A, in notation $B\succ A$, if $B>A$ and for no C, $B>C>A$. For $A,B\in\mathcal{L}$, $A<B$, the interval $[A,B]$ is $\{X\in\mathcal{L} \mid A\leq X\leq B\}$. A finite interval is an interval with a finite number of elements. The symbol $\cong$ between two lat-

tices (ordered sets) indicates isomorphism of lattices (ordered sets). The set of all intervals of $\mathcal{L}$ is denoted by $I(\mathcal{L})$ and the set of all finite intervals is denoted by $I_f(\mathcal{L})$. All rings are associative with 1 and all modules are unitary. A subquotient of a module M is a module of the form B/A with $A \leq B \leq M$, that is a submodule of a quotient module of M.

## 1. Uniserial modules

Let $\mathcal{L}$ be an algebraic totally ordered set, and let $I_f(\mathcal{L})$ be the set of all finite intervals in $\mathcal{L}$. Assume that $\sim$ is an equivalence relation in $I_f(\mathcal{L})$ with the following property:

(*) if $[A,B]$, $[C,D] \in I_f(\mathcal{L})$ and $[A,B] \sim [C,D]$, then $[A,B] \cong [C,D]$ as ordered sets, and if $i:[A,B] \to [C,D]$ is the unique order-isomorphism then $[A_1, B_1] \sim [i(A_1), i(B_1)]$ for every $A_1, B_1 \in \mathcal{L}$ with $A \leq A_1 < B_1 \leq B$.

THEOREM 1. There exist a ring R, a uniserial R-module M, and an order-isomorphism $\Phi:\mathcal{L} \to \mathcal{L}(M)$ such that for all $[A,B]$, $[C,D] \in I_f(\mathcal{L})$, $[A,B] \sim [C,D]$ if and only if the R-modules $\Phi(B)/\Phi(A)$ and $\Phi(D)/\Phi(C)$ are isomorphic.

PROOF. Let $\mathcal{L}_o = \{A \in \mathcal{L} \mid A \succ B$ for some $B \in \mathcal{L}\}$ be the set of all elements that cover some element of $\mathcal{L}$. Define a relation $\equiv$ in $\mathcal{L}_o$: if $A, B \in \mathcal{L}_o$, set $A \equiv B$ if there are only finitely many elements of $\mathcal{L}$ between A and B. Clearly $\equiv$ is an equivalence relation in $\mathcal{L}_o$. Let $\mathcal{L}_o/\equiv$ denote the quotient set of $\mathcal{L}_o$ modulo $\equiv$. Since the equivalence classes modulo $\equiv$ are convex subsets of $\mathcal{L}_o$, the ordering relation $\leq$ on $\mathcal{L}_o$ induces a total ordering relation $\leq$ on $\mathcal{L}_o/\equiv$.

If $A \in \mathcal{L}_o$ and $[A]$ is the equivalence class of A modulo $\equiv$, then $[A]$ is a totally ordered set with the property that there are only a finite number of elements between any two elements of $[A]$. Hence four cases may arise: $[A]$ must be order-anti-isomorphic to either $\mathbb{Z}$, or $\mathbb{Z}^+$, or $\mathbb{Z}^-$, or $[0,n]$, where $\mathbb{Z}$, $\mathbb{Z}^+$, $\mathbb{Z}^-$, $[0,n]$ denote the set of all integers, positive integers, negative integers, integers between 0 and n, respectively. For each $[A]\in\mathcal{L}_0/\equiv$ fix an order-anti-isomorphism $\phi_{[A]}$ of $[A]$ into $\mathbb{Z}$, $\mathbb{Z}^+$, $\mathbb{Z}^-$, $[0,n]$, respectively.

Let K be any field and x be an indeterminate over K. If K[[x]] is the ring of formal power series, define a K[[x]]-module $M_{[A]}$ for each $[A]\in\mathcal{L}_0/\equiv$ in the following way: if $[A]\cong_a\mathbb{Z}$ (anti-isomorphism), define $M_{[A]}=K((x))$, the field of formal Laurent series, which is the field of fractions of K[[x]], viewed as a module over K[[x]]; if $[A]\cong_a\mathbb{Z}^+$, define $M_{[A]}=xK[[x]]$, the maximal ideal of the ring K[[x]]; if $[A]\cong_a\mathbb{Z}^-$, define $M_{[A]}=K((x))/K[[x]]$; finally, if $[A]\cong_a[0,n]$, define $M_{[A]}=K[[x]]/x^{n+1}K[[x]]$. Set $M=\bigoplus_{[A]\in\mathcal{L}_0/\equiv}M_{[A]}$; then M is a K[[x]]-module; in particular it is a vector space over K.

For each $A,B\in\mathcal{L}_0$ such that $A\not\equiv B$ and $[B]<[A]$ and for each $f\in\mathrm{Hom}_K(M_{[A]},M_{[B]})$ let $y_{A,B,f}\in\mathrm{End}_K(M)$ be defined as the endomorphism of M whose restriction to $M_{[A]}$ is equal to f and whose restriction to $M_{[C]}$ is equal to zero for all $[C]\in\mathcal{L}_0/\equiv$, $[C]\neq[A]$.

For each $[A,B]\in I_f(\mathcal{L})$ and each $[C]\in\mathcal{L}_0/\equiv$, let $g_{[A,B],[C]}:K((x))\to K((x))$ be the unique continuous K-linear mapping such that

$$g_{[A,B],[C]}(x^r)=x^s$$

if $r\in\phi_{[C]}([C])$, there exists $D\in\mathcal{L}$ such that $[A,B]\sim[D,\phi_{[C]}^{-1}(r)]$, and $s=\phi(D^+)$, where $D^+$ is the unique element that covers D, and

$$g_{[A,B],[C]}(x^r)=0$$

otherwise. Note that $g_{[A,B],[C]}$ is well-defined because if $D,D'\in[C]$ are such that $[A,B]\sim[D,\phi_{[C]}^{-1}(r)]$ and $[A,B]\sim[D',\phi_{[C]}^{-1}(r)]$, then $[D,\phi_{[C]}^{-1}(r)]\cong[D',\phi_{[C]}^{-1}(r)$ have finite length >0, so that D=D' and D is covered by a unique element $D^+$. Moreover $g_{[A,B],[C]}(x^nK[[x]])\subset x^nK[[x]]$ for $n\geq 0$, so that $g_{[A,B],[C]}$ induces a K-linear morphism $\bar{g}_{[A,B],[C]}:M_{[C]}\to M_{[C]}$. Let $z_{[A,B]}:M\to M$ be the direct sum of all $\bar{g}_{[A,B],[C]}$ where [C] ranges in $\mathcal{L}_0/\equiv$.

Now K[[x]] acts on M, so that there is a ring morphism $\varepsilon:K[[x]]\to\mathrm{End}_K(M)$. Let R be the subring of $\mathrm{End}_K(M)$ generated by $\varepsilon(K[[x]])$, all the endomorphisms

$y_{A,B,f}$ (with $A,B\in\mathcal{L}_o$, $A\neq B$, $[B]<[A]$ and $f\in\mathrm{Hom}_K(M_{[A]}, M_{[B]})$), and all the endomorphisms $z_{[A,B]}$ (with $[A,B]\in I_f(\mathcal{L})$.) Then M is the left R-module we look for.

In fact, it is trivial to verify that M is uniserial, because its nonzero cyclic submodules are exactly those generated by $x^{\phi_{[A]}(A)}\in M_{[A]}$ (if $\phi_{[A]}([A])=\mathbb{Z}$ or $\mathbb{Z}^+$) and by $x^{\overline{\phi_{[A]}(A)}}\in M_{[A]}$ (if $\phi_{[A]}([A])=\mathbb{Z}^-$ or $[0,n]$) for $A\in\mathcal{L}_o$. Hence $\mathcal{L}(M)$ and $\mathcal{L}$ have isomorphic join-semilattices of compact elements (the compact elements of $\mathcal{L}(M)$ are the cyclic submodules of M, and the compact elements of $\mathcal{L}$ are 0 and the elements of $\mathcal{L}_o$). Since any algebraic lattice is isomorphic to the lattice of all ideals of the join-semilattice of its compact elements [5, Th. II.3.13], there is an isomorphism $\Phi:\mathcal{L}\to\mathcal{L}(M)$.

Let us show that if $[A,B],[C,D]\in I_f(\mathcal{L})$ then $[A,B]\sim[C,D]$ if and only if $\Phi(B)/\Phi(A)$ and $\Phi(D)/\Phi(C)$ are R-isomorphic.

If $[A,B]\sim[C,D]$, let $i:[A,B]\to[C,D]$ be the unique order-isomorphism. Note that $\Phi(B)/\Phi(A)$ is canonically K-isomorphic to $\bigoplus_n Kx^n$, where n ranges in $[\phi_{[B]}(B), \phi_{[B]}(A^+)]$. Similarly $\Phi(D)/\Phi(C)\cong\bigoplus_m Kx^m$, with $m\in[\phi_{[D]}(D), \phi_{[D]}(C^+)]$. Define $\psi:\Phi(B)/\Phi(A)\to\Phi(D)/\Phi(C)$ as $\psi(x^n)=x^{\phi_{[D]}(i(\phi_{[B]}^{-1}(n)))}$. Showing that $\psi$ is an R-isomorphism is a routine verification. We leave the details to the reader.

Conversely, suppose that $\Phi(B)/\Phi(A)\cong\Phi(D)/\Phi(C)$. Now $\Phi(B)/\Phi(A)\cong\bigoplus_n Kx^n$ with n in $[\phi_{[B]}(B), \phi_{[B]}(A^+)]$ and $z_{[A,B]}x^{\phi_{[B]}(B)}=\overline{g}_{[A,B],[B]}x^{\phi_{[B]}(B)}=x^{\phi_{[B]}(A^+)}\neq 0$. Therefore $z_{[A,B]}$ does not annihilate $\Phi(B)/\Phi(A)\cong\Phi(D)/\Phi(C)$. Hence $z_{[A,B]}$ does not annihilate one of the $x^m$, $\phi_{[D]}(D)\leq m\leq\phi_{[D]}(C^+)$. But it is easy to check that $z_{[A,B]}x^m=0$ for $\phi_{[D]}(D)<m\leq\phi_{[D]}(C^+)$, so that $z_{[A,B]}x^{\phi_{[D]}(D)}\neq 0$. It follows that $[A,B]\sim[C',D]$ for some C'. But $[C,D]$, $[A,B]$ and $[C',D]$ have the same length, so that C=C' and $[A,B]\sim[C,D]$. This concludes the proof of the theorem.

Theorem 1 essentially says that we may construct uniserial modules with arbitrarily fixed isomorphism among the subquotients of finite length (arbitrarily, apart from the obvious condition (*)). As soon as we put some extra conditions on the ring R the situation changes radically. A trivial example is the following: suppose we want the ring R to be a chain ring (i.e., its left ideals to be totally ordered with respect to inclusion). If R is a chain ring, any two uniserial R-modules of the same finite length are isomorphic. Therefore there is no freedom of choice of $\sim$ in this trivial case.

A more natural and less trivial condition is the commutativity of R. It is possible to show that the lattices isomorphic to lattices of uniserial modules over commutative rings are exactly the lattices isomorphic to augmented ideal lattices of convex subsets of totally ordered abelian groups (For the terminology see [5]. For a similar result see [10]). If M is a uniserial module over a commutative ring R, any two subquotients of M of the same finite length are isomorphic, as the next proposition shows. Therefore there is no freedom of choice of $\sim$ again.

PROPOSITION 2. Let M be a uniserial module over a commutative ring R. Assume that $A \leq B$, $C \leq D$ are submodules of M and B/A, D/C have finite length. Then B/A and D/C are isomorphic R-modules if and only if [A,B], [C,D] are isomorphic ordered sets, i.e., if and only if they have the same length.

PROOF. Suppose [A,B],[C,D] have the same finite length $\geq 1$. Then B and D are cyclic, and without loss of generality we may suppose $B \leq D$. If B=Rb and D=Rd, then b=rd for some $r \in R$. The multiplication by r is an endomorphism of M, which induces an isomorphism $B/\mathrm{Ann}_M(r) \cong D$. Since [A,B] and [C,D] have the same length, $A/\mathrm{Ann}_M(r)$ corresponds to C via this isomorphism. Hence B/A and D/C are isomorphic. The converse is obvious.□

We conclude this section by pointing out the close connection between uniserial modules and rings of triangular matrices. Rings strongly related with rings of triangular matrices have been considered in the study of Noetherian hereditary semi-perfect prime rings [8]. Block lower triangular matrix rings are exactly indecomposable hereditary serial rings [9]. There are other applications of rings of upper triangular matrices in Warfield [11] and Haack [6]. A careful examination of the ring R constructed in the proof of Theorem 1 reveals that R is a subring of a ring of triangular matrices. The next result and its corollary show that this was almost necessary.

THEOREM 3. Let R be a ring, M be a faithful Artinian uniserial left R-module. If $A=\mathrm{End}_R(M)$ is a division ring, then R is isomorphic to a subring of a ring of block upper triangular column finite matrices over A.

PROOF. Note that every submodule of ${}_RM$ is fully invariant, that is, it is a subbimodule of ${}_RM_A$. In fact, if $\alpha \in A$, then either $\alpha=0$ or $\alpha$ is an automorphism of ${}_RM$ (because A is a division ring). In this second case $\alpha$ induces an order-automorphism $\mathcal{L}({}_RM) \to \mathcal{L}({}_RM)$. But $\mathcal{L}({}_RM)$ is an Artinian totally ordered set, i.e., it is well-ordered, and a well-ordered set has no automorphism different from the identity. Therefore $\alpha$ induces the identity on $\mathcal{L}({}_RM)$, i.e. every submodule of ${}_RM$ is $\alpha$-invariant.

Since M is faithful, there is a canonical embedding $\omega: R \to \mathrm{End}(M_A)$. We have just seen that the R-submodules of M are (R-A)-subbimodules and are well-ordered with respect to inclusion; if $\{M_\gamma \mid \gamma \leq \alpha\}$ are all these submodules (with $\alpha$ an ordinal and $0=M_0 \leq M_1 \leq \ldots \leq M_\alpha = M$), and for all $\gamma<\alpha$ $C_\gamma$ is a complement, i.e. $M_{\gamma+1} = M_\gamma \oplus C_\gamma$ as right vector spaces over A, then $M_\delta = \bigoplus_{\gamma<\delta} C_\gamma$ for all $\delta \leq \alpha$ as right vector spaces (an easy induction on $\delta$). The conclusion follows immediately. □

COROLLARY 4. Let R be a ring, M a faithful Artinian uniserial left R-module. Assume that the socle of M is not isomorphic to any other subquotient of M. Then $A=\mathrm{End}_R(M)$ is a division ring, and R is isomorphic to a subring of a ring of block upper triangular column finite matrices over A.

PROOF. We must show that if $\alpha \in A = \mathrm{End}_R(M)$ and $\alpha \neq 0$, then $\alpha$ is a bijection. Now $\alpha: M \to M$ induces a monomorphism $\overline{\alpha}: M/\ker\alpha \to M$ with $M/\ker\alpha \neq 0$, and therefore it induces an isomorphism between $\mathrm{Soc}(M/\ker\alpha)$ and $\mathrm{Soc}(M)$. By our assumptions $\ker\alpha = 0$, i.e., $\alpha$ is injective. It follows that $\alpha$ induces an order-monomorphism $\mathcal{L}(M) \to \mathcal{L}(M)$. But $\mathcal{L}(M)$ is well-ordered and a well-ordered set cannot be isomorphic to a proper initial interval. Thus the order-monomorphism $\mathcal{L}(M) \to \mathcal{L}(M)$ induced by $\alpha$ is surjective. It follows that $\alpha$ is also surjective.

2. Semisimple modules

Von Neumann's celebrated Coordinatization Theorem may be stated in the following way: "Every atomic arguesian lattice of finite length is isomorphic to the lattice of all right ideals of a suitable semi-simple ring." [2, Ch. IV, Th. 17]. We need the Coordinatization Theorem in the following form, which characterizes the lattices of semisimple modules.

THEOREM 5. Let $\mathcal{L}$ be a lattice. The following statements are equivalent:

i) $\mathcal{L}$ is isomorphic to $\mathcal{L}(M)$ for a semisimple module M over a ring R;

ii) $\mathcal{L}$ is complemented and is isomorphic to $\mathcal{L}(M)$ for a module M over a ring R;

iii) $\mathcal{L}$ is a complemented arguesian algebraic lattice, and every interval [A,B] in $\mathcal{L}$ of length two has either an infinite number of atoms or $p^n+1$ atoms for some prime p and some $n \geq 0$.

PROOF. The equivalence of i) and ii) is trivial, because a module M is semisimple if and only if every submodule of M is a direct summand, that is, if and only if $\mathcal{L}(M)$ is complemented.

i)$\Rightarrow$iii). It is well-known that $\mathcal{L}(M)$ is an arguesian algebraic lattice. We must prove the statement about the intervals of length two. By taking a subquotient of M we may suppose that M is a semisimple module with $\mathcal{L}(M)$ of length two. Then M is the direct sum of two simple modules, $M=S_1\oplus S_2$. If $S_1$ and $S_2$ are not isomorphic, then $\mathcal{L}(M)=\{0, S_1, S_2, M\}$ has two atoms. If $S_1\cong S_2$, we may suppose M a faithful R-module (by passing to $R/\mathrm{Ann}_R M$). In this case $S=S_1$ is a faithful simple module, and therefore R is a left primitive ring. By the Density Theorem $D=\mathrm{End}(_R S)$ is a division ring, and R is dense in $\mathrm{End}(S_D)$. Moreover the left R- and $\mathrm{End}(S_D)$-submodules of $S\oplus S$ coincide. In fact, if $f\in\mathrm{End}(S_D)$ and $(s_1,s_2)\in S\oplus S$, then $f(s_1,s_2)=(f(s_1),f(s_2))=(rs_1,rs_2)=r(s_1,s_2)$ for some $r\in R$ because R is dense in $\mathrm{End}(S_D)$. Therefore the number of atoms in $\mathcal{L}(M)$ is equal to the number of atoms in $\mathcal{L}(S\oplus S)$ considered as an $\mathrm{End}(S_D)$-module. If we compute the cyclic $\mathrm{End}(S_D)$-submodules of $S\oplus S$ we see that $(s_1,s_2)$ generates $S\oplus S$ if $s_1$ and $s_2$ are linearly independent over D and generates $\{(s, s\lambda) \mid s\in S_D\}$ if $s_2=s_1\lambda$ for $\lambda\in D$, say. Therefore the atoms of $\mathcal{L}(M)$ are $|D|+1$. The conclusion follows immediately.

iii)⇒i). We leave the proof to the reader. Recall that a complemented modular algebraic lattice is a direct product of subdirectly irreducible lattices [3, 11.10], and every complemented arguesian algebraic subdirectly irreducible lattice of length 3 or more is isomorphic to the lattice of a vector space over some division ring. The hypothesis on the number of atoms is used for the subdirectly irreducible direct factors of length two. □

THEOREM 6. Let $M_i$ be a semisimple $R_i$-module, i=1,2, and let $\Phi:\mathcal{L}(M_1)\to\mathcal{L}(M_2)$ be a lattice isomorphism. Then for all $A\leq B$, $C\leq D$ submodules of $M_1$, the $R_1$-modules B/A and D/C are isomorphic if and only if the $R_2$-modules $\Phi(B)/\Phi(A)$ and $\Phi(D)/\Phi(C)$ are isomorphic.

PROOF. The lattices $\mathcal{L}(M_1)$, $\mathcal{L}(M_2)$ are direct products of subdirectly irreducible lattices in a unique way, and these decompositions correspond to the direct decompositions of $M_1$, $M_2$ into direct sums of isotopic components. Therefore $M_i=\sum_\lambda M_{i,\lambda}$ and $\mathcal{L}(M_i)\cong\prod_\lambda\mathcal{L}(M_{i,\lambda})$, where $M_{i,\lambda}$ is a sum of isomorphic simple modules, and no nonzero submodules of $M_{i,\lambda}$, $M_{i,\mu}$ are isomorphic for $\lambda\neq\mu$. Thus $\Phi=\prod_\lambda \Phi_\lambda$, where $\Phi_\lambda:\mathcal{L}(M_{1,\lambda})\to\mathcal{L}(M_{2,\lambda})$ is an isomorphism. Now $A=\bigoplus_\lambda(A\cap M_{1,\lambda})$, and similarly for B,C,D. Therefore $B/A\cong D/C$ if and only if $\bigoplus_\lambda(B\cap M_{1,\lambda}/A\cap M_{1,\lambda})\cong$ $\cong\bigoplus_\lambda(D\cap M_{1,\lambda}/C\cap M_{1,\lambda})$, i.e., if and only if (**) $B\cap M_{1,\lambda}/A\cap M_{1,\lambda}\cong D\cap M_{1,\lambda}/C\cap M_{1,\lambda}$ for all $\lambda$. Since these are isotopic semisimple modules, this holds if and only if the two members of (**) are direct sums of equipotent families of simple modules, that is, if and only if the intervals $[A\cap M_{1,\lambda}, B\cap M_{1,\lambda}]$ and $[C\cap M_{1,\lambda}, D\cap M_{1,\lambda}]$ are isomorphic for all $\lambda$. It is now clear that the structure of the lattice completely determines the isomorphisms between the subquotients, and the Theorem follows easily. □

We conclude with a remark. The Coordinatization Theorem has been generalized in various ways. One of the best generalizations is due to Jónsson and Monk and concerns primary arguesian lattices [7]. This situation had also been considered by Baer [1]. Every primary lattice is either totally ordered or simple [7, Th. 6.2]. In Section 1 we already considered the case of totally ordered lattices in detail. If a lattice is primary and simple, then any two intervals of length one are projective. It follows that if the module M is not uniserial and $\mathcal{L}(M)$ is a primary lattice, then all simple subquotients of M must be isomorphic.

References

1. Baer, R. Kollineationen primärer Praemoduln, in *Studies on Abelian Groups*, (Symposium, Montpellier, 1967), pp. 1-36, Springer, Berlin, 1968.

2. Birkhoff, G., *Lattice theory*, Third edition, Amer. Math. Soc., Providence R. I., 1967.

3. Crawley, P. and Dilworth, R. P., *Algebraic theory of lattices*, Prentice-Hall, Englewood Cliffs, N. J., 1973.

4. Fuchs, L., Torsion preradicals and the ascending Loewy series of modules

*J. reine angew. Math.* 239/240 (1969) 169-179.

5. Grätzer, G., *General lattice theory*, Birkhäuser Verlag, Basel, 1978.

6. Haack, J. K., Serial rings and subdirect products, *J. Pure Appl. Algebra* 30 (1983) 157-165.

7. Jónsson, B. and Monk, G. S., Representations of primary arguesian lattices, *Pacific J. Math.* 30 (1969) 95-139.

8. Michler, G. O., Structure of semi-perfect hereditary noetherian rings, *J. Algebra* 13 (1969) 327-344.

9. Robson, J. C., Idealizers and hereditary noetherian prime rings, *J. Algebra* 22 (1972) 45-81.

10. Shores, T. S., On generalized valuation rings, *Michigan Math. J.* 21 (1975) 405-409.

11. Warfield, R. B., Serial rings and finitely presented modules, *J. Algebra* 37 (1975) 187-222.

# A MODULE-THEORETICAL APPROACH TO VECTOR SPACE CATEGORIES

DANIEL SIMSON
NICHOLAS COPERNICUS UNIVERSITY
87-100 TORUŃ, POLAND

## 1. Introduction

The concepts of a vector space category over an algebraically closed field and a subspace category were introduced by Nazarova and Rojter [9] in a connection with the second Brauer-Thrall conjecture. In [11, 12] Ringel presents a nice categorical explanation of these concepts and of their use. In the present note we want to give a brief introduction to the socle projective modules technique in the study of vector space categories and indecomposable modules over artinian rings. This approach was introduced in [15, 18] as a generalization of the Gabriel's I-spaces technique [5] and of the Coxeter type arguments by Drozd [4] applied to matrix representations of posets introduced by Nazarova and Rojter in [8].

Let F be a division ring. We recall from [9, 11] that a vector space category $\mathbb{K}_F$ is an additive category $\mathbb{K}$ together with a faithful additive functor $|-|:\mathbb{K} \longrightarrow \mathrm{mod}\, F$ where by mod R we mean the category of finitely generated right R-modules. A subspace category $U(\mathbb{K}_F)$ of $\mathbb{K}_F$ consists of triples $\underline{C} = (V_F, X, t)$ where $V_F$ is in mod F , X is an object of $\mathbb{K}$ and $t: V_F \longrightarrow |X|_F$ is a linear map. A map from $\underline{C}$ to $\underline{C}'$ in $U(\mathbb{K}_F)$ is a pair (u,h) where $u \in \mathrm{Hom}_F(U, U')$ and $h: X \to X'$ is a map in $\mathbb{K}$ such that $|h|t = t'u$. The factor space category $V(\mathbb{K}_F)$ of $\mathbb{K}_F$ consists of triples $\underline{C} = (V_F, X, t)$ where $t: |X|_F \to V_F$ is a map in mod F. We define a pair of additive functors

$$U(\mathbb{K}_F) \underset{S^+}{\overset{S^-}{\rightleftarrows}} V(\mathbb{K}_F)$$

by taking the cokernel and the kernel of t, respectively. Given a full subcategory $\underline{\underline{B}}$ of $\underline{\underline{A}}$ we denote by $[\underline{\underline{B}}]$ the two sided ideal in $\underline{\underline{A}}$ generated by all maps in $\underline{\underline{B}}$. We denote by $\underline{\underline{A}}/[\underline{\underline{B}}]$ the factor category of $\underline{\underline{A}}$ modulo $[\underline{\underline{B}}]$.

One of the aims of this note is to give a functorial connection between categories $U(\mathbb{K}_F)$, $V(\mathbb{K}_F)$ and categories $\mathrm{mod}_{sp} R$ of socle projective modules over right peak rings R. A ring R is called right peak ring if R is semiperfect, $\mathrm{soc}\, R_R$ is essential in R and it is a finite direct sum of copies of a simple projective right ideal P of R called a right peak of R. A module $X_R$ is socle projective if $\mathrm{soc}\, X_R$ is a projective and essential submodule of $X_R$. The category $\mathrm{mod}_{ti} R$ of top injective modules is defined analogously.

## 2. BASIC REDUCTIONS

Throughout we suppose for simplicity that $\mathbb{K}_F$ is a Krull-Schmidt vector space category with a finite number of pairwise nonisomorphic indecomposable objects $K_1,\dots,K_n$. The general case is discussed in [19]. We associate to $\mathbb{K}_F$ the right peak ring

$$(2.1') \quad R_{\mathbb{K}} = \begin{bmatrix} E & {}_EK_F \\ 0 & F \end{bmatrix} = \begin{bmatrix} F_1 & {}_1K_2 & \cdots & {}_1K_n & {}_1K_{n+1} \\ {}_2K_1 & F_2 & \cdots & {}_2K_n & {}_2K_{n+1} \\ \vdots & & \ddots & \vdots & \vdots \\ {}_nK_1 & {}_nK_2 & \cdots & F_n & {}_nK_{n+1} \\ 0 & 0 & \cdots & 0 & F_{n+1} \end{bmatrix}$$

where $E = \mathrm{End}(K_1 \oplus\dots\oplus K_n)$, ${}_EK_F = {}_E|K_1 \oplus\dots\oplus K_n|_F$, $F_j = \mathrm{End}\, K_j$ for $j \leqslant n$, $F_{n+1} = F$, ${}_iK_j = \mathbb{K}(K_j, K_i)$ for $i,j \leqslant n$, ${}_iK_{n+1} = {}_{F_i}|K_i|_F$ and the multiplication is defined by bimodule maps $c_{ijs}: {}_iK_j \otimes {}_jK_s \longrightarrow {}_iK_s$ defined in a natural way [18]. We suppose that all $F_j$ are local rings. Any right $R_{\mathbb{K}}$-module $X$ will be identified with a triple $X = (X'_E, X''_F; \varphi)$ where $X'_E$ is in mod $E$, $X''_F$ is in mod $F$ and $\varphi: X'_E \otimes_E K_F \to X''_F$ is a linear map. $X$ is socle projective if and only if the map $\overline{\varphi}$ adjoint to $\varphi$ is injective. It is easy to check that there is a commutative diagram

$$\begin{array}{ccc} \mathbb{K} & \xrightarrow{\;|-|\;} & \mathrm{mod}\, F \\ \big\downarrow w & \nearrow & -\otimes_E K_F \\ \mathrm{pr}(E) & & \end{array}$$

where pr(E) is the category of finitely generated projective right E-modules and w is the Yoneda equivalence. Following an idea of Drozd [4] we define the functor $H:V(\mathbb{K}_F) \to \mathrm{mod}_{sp} R_{\mathbb{K}}$ as the composition $V(\mathbb{K}_F) \xrightarrow{w^+} \mathrm{mod}(R_{\mathbb{K}}) \xrightarrow{\breve{\Theta}} \mathrm{mod}_{sp} R_{\mathbb{K}}$ where $w^+$ is the full embedding induced by w and $\breve{\Theta}(X) = (\check{X}'_E, X''_F; \check{\varphi})$ with $\check{X}'_E = \mathrm{Im}\,\overline{\varphi}$, and $\check{\varphi}$ is the map adjoint to the inclusion $\check{X}'_E \hookrightarrow \mathrm{Hom}_F({}_E K_F, X''_F)$.

THEOREM 2.1. The functor H is full and dense. $H(\underline{C}) = 0$ if and only if $\underline{C} = (0,X,0)$. Moreover, $\mathrm{Ker}\, H = [(0,K_1,0) \ldots (0,K_n,0)]$

Proof. If $\underline{C}$ is indecomposable with $V_F \neq 0$ and $w^+(\underline{C}) = (P_E, V_F, \varphi)$ then $H(\underline{C}) = (\check{P}_E, V_F, \check{\varphi})$ and it is easy to see that $\varphi : P_E \to \check{P}_E$ is a projective cover. Hence the theorem follows by standard projective cover arguments [18, Theorem 3.3].

Now we denote by $\mathbb{K}^*_F$ the category $\mathbb{K}^{op}$ together with the composed functor $\mathbb{K}^{op} \xrightarrow{|-|^{op}} (\mathrm{mod}\ F)^{op} \xrightarrow{(-)^*} \mathrm{mod}(F^{op})$. It is easy to see that $R_{\mathbb{K}^*} = \begin{pmatrix} E^{op} & K^* \\ 0 & F^{op} \end{pmatrix}$, $R^{op}_{\mathbb{K}} = \begin{pmatrix} F^{op} & K \\ 0 & E^{op} \end{pmatrix}$ and if $R_{\mathbb{K}}$ is an artinian PI-ring then there is an equivalence of categories

$$\mathrm{mod}_{sp} R_{\mathbb{K}^*} \xrightarrow{\nabla} \mathrm{mod}_{ti} R^{op}_{\mathbb{K}} \xrightarrow{D} (\mathrm{mod}_{sp} (R_{\mathbb{K}})_*)^{op}$$ where

$\nabla(X'; X''; \varphi) = (Y'; Y''; \Psi)$, $Y' = X''$; $Y'' = \mathrm{Coker}\,\overline{\varphi}$ and $\Psi$ is the composition of the cokernel map with the natural isomorphism $Y' \otimes_{F^{op}} K \cong \mathrm{Hom}_{F^{op}}(K^*, Y')$; D is a Morita duality and $(R_{\mathbb{K}})_*$ is a ring Morita dual to $R_{\mathbb{K}}$. If $R_{\mathbb{K}}$ is an Artin algebra then $(R_{\mathbb{K}})_* \cong R_{\mathbb{K}}$.

From Theorem 2.1 immediatelly follows

COROLLARY 2.2. If $(-)^*:U(\mathbb{K}_F) \longrightarrow V(\mathbb{K}_F^*)^{op}$ is the natural duality then the functor $H^* = (-)^*H^{op}:U(\mathbb{K}_F) \to (\mathrm{mod}_{sp}R_{\mathbb{K}^*})^{op}$ is full, dense and $\mathrm{Ker}\, H^* = [\{(0,K_1,0),\dots,(0,K_n,0)\}]$.

Consider the second reduction functor

$G:U(\mathbb{K}_F) \longrightarrow \mathrm{mod}_{sp}(R_{\mathbb{K}})_*$ which is the composed functor

$$U(\mathbb{K}_F) \xrightarrow{(-)^*} V(\mathbb{K}_F^*)^{op} \xrightarrow{H^{op}} (\mathrm{mod}_{sp}R_{\mathbb{K}^*})^{op} \xrightarrow{D\nabla} \mathrm{mod}_{sp}(R_{\mathbb{K}})_*.$$

From Theorem 2.1 immediately follows

THEOREM 2.3. G is full, dense and $\mathrm{Ker}\, G = \mathrm{Ker}\, H^*$.

Let us describe H and G in a special case. We say that $\mathbb{K}_F$ is of the <u>poset type</u> if $\mathbb{K}_F$ is schurian (i.e. $F_i$ is a division ring for any i) and $\dim_{F_j}|K_j|_F = \dim |K_j|_F = 1$ for all j. In this case we define a partial order on the set $I = I(\mathbb{K}_F) = \{1,\dots,n,n+1\}$ by $i<j$ iff ${}_iK_j \neq 0$. It is clear that I is a partially ordered set (abbr. poset) with a unique maximal element n+1.

Let us recall from [5] that an I-<u>space</u> is a system $(V, V_j)$, $j\in I$, where V is in mod F and $V_j \subseteq V$ is a subspace such that $V_i \subseteq V_j$ provided $i \leqslant j$. We denote by I-sp the category of all I-spaces.

COROLLARY 2.4. If $\mathbb{K}_F$ is of the poset type then

(1) $I = I(\mathbb{K}_F)$ is a poset, $R_{\mathbb{K}}$ is isomorphic to the incidence ring FI of I with coefficients in F, $\mathrm{mod}_{sp}R_{\mathbb{K}} \cong \hat{I}\text{-sp} \cong \mathrm{mod}_{sp}(R_{\mathbb{K}})_*$ and $\mathrm{mod}_{sp}R_{\mathbb{K}^*} \cong \hat{I}^{op}\text{-sp}$ where $\hat{I} = I - \{n+1\}$.

(2) The functors H and G induce functors

$$\mathcal{H}: V(\mathbb{K}_F) \longrightarrow I\text{-sp}, \quad \mathcal{G}: U(\mathbb{K}_F) \longrightarrow I\text{-sp}$$

having the following properties. If $Y \cong K_1^{s_1} \oplus \dots \oplus K_n^{s_n}$, $\underline{C} = (V, Y, t)$, $\mathcal{H}(\underline{C}) = (U, U_j)$ and $\mathcal{G}(\underline{C}) = (U', U'_j)$ then

(a) $U = V$, $U' = V$ and $U_j = \operatorname{Im}(\bigoplus_{i \leq j} |K_i^{s_i}|_F \hookrightarrow |Y|_F \xrightarrow{t} V_F)$, $U'_j = \operatorname{Ker}(V_F \xrightarrow{t} |Y|_F \xrightarrow{\pi_j} \bigoplus_{i > j} |K_i^{s_i}|_F)$ where $\pi_j$ is the j-projection.

(b) If $\underline{C}$ is indecomposable and $V \neq 0$ then

$$s_j = \dim(U_j / \sum_{i<j} U_i)_F = \dim(\bigcap_{i>j} U'_i / U'_j)_F.$$

Proof. (1) Since $\mathbb{K}_F$ is of the poset type then $F_i \cong F$ and there are bimodule isomorphisms ${}_iK_{n+1} \cong {}_FF_F$. Hence without loss of generality we can suppose $F_i = F$, ${}_iK_{n+1} = {}_FF_F$ in the matrix form (2.1'). For any i,j we have a bimodule isomorphism $\bar{c}_{ijn+1}: {}_iK_j \to \operatorname{Hom}_F({}_jK_{n+1}, {}_iK_{n+1})$. Let ${}_ie_j$ be the element in ${}_iK_j$ corresponding to id via $\bar{c}_{ijn+1}$. It is easy to see that $c_{ijt}({}_ie_j \otimes {}_je_t) = {}_ie_t$ and $r\,{}_ie_j = {}_ie_j r$ for all $r \in F$ and all i,j,t. Then the standard bimodule isomorphisms ${}_iK_j \cong {}_FF_F$, ${}_ie_j \mapsto 1$, induce a ring isomorphism $R_{\mathbb{K}} \cong FI$. The remaining statements in (1) easily follow from [18; 2.4].

(2) The statements (a),(b) for $\mathcal{H}$ can be derived directly from the definition of $H(\underline{C})$ by the projective cover arguments (see [14;Thm1.5], [18;Thm 3.3]). In order to prove (2) for $\mathcal{G}$ we note that D is the usual F-duality whereas $\nabla: I^{op}\text{-sp} \to I^{op}\text{-fsp} \subseteq \operatorname{mod} R_{\mathbb{K}}^{op}$ is given by $\nabla(V, V_j) = (V, U_j)$ where $U_j = V/V_j$. Here fsp means factor spaces [18; 2.9]. Then (2) follows by the duality and the snake lemma.

## 3. A COXETER SCHEME

Let $\mathbb{K}_F$ be a vector space category such that $R = R_{\mathbb{K}}$ is a finite dimensional algebra over a field k. A module in $\mathrm{mod}_{sp}R$ is called sp-<u>injective</u> if it is injective with respect to those monomorphisms in $\mathrm{mod}_{sp}R$ whose cokernels are in $\mathrm{mod}_{sp}R$. The <u>Coxeter scheme</u> of $\mathbb{K}_F$ is the following diagram

$$\begin{array}{ccc} V(\mathbb{K}_F) & \underset{S^-}{\overset{S^+}{\rightleftarrows}} & U(\mathbb{K}_F) \\ \downarrow H & & \downarrow G \\ \mathrm{mod}_{sp}R & \underset{\Delta^-}{\overset{\Delta^+}{\rightleftarrows}} & \mathrm{mod}_{sp}R \end{array}$$

where $\Delta^+$, $\Delta^-$ are maps defined on modules by the formulas $\Delta^+N = D\,tr\widetilde{N}$, $\Delta^-N = \tilde{\Theta}trDN$ with $\widetilde{N} = P/\mathrm{soc}\,\mathrm{Ker}(u)$ where $u: P \to N$ is the projective cover(comp.[3]). Here tr is the Auslander's transpose and D is the k-duality. It follows from our results below that the Coxeter scheme can be used in computing indecomposables in $U(\mathbb{K}_F)$ in a similar way as the partial Coxeter functors [1, 4, 13]. We have the following[18]

PROPOSITION 3.1. If $\underline{C}$ is an indecomposable object in $V(\mathbb{K}_F)$ or in $U(\mathbb{K}_F)$ then $\Delta^+H\underline{C} \cong GS^+\underline{C}$ and $\Delta^-G\underline{C} \cong HS^-\underline{C}$ if the terms are nonzero. If N is an indecomposable nonprojective module in $\mathrm{mod}_{sp}R$ and L is an indecomposable non-sp-injective module in $\mathrm{mod}_{sp}R$ then there are almost split sequences

$$0 \to \Delta^+N \to X \to N \to 0 \text{ and } 0 \to L \to Y \to \Delta^-L \to 0.$$

Let $R = P_1 \oplus \ldots \oplus P_{n+1}$ where $P_j$ is the $j^{th}$ row right ideal in the form (2.1′); $P_{n+1}$ is the right peak of R.

A module N in $mod_{sp}R$ is <u>exact</u> if the projective cover of N has the form $P_1^{s_1} \oplus \ldots \oplus P_n^{s_n}$ and $s_j \neq 0$ for all j. We put $cdn(N) = (s_1, \ldots, s_n, s_{n+1})$ if $soc(N) \cong P_{n+1}^{s_{n+1}}$. If $R = R_{\mathbb{K}}$ is schurian we consider the rational quadratic form

$$q(x) = \sum_{i=1}^{n+1} x_i^2 f_i + \sum_{i,j=1}^{n} x_i x_j f_i d_{ij} - (\sum_{i=1}^{n} x_i f_i d_{in+1}) x_{n+1}$$

where $x = (x_1, \ldots, x_{n+1})$, $d_{ij} = \dim_{F_i}({}_iK_j)$ and $f_i = \dim_k F_i$. Applying arguments of Drozd [4] one can prove [18, Thm 3.11]

THEOREM 3.2. $U(\mathbb{K}_F)$ is of finite type if and only if q is weakly positive i.e. $q(x) > 0$ for $x \neq 0$ with nonnegative coordinates. If q is weakly positive and $\delta = \delta_1 \ldots \delta_{n+1}$, where $\delta_j$ is the $j^{th}$ reflection defined by the symmetric bilinear form of q, then

(a) If X is indecomposable in $mod_{sp}R$ then $\mathrm{End}X \cong F_j$ for some j. If in addition X and $\Delta^+X$ are exact then $cdn(\Delta^+X) = \delta(cdn(X))$ and $\mathrm{Hom}_R(N,X) \cong \mathrm{Hom}_R(\Delta^+N, \Delta^+X)$ for any indecomposable module N in $mod_{sp}R$ with $\Delta^+N \neq 0$.

(b) Let X be an indecomposable module in $mod_{sp}R$ with $\mathrm{End}X \cong F_i$. Then $q(cdn(X)) = f_i$ and X is uniquely determined by its composition factors. Moreover, there exists an integer j such that $(\Delta^+)^j X$ is either projective or non-exact.

## 4. APPLICATIONS

Now suppose A is an artinian PI-ring of the form

$$A = \begin{bmatrix} F & {}_FM_T \\ 0 & T \end{bmatrix}$$

where F is a division ring and T is a ring of finite representation type. Then $\mathbb{K}_F = \mathrm{Hom}_T({}_F M_T, \mathrm{mod}T)$ is a vector space category [18;4.7] and by [12;2.5] we have a functor $\Phi:\mathrm{mod}A \longrightarrow U(\mathbb{K}_F)$, Let us denote by

$$H_+^*:\mathrm{mod}A \longrightarrow (\mathrm{mod}_{sp}R_{\mathbb{K}^*})^{op} \text{ and } G_+:\mathrm{mod}A \longrightarrow \mathrm{mod}_{sp}(R_{\mathbb{K}})_*$$

the composed functors $H^*\Phi$ and $G\Phi$. From 2.1 - 2.4 we get

THEOREM 4.1.(a) $H_+^*$ and $G_+$ are full and dense; $\mathrm{Ker}\, H_+^* = \mathrm{Ker}\, G_+ = [\mathrm{mod}T]$ and there are equivalences of categories

$$\mathrm{mod}A/[\mathrm{mod}T] \cong (\mathrm{mod}_{sp}R_{\mathbb{K}^*})^{op}, \quad \mathrm{mod}A/[\mathrm{mod}T] \cong \mathrm{mod}_{sp}(R_{\mathbb{K}})_*.$$

(b) If $\mathbb{K}_F$ is of the poset type, $X = (X_F', X_T'', \varphi)$ is an indecomposable A-module, $X_T'' \cong L_1^{s_1} \oplus \ldots \oplus L_n^{s_n}$, $L_j$ are pairwise nonisomorphic indecomposable, $K_j = \mathrm{Hom}_T({}_F M_T, L_j)$, $X_F' \neq 0$ and $H_+X = (U, U_j) \in I^{op}\text{-sp}$, $G_+X = (V, V_j) \in I\text{-sp}$, $I = I(\mathbb{K}_F)$, then $X_F' \cong U^* \cong V$ and $s_j = \dim(U_j/\sum_{i>j} U_i)_F = \dim(\bigcap_{i>j} V_i/V_j)_F$.

A right peak ring R is said to be sp-<u>representation finite</u> if $\mathrm{mod}_{sp}R$ is of finite representation type. We denote by $\mathrm{mod}_{sp}^{-}R$ the full subcategory of $\mathrm{mod}_{sp}R$ consisting of modules having no injective summands. A counterpart of Theorem 4.1 for socle projective modules allows us to prove the following structure theorem.

THEOREM 4.2. Let R be a right peak nonsemisimple artin algebra. The following conditions are equivalent:

(a) R is sp-representation finite.

(b) There is a chain $0 = \underline{\underline{B}}_0 \subsetneq \underline{\underline{B}}_1 \subseteq \ldots \subseteq \underline{\underline{B}}_m = \mathrm{mod}_{sp} R$ of full subcategories of $\mathrm{mod}_{sp} R$ such that $\underline{\underline{B}}_i / [\underline{\underline{B}}_{i-1}] = \mathrm{mod}^-_{sp} R_i$ where $R_i = \begin{bmatrix} D_i & N_i \\ 0 & F_i \end{bmatrix}$, $D_i$ and $F_i$ are division algebras and $(\dim_{D_i}(N_i))(\dim(N_i)_{F_i}) \leqslant 3$.

We note that Theorems 3.2 and 4.2 give us a method for solving schurian vector space categories of finite type and for calculating their indecomposable subspaces. In particular, Theorem 4.2 allows us to reduce in a finite number of steps the classification of indecomposables in $U(\mathbb{K}_F)$ to the well-known classification of indecomposable modules over hereditary algebras of the Dynkin types $\circ \xrightarrow{(d,d')} \circ$ with $dd' \leqslant 3$. Moreover, Theorems 4.1 and 4.2 give us an algorithm for the classification of indecomposable modules over a a large class of triangulated algebras. For applications of the technique presented here the reader is referred to [7,10,15-19]. Let us mention that in [7] a diagrammatic characterization of schurian vector space PI-categories is given. In [18] the differentiation algorithm [8, 2] is extended to right peak rings.

Most of the results presented in this note can be extended to the case when instead of the division ring F in $\mathbb{K}_F$ we take an arbitrary direct sum of division rings or even an arbitrary ring. This leads to the notions of a right multipeak ring ( in general without identity)[18;Sec.6],[19] and a moduled category [20]. An advantage of this generalization is that we can apply the Gabriel's covering techniqe in the study of vector space categories via right peak rings. Applying the same type of arguments as in [6] we get the following useful result

THEOREM 4.3. If $T:\tilde{R} \longrightarrow R$ is a Galois covering functor in the sense of [6] and R is a locally finite dimensional basic right multipeak algebra then $\tilde{R}$ is a right multipeak algebra, the push-down functor $T_\lambda:\mathrm{Mod}\tilde{R} \longrightarrow \mathrm{Mod}R$ and the pull up functor $T_\bullet:\mathrm{Mod}R \longrightarrow \mathrm{Mod}\tilde{R}$ induce functors

$$\mathrm{Mod}_{sp}\tilde{R} \underset{T_\bullet}{\overset{T_\lambda}{\rightleftarrows}} \mathrm{Mod}_{sp}R$$

having properties analogous to those in [6; 3.2-3.9].

We hope that a counterpart of Theorem 4.1 for $\mathrm{mod}_{sp}R$ together with the differentiation algorithm and the covering technique are sufficient tools for solving arbitrary vector space PI-categories of finite type.

REFERENCES

1. Auslander M., Platzeck M.I., Reiten I., Coxeter functors without diagrams, Trans.Amer.Math.Soc. 250(1979), 1-46.

2. Bautista R., Simson D., Torsionless modules over 1-hereditary 1-Gorenstein artinian rings, Comm. Algebra 12, 1984, to appear.

3. Bünermann, Auslander-Reiten quivers of exact one-parame-

ter partially ordered sets, Lecture Notes in Math., No. 903, pp.55-61.

4. Drozd Ju.A., Coxeter transformations and representation of partially ordered sets, Funkc.Anal i Priložen., 8(1974), 34-42 (in Russian).

5. Gabriel,P., Indecomposable representations II, "Symposia Mathematica", Vol. XI, pp.81-104, 1973.

6. Gabriel P., The universal covering of a representation finite algebra, Lecture Notes in Math., 903, pp. 66-105.

7. Klemp B., Simson D., A diagrammatic characterization of schurian vector space PI-categories of finite type, Bull.Pol.Ac.Math., 32(1984), to appear.

8. Nazarova L.A., Rojter A.V., Representations of partially ordered sets, Zap.Naučn.Sem. LOMI 28(1972), 5-31.

9. Nazarova L.A., Rojter A.V., Kategorielle Matrizen-Probleme und die Brauer-Thrall-Vermutung, Mitt.Math.Sem. Giessen 115(1975), 1-153.

10. Pogorzały Z., Representation-finite special algebras, Commentationes Math., to appear.

11. Ringel C.M., Reports on the Brauer-Thrall conjectures, Lecture Notes in Math., Vol.831, 1980, pp. 104-136.

12. Ringel C.M., Tame algebras, Lecture Notes in Math., Vol, 831, 1980, 137-287.

13. Simson D. Partial Coxeter functors and right pure semisimple hereditary rings, J.Algebra 71(1980), 195-218.

14. Simson D., Special schurian vector space categories and 1-hereditary right QF-2 rings, Commenationes Math. 24(1984), to appear.

15. Simson D., On methods for the computation of indecomposable modules over artinian rings, in Reports of 28$^{th}$ Symposium on Algebra:"Ring Theory and Algebraic Geometry", University of Chiba(Japan), 1982, pp. 143-170.

16. Simson D., Indecomposable modules over one-sided serial local rings and right pure semisimple PI-rings, Tsukuba J.Math., 7(1983), 87-103.

17. Simson D., Right pure semisimple 1-hereditary PI-rings, Rend.Sem.Mat.Univ.Padova, 71(1984),1-35.

18. Simson D., Vector space categories, right peak rings and their socle projective modules, J.Algebra, to appear

19. Simson D., Socle reductions and socle projective modules, to appear.

20. Simson D., Representations of partially ordered sets, vector space categories and socle projective modules, Paderborn, 1983.

# THE REPRESENTATION TYPE OF GROUP ALGEBRAS

ANDRZEJ SKOWROŃSKI
NICHOLAS COPERNICUS UNIVERSITY
87-100 TORUŃ, POLAND

## 1. INTRODUCTION

Throughout this paper we use the term algebra to mean finite-dimensional algebra over a fixed algebraically closed field K and the term module to mean finitely generated right module. Algebras, as is usual in representation theory, are assumed to be basic. For any algebra A we will denote by mod A the category of finitely generated A-modules and by ind A the full subcategory of mod A consisting of all indecomposable modules.

Recall that an algebra A is called wild provided mod A has a full subcategory representation equivalent to the category of finite-dimensional, as K-vector spaces, modules over the free associative algebra K⟨x,y⟩ in two (non-commuting) variables x and y. Further, A is of finite type if ind A has only finitely many nonisomorphic objects. If A is neither wild nor of finite type, then A is said to be tame. Drozd showed in [1,2] that an algebra A is tame if and only if A is not of finite type and for any dimension d, there exists a finite family of functors $F_i$: mod $R_i \longrightarrow$ mod A, $i = 1,\dots,n_d$, where $R_i$ = K or $R_i$ is a rational algebra $K[T]_f$ of dimension 1, satisfying the conditions:

(a) For each i, $1 \leqslant i \leqslant n_d$, $F_i = ? \otimes_{R_i} Q_i$ where $Q_i$ is an $R_i$-A-bimodule being a finitely generated free right $R_i$-module,

(b) Every indecomposable A-module M of dimension d is of the form $M \cong F_i(S)$ for some i and a simple $R_i$-module S.

Let A be an algebra and let G be a finite group. We are concerned with the problem of determining the representation type of the group algebra **AG**. A characterization of group algebras AG of finite type has been obtained in the case A = K by Higman [3] and in the general case by the author and Meltzer [4,5]. Krugliak showed in [6] that if K is of charakteristic p> 2, then the group algebras KG of noncyclic p-groups are wild, and Brenner [7] has shown that for p = 2, the group algebras KG of all 2-groups beside the cyclic, dihedral, semidihedral and quaternion groups are wild. Bondarenko [8] and Ringel [9] have indepedently shown that in characteristic 2 the group algebras of dihedral 2-groups are tame. Finally, Drozd and Bondarenko [10] have also shown that the group algebras of semidihedral and

and quaternion 2-groups are tame in characteristic 2 (see also [11]).

In this paper we give necessary and sufficient conditions for AG to be tame where A is an arbitrary algebra and G a finite group.

In order to state the main theorem we need some notations. For any positive integer n we will denote by $X_n$ the quiver

$$X_n : \quad 1 \xrightarrow{\alpha_1} 2 \xrightarrow{\alpha_2} \ldots \to n-1 \xrightarrow{\alpha_{n-1}} n$$

and by $Y_n$ the oriented cycle

$$Y_n : \quad \begin{array}{c} n \xrightarrow{\alpha_n} 1 \\ \alpha_{n-1} \nearrow \qquad \searrow \alpha_1 \\ n-1 \qquad\qquad 2 \\ \uparrow \qquad\qquad \downarrow \\ \bullet \; \bullet \qquad \bullet \; \bullet \end{array}$$

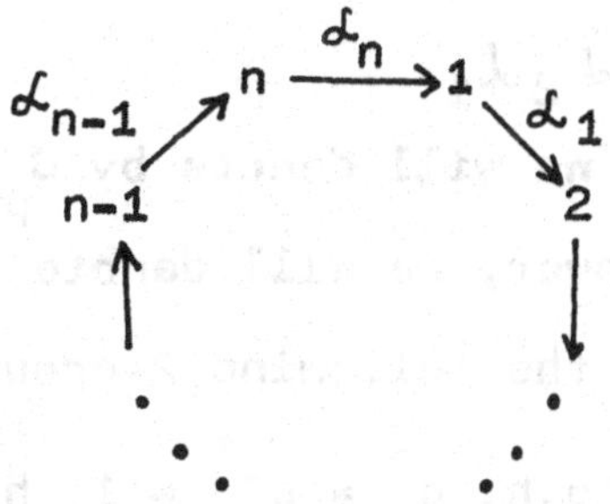

For $n \geqslant 5$ and any sequence of positive integers $n_1, \ldots, n_s$ satisfying the conditions: $s \geqslant 2$, $n_1 > 1$, $n_s < n$, and $n_i + 1 < n_{i+1}$, $i = 1, \ldots, s-1$, we will denote by $A^n_{(n_1, \ldots, n_s)}$ the bounden quiver algebra $KX_n / I^n_{(n_1, \ldots, n_s)}$ (in the sense of [12]) where $I^n_{(n_1, \ldots, n_s)}$ is the ideal in the quiver algebra $KX_n$ generated by the composed arrows $\alpha_i \alpha_{i-1}$, $i \neq n_1$, $\ldots, n_s$, $i = 2, \ldots, n-1$. For $n \geqslant 2$ and any sequence of positive integers $n_1, \ldots, n_s$ satisfying the conditions: $s \geqslant 1$, $n_1 > 1$, $n_s \leq n$, $n_i + 1 < n_{i+1}$, $i = 1, \ldots, s-1$, we will denote by $B^n_{(n_1,\dots,n_s)}$ the bounden quiver algebra $KY_n/J^n_{(n_1,\dots,n_s)}$ where $J^n_{(n_1,\dots,n_s)}$ is the ideal in the quiver algebra $KY_n$ generated by the composed arrows $\alpha_i\alpha_{i-1}$, $i \neq n_1,\dots,n_s$, $i = 1,\dots,n$, and $\alpha_0 = \alpha_n$. Moreover, for $n \geqslant 1$, we will denote by $B^n$ the bounden quiver algebra $KY_n/P_n$ where $P_n$ is the ideal in $KY_n$ generated by the composed arrows $\alpha_i\alpha_{i-1}$, $i = 1,\dots,n$, and $\alpha_0 = \alpha_n$. Finally, we will denote by $R_1$ the quiver algebra $KX_4$, by $R_2$ the bounden quiver algebra $KX_4/I_2$ where $I_2$ is generated by $\alpha_3\alpha_2\alpha_1$, by $R_3$ the quiver algebra $KX_2$, and by $R_4$ the bounden quiver algebra $KX_3/I_3$ where $I_3$ is generated by $\alpha_2\alpha_1$.

For a prime number p we will denote by $C_{p^r}$ the cyclic p-group of order $p^r$. Moreover, we will denote by $D_m$, $m \geqslant 1$, $S_m$, $m \geqslant 3$, and $Q_m$, $m \geqslant 2$, the following 2-groups:

the dihedral groups $D_m = \langle g,h;\ g^2 = h^{2^m} = 1,\ hg = gh^{-1} \rangle$

the semidihedral groups $S_m = \langle g,h;\ g^2 = h^{2^m} = 1,\ hg = gh^{2^{m-1}-1} \rangle$

the quaternion groups $Q_m = \langle g,h;\ g^2 = h^{2^{m-1}},\ g^4 = 1,\ hg = gh^{-1} \rangle$.

Then using diagrammatic methods, vector space category methods, Auslander-Reiten theory of almost split sequences and coverings techniques we prove the following theorem.

THEOREM 1. Let G be a finite group and let A be a connected finite-dimensional basic algebra over an algebraically closed field K of characteristic p. Then AG is tame if and only if one of the following cases holds.

(i) p does not divide the order of G and A is tame.

(ii) p divides the order of G and one of the following holds:

(1) $p = 3$, a 3-Sylow subgroup of G is isomorphic to $C_3$, and A is isomorphic to $R_4$,

or (2) $p = 2$ and one of the following holds:

(a) $A = K$ and a 2-Sylow subgroup of G is isomorphic to one of the groups $D_m$, $S_m$, or $Q_m$,

(b) a 2-Sylow subgroup of G is isomorphic to $C_4$ and A is isomorphic to $R_3$,

(c) a 2-Sylow subgroup of G is isomorphic to $C_2$ and A is isomorphic to one of the algebras $R_1$, $R_2$, $A^n_{(n_1,..,n_s)}$, $B^n_{(n_1,\dots,n_s)}$, or $B^n$.

## 2. GALOIS COVERINGS

In the proof of Theorem 1 one of the main role is played by coverings techniques introduced by Bongartz-Gabriel [13], Gabriel [14], and Riedtmann [15] for the investigation of algebras of finite type, and developed recently by the author and Dowbor [16,17] for the research of modules over arbitrary algebras. Here we shall formulate some results from [16,17] needed in the proof of Theorem 1.

Let R be a locally bounded K-category [14] and let H be a group of K-authomorphisms of R. We will identify

R-modules with K-linear functors from $R^{op}$ to Mod K where Mod K is the category of K-vector spaces. The category of all R-modules we will denote by Mod R. For each R-module M and $h \in H$, we denote by ${}^hM$ the translated R-module $M(h^{-1}(?))$. We will assume that H acts freely on the isoclasses of indecomposable finite-dimensional R-modules, that is, we have ${}^hM \not\cong M$ for each $M \in$ ind R and each $h \neq 1$ in H. Following Gabriel [14] we have the Galois covering functor $F : R \longrightarrow R/H$ which assigns to each object x from R its H-orbit and the push-down functor $F_\lambda$ : Mod R $\longrightarrow$ Mod R/H which assigns to each R-module M the R/H-module $F_\lambda(M)$ with $F_\lambda(M)(a) = \bigoplus_{x/a} M(x)$ where x ranges over all objects x of R such that $F(x) = a$. We denote by $\mathrm{Ind}^H R$ the full subcategory of Mod R formed by chosen representatives of the isoclasses of all indecomposable R-modules M satisfying the conditions $\dim_K M(x) < \infty$ for each object x in R and there exists an element $h \neq 1$ in H such that $\mathrm{supp}({}^hM) = \mathrm{supp}(M)$, where supp(M) denotes the support of M. From our assumption, the support of any module from $\mathrm{Ind}^H R$ is infinite. We say that a module M from $\mathrm{Ind}^H R$ is of <u>type $A_\infty^\infty$</u> if the support of M is of the form

$$A_\infty^\infty \quad : \qquad \cdots - \bullet - \bullet - \bullet - \bullet - \cdots$$

where $\bullet - \bullet$ denotes $\bullet \longrightarrow \bullet$ or $\bullet \longleftarrow \bullet$.

Assume that M is an R-module of type $A_\infty^\infty$ from $\mathrm{Ind}^H R$. Then it is easy to see that $H_M = \{h \in H \; ; \; \mathrm{supp}({}^hM) = \mathrm{supp}(M)\}$ is an infinite cyclic group. Hence, since the group $H_M$ acts in a natural way on $F_\lambda(M)$, $F_\lambda(M)$ has a structure of $K[T,T^{-1}]$-R/H-bimodule. We denote by $F_M$ the functor

$$? \otimes F_\lambda(M) \; : \; \mathrm{Mod}\; K[T,T^{-1}] \longrightarrow \mathrm{Mod}\; R/H \; .$$

It is shown in [17] that $F_M$ restricted to mod $K[T,T^{-1}]$ preserves indecomposables and reflects isomorphisms.

A locally bounded K-category R is said to be <u>tame</u> if any finite full subcategory of R is tame [18,19].

We have the following theorem being a special case of a more general theorem proved in [17].

THEOREM 2. Let R be a locally bounded K-category and let a group H of K-authomorphisms of R acts freely on the isoclasses of finite-dimensional indecomposable R-modules. Assume that any R-module from $\mathrm{Ind}^H R$ is of type $A_\infty^\infty$ and let $\underline{O}$ be a set of representatives of the H-orbits in $\mathrm{Ind}^H R$. Let $F_\lambda$ : Mod R $\longrightarrow$ Mod R/H be the push-down functor associated with the Galois covering $F : R \longrightarrow R/H$ and, for each M from $\underline{O}$, let $F_M$ : Mod $K[T,T^{-1}] \longrightarrow$ Mod R/H be the functor defined above. Then each indecomposable R/H-module L from mod R/H is isomorphic to some $F_\lambda(U)$, $U \in$ ind R, or to some $F_M(V)$ where $M \in \underline{O}$ and V is an indecomposable finite-dimensional $K[T,T^{-1}]$-module. Moreover, R is tame if and only if R/H is tame.

Following [16] a locally bounded K-category R is called locally suport-finite provided for each object x of R, the full subcategory of R formed by the sum of supports supp(M) of all indecomposable finitely generated R-modules M such that $M(x) \neq 0$ is finite. Then we have the following consequence of the above theorem.

COROLLARY 1. Let R be a locally support-finite K-category and let a group H of K-authomorphisms of R acts freely on the isoclasses of indecomposable finitely generated R-modules. Then the push-down functor $F_\lambda$ : Mod R $\longrightarrow$ Mod R/H associated with the Galois covering F : R $\longrightarrow$ R/H induces a bijection between the H-orbits of isoclasses of finitely generated indecomposable R-modules and the isoclasses of finitely generated indecomposable R/H-modules.

## 3. PROOF OF THEOREM 1

Let G be a finite group and let A be a connected basic algebra over an algebraically closed field K of characteristic p. As in [10,Proposition 3] one proves that, if the order of G is invertible in A, then AG is tame if and only if A is. Moreover, if p divides the order of G and F is a p-Sylow subgroup of G, then AG is tame if and only if AF is. Hence we can assume that G is a p-group. If A = K, then from [6-10] KG is tame if and only if p = 2 and G is one of the groups $D_m$, $S_m$, or $Q_m$. Suppose that A is not semisimple,

G is not cyclic and AG is tame. Then, since KG is a factor algebra of AG, $p = 2$ and G is one of the groups $D_m$, $S_m$, or $Q_m$. Using [4, Proposition 1] it is not hard to show that the separated quiver [20] of AG contains a subquiver of the form

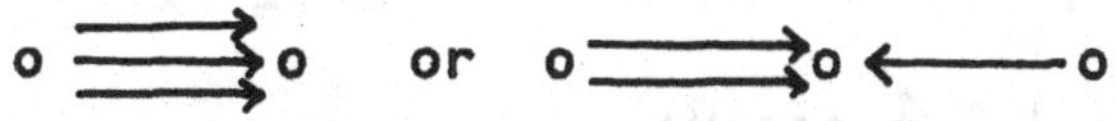

and by [21], AG is wild. Consequently, if G is not cyclic and AG is tame, then $A = K$, $p = 2$, and $G = D_m$, $S_m$, or $Q_m$. Now, for our aim, we can assume that A is nonsemisimple and G is a cyclic p-group, say of order $p^r$. Then, using diagrammatic methods [20,21] and [4, Proposition 1] again, one can prove that if AG is tame, then A is isomorphic to a bounden quiver algebra KQ/I where Q is an oriented cycle $Y_n$ or a linear quiver $Z_n$

$$Z_n : \quad 1 \xrightarrow{\alpha_1} 2 \text{---} \cdots \text{---} n-1 \xrightarrow{\alpha_{n-1}} n .$$

Then as in [4, Lemma 3] one proves the following proposition

PROPOSITION 1. Under the above assumptions and notation, if AG is tame, then AG is isomorphic to the bounden quiver algebra $KQ'/I'$ where $Q'$ is the quiver

$Y'_n$ :

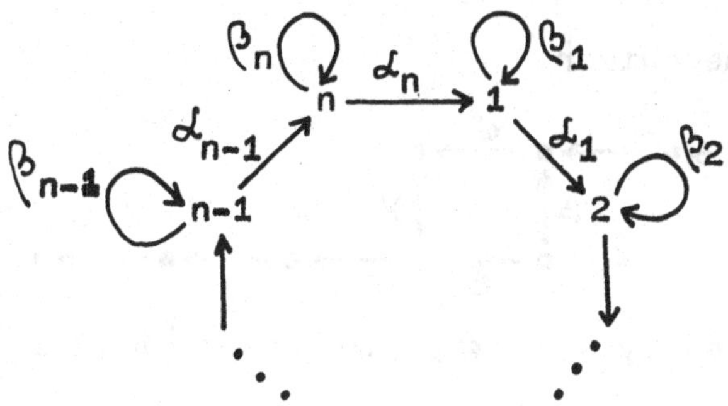


if $A = KY_n/I$

or

$Z'_n$ : 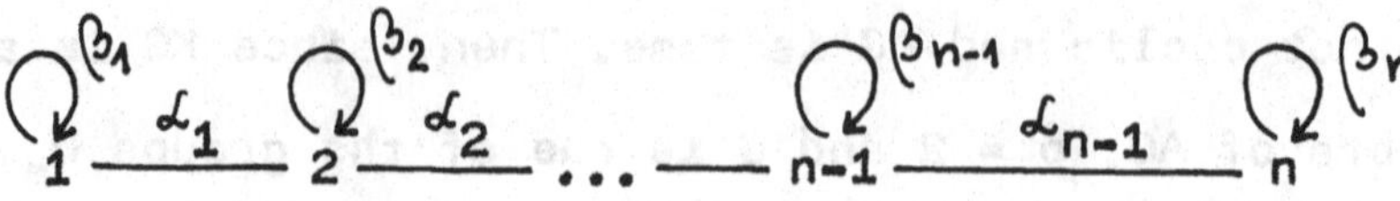

if $A = KZ_n/I$, and $I'$ is generated by $I$, $\beta_i^r$, $i = 1,\dots,n$, $\alpha_i\beta_i - \beta_{i+1}\alpha_i$ if $i \xrightarrow{\alpha_i} i+1$ and $\beta_i\alpha_i - \alpha_i\beta_{i+1}$ if $i \xleftarrow{\alpha_i} i+1$, $i = 1,\dots,n$.

In order to obtain other necessary conditions for AG to be tame, we use Galois coverings. Applying constructions of the universal Galois coverings of bounden quivers [14,22] we get a Galois covering functor $F : K\tilde{Q}'/\tilde{I}' \longrightarrow KQ'/I'$ with a group H being an infinite cyclic group in the case $Q = Z_n$ and a free abelian group of rank 2 in the case $Q = Y_n$. Now let $AG = KQ'/I'$ be of infinite type and nonisomorphic to one of the algebras stated in Theorem 1. Then, using the description of group algebras of finite type given in [4,5] it is not hard to show that $K\tilde{Q}'/\tilde{I}'$ contains a finite full subcategory of wild type, and hence AG is wild. For instance, if A is the quiver algebra of

$$Q : \bullet \longrightarrow \bullet$$

and $G = C_p$, $p \geqslant 5$, then $K\tilde{Q}'$ contains the bounden quiver algebra KT/J of the quiver

T : 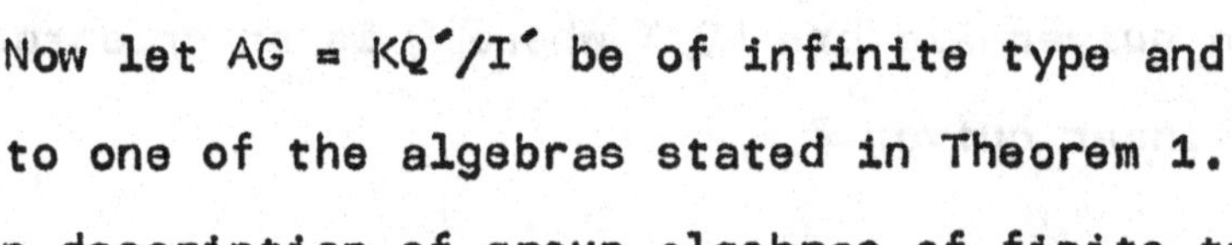

where J is generated by $\alpha\beta - \gamma\sigma$, which from [18,2.3] is wild.

In the proof of the sufficiency we apply Theorem 2, Corollary 1, Auslander-Reiten theory of almost split sequences [12,23,24] and the vector space category methods introduced by Nazarova and Rojter [25] in a connection with the second Brauer-Thrall conjecture, and developed by Ringel [18] and Simson [26-28]. We prove that, if AG is one of the algebras $A^n_{(n_1,\ldots,n_s)}C_2$ , $R_1C_2$, $R_2C_2$, $R_3C_4$ and $R_4C_3$, then the universal Galois covering $\widetilde{AG}$ of AG is a locally support-finite tame K-category, and consequently from Corollary 2, AG is tame. For instance, $\widetilde{R_1C_2}$ can be obtained by one-point extension procedures from the locally representation-finite K-category $\widetilde{RC_2}$, where R is the quiver algebra $KX_3$, and the indecomposable $\widetilde{R_1C_2}$-modules which are not in mod $\widetilde{RC_2}$ are indecomposable representations of the vector space category

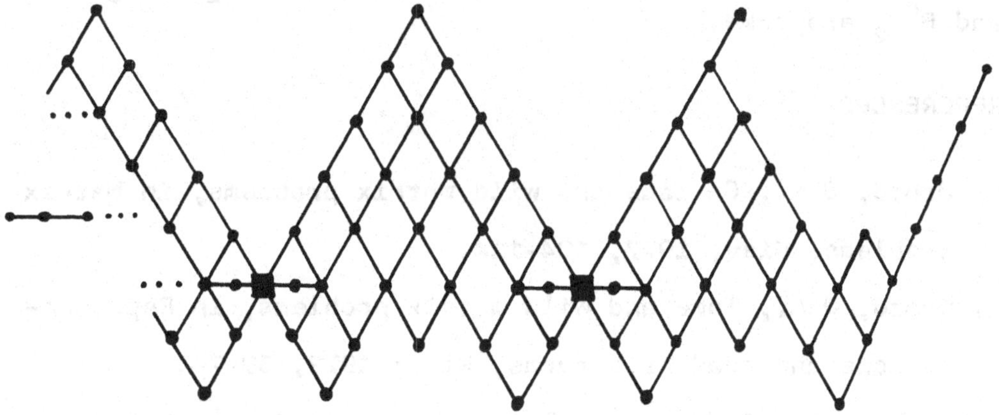

Applying the differential algorithms from [26-29], we can reduce this category, preserving the representation type, to the Ringel's pattern $(\tilde{E}_8,5)$ [18]. Hence we deduce that $\widetilde{R_1C_2}$ is a locally support-finite tame K-category. Similarly the corresponding vector space categories for $\widetilde{R_3C_4}$, $\widetilde{R_4C_3}$, and $\widetilde{A^n_{(n_1,\ldots,n_s)}C_2}$ can be reduced to the Ringel's patterns $(\tilde{E}_8,5)$, $(\tilde{E}_7,3)$, and $(\tilde{D}_m,\frac{m-2}{m-2})$. Moreover, $R_2C_2$ as a factor algebra of $R_1C_2$ is also tame. Using properties of almost split sequences [12,23,24] and theorems on the structure of indecomposable representations of vector space categories considered in [18,29,30], one proves that $\widetilde{B^n_{(n_1,\ldots,n_s)}C_2}$ and $\widetilde{B^nC_2}$ are tame categories and their indecomposable finite-dimensional representations derive from the indecomposable finite-dimensional representations of Dynkin quivers of type $A_m, D_m$ and extended Dynkin quivers of type $\tilde{D}_m$. Then a simple analysis shows that the assumptions of Theorem 2 are satisfied and consequently the algebras $B^n_{(n_1,\ldots,n_s)}C_2$ and $B^nC_2$ are tame.

REFERENCES

1. Drozd, J.A., On tame and wild matrix problems, in Matrix problems, Kiev, 1977, 104-114.

2. Drozd, J.A., Tame and wild matrix problems, in Representations and quadratic forms, Kiev, 1979, 39-74.

3. Higman, D., Indecomposable representations at characteristic p, Duke Math. J. 21,377-381, 1954.

4. Meltzer, H., Skowroński, A., Group algebras of finite

representation type, Math. Z. 182, 129-148, 1983

5. Meltzer, H., Skowroński, A., Correction to " Group algebras of finite representation type ", Math. Z., to appear.

6. Krugliak, S.A., Representations of the (p,p) group over a field of charakteristic p, Dokl. Akad. Nauk SSSR 153, 1253-1256, 1963.

7. Brenner, S., Modular representations of p groups, J. Algebra 15, 89-102, 1970.

8. Bondarenko, V.M., Representations of dihedral groups over a field of characteristic 2. Math. Sbornik 96, 63-74, 1975.

9. Ringel, C.M., The indecomposable representations of dihedral 2-groups, Math. Ann. 214, 19-34, 1975.

10. Bondarenko, V.M., Drozd, J.A., Representation type of finite groups, Zap. Naučn. Sem. LOMI 71, 24-41, 1977.

11. Ringel, C.M., Tame group algebras, in Darstellungstheorie, Endlich Dimensionaler Algebren, Oberwolfach, 1977, 92-95.

12. Gabriel, P., Auslander-Reiten sequences and representation-finite algebras, Lecture Notes in Math. 831, 1980, 1-71.

13. Bongartz, K., Gabriel, P., Covering spaces in representation theory, Invent. math. 65, 331-378, 1982.

14. Gabriel, P., the universal cover of a representation-

-finite algebra, Lecture Notes in Math. 903, 1981, 68-105.

15. Riedtmann, Chr., Algebren, Darstellungsköcher, Überlagerungen and zurück, Comment. Math. Helv. 55, 199-224, 1980.

16. Dowbor, P., Skowroński, A., On Galois coverings of tame algebras, to appear.

17. Dowbor, P., Skowroński, A., Galois coverings of algebras of infinite representation type, to appear.

18. Ringel, C.M., Tame algebras, Lecture Notes in Math. 831 1980, 137-287.

19. Dowbor, P, Skowroński, A., Some remarks on the representation type of locally bounded categories, to appear

20. Gabriel, P., Indecomposable representations II, Symposia Math. 11, 81-104, 1973.

21. Dlab, V., Ringel, C.M., Indecomposable representations of graphs and algebras, Mem. Amer. Math. Soc. 173, 1976

22. Waschbüsch, J.,Universal coverings of selfinjective algebras, Lecture Notes in Math. 903, 1981, 331-349.

23. Auslander, M., Reiten, I., Representation theory of artin algebras III, Almost split sequences, Comm. Algebra 3, 239-294, 1975.

24. Auslander, M., Reiten, I., Representation theory of artin algebras IV, Invariants given by almost split sequences, Comm. Algebra 5, 443-518.

25. Nazarova, L.A., Rojter, A.V., Kategorielle Matrizen-

-Probleme und die Brauer-Thrall-Vermuttung, Mitt. Math. Sem. Giessen 115, 1975, 1-153.

26. Simson, D., Vector space categories, right peak rings and their socle projective modules, J. Algebra, to appe ar.

27. Simson, D., Representations of partially ordered sets, vector space categories and socle projective modules, Paderborn, 1983.

28. Simson, D., A module-theoretical approach to vector space categories, to appear.

29. Nazarova, L.A., Rojter, A.V., Representations of partially ordered sets, Zap. Naučn. Sem. LOMI 28, 1972, 5-31.

30. Pogorzały, Z., Skowroński, A., On algebras whose indecomposable modules are multiplicity-free, Proc. London Math. Soc. 47, 463-479, 1983.

Printed in the United States
By Bookmasters